How Life Emerges from Inanimate Matter

How Life Emerges from Inanimate Matter:

From Atoms to a Living Individual

By

Bernard Korzeniewski

Cambridge Scholars Publishing

How Life Emerges from Inanimate Matter:
From Atoms to a Living Individual

By Bernard Korzeniewski

This book first published 2023

Cambridge Scholars Publishing

Lady Stephenson Library, Newcastle upon Tyne, NE6 2PA, UK

British Library Cataloguing in Publication Data
A catalogue record for this book is available from the British Library

ISBN (10): 1-5275-1980-5
ISBN (13): 978-1-5275-1980-0

I dedicate this book to Stanisław Lem, the man who has had the greatest impact on my intellectual development.

TABLE OF CONTENTS

INTRODUCTION

Let us consider a complex mechanism (a functional system) built by a person, such as a car. We do not have the slightest problem in answering the question: what purpose does the car serve? Of course, to reach a designated place fast, comfortably, and safely. We also know perfectly well the role of particular parts and the sub-systems of the car. The engine propels, the steering wheel controls the direction, the wheels turn, the car body protects other parts/sub-systems, the crankshaft transfers the drive from the engine to the wheels, and the gas pedal decides the speed. The structure and function of the car are organized hierarchically, which means larger parts/sub-systems are composed of smaller parts/sub-systems. For example, the cylinders, fuel injectors, and spark plugs control the engine. Those, in turn, are composed of still smaller elements. The car and its parts/sub-systems have been purposefully designed to fulfill the tasks and functions subordinate (ancillary) in relation to it. It is easy to understand all of this; alternatively, comprehending particular details and aspects of the construction and operation of the car requires some technical knowledge. Each element of the car's structure and function possesses its clearly specified purpose. In other words, we know perfectly well the car's purpose.

Let us now consider another complex system, which originated from billions of years of biological evolution—the frog. It is organized hierarchically, executes various tasks, and contains numerous systems and sub-systems. So, we have here organic macromolecules (first of all, proteins and nucleic acids: DNA and RNA), cellular organelles (e.g., mitochondria), whole cells (together with their components and complex structures), tissues, organs, and organ systems. Thanks to progress in biology, we know a lot about how they are built, how they function, and what purpose they serve. Thus, the heart pumps blood that delivers oxygen and nutrients to all body cells while receiving carbon dioxide and the waste products of metabolism. Muscles serve locomotion, vocalization (producing sounds), and catching insects (tongue, jaw muscles). The brain directs the whole frog's behavior, integrates stimuli from the surroundings by sense organs, undertakes decisions concerning actions and implements them (through stimulation of muscle contraction), and accumulates memory. Sense organs (receptors) collect data from the surroundings, which the peripheral nervous system transfers to the brain. On the other hand, skeletal muscles (effectors), excited through peripheral motor nerves, implement the brain's directives. The digestive system serves the intake, digestion, and assimilation (absorption into the blood circulation system) of food (nourishment). The skin insulates and protects the internal organs, and also, as the frog is an amphibian, it participates significantly in respiration. The kidneys extract and densify the waste products of metabolism, especially nitrogen metabolism (urea in adults, ammonia in tadpoles), which are then expelled. Particular organ systems and organs are composed of parts and sub-systems that perform service and subordinate functions—it is enough to see an anatomical section of the kidney or brain to become familiarized with the mechanisms of their operation. The structure and function of all organs are based on living cells—specific for particular organs and tissues (frequently as different as epithelium cells, neurons, and muscle cells). Most cells contain organelles—a nucleus, mitochondria, endoplasmic reticulum, and macromolecule complexes, such as a cytoskeleton or ribosomes serving protein synthesis.

All parts/sub-systems of the organism of a frog support and mutually serve each other. The heart, together with the circulatory system, enables the functioning of the brain, muscles, and digestive system, the digestive system sources nutrients for all organs, the muscles enable locomotion directed at gaining food and fleeing from a predator, and the brain adequately supervises the behavior of the frog in response to stimuli from the surroundings, enabling it to survive together with all its organs. The same reasoning can be applied to organs and organ systems and every cell (of one of several hundred types). Ultimately, all parts/sub-systems of an individual frog serve all other parts/sub-systems. This property is called *causal closure*.

All of this seems simple and understandable. A problem appears only when we ask what the role, function, or purpose of a living individual in general and our frog in particular is. Certainly, this superior, long-term aim is not the survival of itself, as every complex organism eventually dies, either because of senility, in the jaws of a predator, or due to illness (e.g., infection by parasites), environmental change (e.g., drying out of a water reservoir) or by accident (only simple organisms such as bacteria can potentially be "forever young" and divide *ad infinitum*). Of course, living individuals strive to survive "until a certain time." But what time? Many organisms die "voluntarily" just after reproduction, for which they use all accessible energy resources, e.g., salmon, after spawning (preceded by an exhausting journey upstream against the river flow), or one-day ephemera after copulation and lying fertilized eggs (they spend their whole life in water in the larval stage and as adults do not even have a functional digestive system). It "pays" for the mantis male to be eaten by the female after copulation, as he supplies building material and energy to his progeny in this way.[1] Certainly, the frog is not purposefully directed to be an element of the ecosystem in which it lives, for instance, to serve as food for storks or to eat insects to control their number. Just the opposite, it struggles to avoid being eaten by other organisms, either by hiding or fleeing, and it is not interested in controlling the number but in the abundance of insects. In turn, insects do what they can to not be consumed by a frog. The opinion that living organisms tamely accept the role of an element of an ecosystem is extremely naïve (leaving aside the fact that it implies ignorance and a lack of understanding of biology). The truth is completely different: living individuals struggle for the best possible place in the ecosystem through competition and evolutionary optimization. Every individual is selfishly directed to itself and not for "the good of the ecosystem" or "the good of the species," as nothing like this exists. Also, the frog does not serve to be a supplier of legs for the French. Leaving aside an extreme anthropomorphism of such a point of view, frogs (and their kin) evolved for hundreds of millions of years before the French (and storks) appeared, and most frog species live in tropical forests nowadays, far away from humans. In the face of this, what is the superior purpose of this incredibly complex ensemble of cells/organs/organ systems entering into the composition of a living individual? What does such an individual, for instance, the frog, strive toward? What is the rationale for its existence in its whole complexity of structure and function? In other words: ***what purpose does the frog serve?***

We will come to the answer to this question step by step. We will see how the phenomenon of life, in particular its fundamental unit called a living individual, gradually emerges from the functioning of inanimate elements of the physical reality when we pass from single atoms (and earlier—elementary particles), through more and more complex configurations of atoms, to the enormously complicated structure and function of living organisms. This transition should be understood in two ways. First, it is about the biological evolution of life from the stage of spontaneously originated organic compounds and their complexes, through initially very simple, living systems, to all biological organisms appearing presently (and in past geological epochs) on the terrestrial globe and, most probably, on many other planets in the universe. Second, it is about how the properties of single atoms, constituting the building material of biological organisms, lead to, after their appropriate arrangement in a certain superordinate functional structure, the creation of a completely new quality absent in the

[1] Of course, it does not pay for the male to be eaten before copulation.

purely physical (and chemical) world, namely the phenomenon of life, based on living individuals, characterized by intrinsic purposefulness and a countless richness of forms. As we are one of the forms of life and (also) living individuals, the solution to this conundrum could help us understand the place of humanity in the (physical) universe.

This book is not addressed to any determined, easy-to-separate circle of readers, e.g., biology students or adepts in the philosophy of nature, and therefore it does not constitute a handbook of the problems dealt within it. Emphasis is put on understanding problems rather than an enumeration of possibly many facts, and before the question of "What?" the priority is given to the question of "How?" More than dazzling with erudition, I prefer a clear (although not always easy to receive) answer to the fundamental questions. This is, first of all, a book written from my point of view, containing what I regard as worth presenting on the considered topic, addressed to all thinking people curious about the world, especially those interested in biology and other natural sciences. Although deeply anchored in contemporary scientific knowledge, it certainly cannot be considered a strictly popular-science item, as I propose in it, especially in Chapter 6, my original interpretation (vision) of the phenomenon of life and a living individual, which has already been partly discussed in my previous works. The cybernetic definition of life (a living individual) is further developed, justified, explained, and confronted with the realities of the physical world. In all earlier chapters, the popular-science motif decidedly predominates, so even if my personal opinion about the essence of life seems too difficult to grasp or is insufficiently convincing, I hope the remaining part of the book will be interesting and deliver some intellectual satisfaction. Finally, in Chapter 7, I shortly summarize my opinions on the essence of mind, subjective psyche, and self-consciousness and their relation to the phenomenon of life. I describe the cybernetic model of the psyche-bearing individual and the formal similarities between them and biological individuals. Generally, I try to justify the theory that the complexity (and, first of all, the complexity at the basis of the mentioned individuals) exists equally realistically and objectively in the external physical world, as does matter, space, and time.

Numerous fundamental questions lie at the basis of the present elaboration. How is the existence of any structure in our universe possible at all? Why, in some cases, does the structure originate? What conditions must be fulfilled in order for this to happen? How can the degree of structural complexity and amount of related information spontaneously increase over long periods of time? What is the specificity of the structure and function of living individuals? What fundamental mechanisms were responsible for their origin, evolution, perfecting, and complications? What is this, the living individual, and what relationships associate it with the phenomenon of life in general? What constitutes the subject of biological evolution? What is the relation between the living individual and the psychical individual? These are just a few of them.

Several major concepts contribute to the leading motif of this book. First, life appears as an emergent property characterized by a determined kind of complexity from the functioning of its elements at a lower physical level. No special factor or agential force is needed for life to come into being. This, which causes the feeling of complete dissimilarity of living systems from inanimate physical systems, and therefore is responsible for the emergence of the biological level, is not so much the degree but rather the *specific kind* of functional complexity of matter constituting the building material of living organisms. For purely physical reasons, living organisms must be dynamic dissipative structures (generally, systems displaced from thermodynamic equilibrium) whose existence is sustained by thermodynamic gradients, unceasing flow of matter, dissipation of energy, and production of entropy. The fundamental unit of life is living individuals that strive to survive in good conditions to the reproduction period, to reproduce and thus transfer their identity to progeny. They (in fact, their identity) can evolve from generation to generation and constitute the unit of natural selection and the subject of Darwinian evolution. The living individual can be best defined in the frame of the cybernetic paradigm as a network of purposeful regulatory mechanisms (first of all, negative feedbacks) supervising the embryonal development (morphogenesis), cognitive functions, behavior, and constancy of internal conditions, directed only and exclusively on the multiplication of itself in a possibly great number of copies, that is reproduction, representing positive feedback. It constitutes an example of the relation of self-applicability (self-reference, recurrence) that is, in my opinion, crucial for understanding not only the phenomenon of life but also of self-consciousness.

The living individual can be conceptualized in a slightly simplified way as a self-controlling, self-copying system; that is, a dynamic *homeostatic replicator*. Life emerges when elements of inanimate matter undergo organization into an appropriately composed SYSTEM, corresponding to the above characteristics.

I would like to thank Andrzej Joachimiak and Ireneusz Ślesak cordially for their critical reading of the manuscript and for detecting several mistakes and obscurities.

CHAPTER 1

WHY IS STRUCTURE POSSIBLE AT ALL IN THE PHYSICAL WORLD?

Structure in physics and chemistry

This book should start from the most elementary problems. An inseparable property of living organisms is their complex, hierarchically-organized structure (and function based on it). The fundamental question is why structure, in general, is possible in our physical universe. However, should this question not be considered completely pointless, as it refers to an obvious and understandable matter? After all, we can easily create, in a sense, naturally, various sorts of structures. We can build different constructions from LEGO playing blocks. We can connect plastic balls of various sizes and colors with plastic sticks to obtain models of even very complicated chemical molecules, for instance, proteins or the DNA double helix. We produce computers, cars, skyscrapers, and toothbrushes. Besides, it is enough to take any product of nature, such as a stone, to see that it possesses a determined shape and internal structure (composed of various crystals, minerals, grains, and strands). Then, is there a problem associated with structure? Is it not enough to shape and process components and then appropriately connect them? In the case of a stone, nature created this through certain geological processes.

Unfortunately, it is not so simple. All enumerated, having structure, macroscopic objects are composed of atoms of elements and molecules of chemical compounds. On a microscopic level, where quantum mechanics absolutely rules, there are no "shaped elements" and appropriate "connectors" between them, such as, for instance, plastic balls and sticks in macroscopic models of chemical molecules. Instead, there are only some objects (elementary particles, atoms), and the only thing between them is various interactions (forces) that can be attractive or repulsive. On the other hand, they lack any "rubber bands," "connectors," or "sticks." (Leaving aside the fact that in the microworld, the distinction between objects and the forces generated by them and acting on them is not fully legitimate, which is discussed more broadly at the end of this chapter). How is it possible in such a situation to get any structure?

Well, a fundamental, far from trivial, answer to the question about the possibility of structure coming into existence in the real world is the statement that the key role is played here by the presence of attractive and repulsive forces between physical objects (e.g., elementary particles or atoms), the forces which counterbalance each other when the considered objects are located at a determined distance from each other. Just this distance determines the characteristic size of a given elementary structure.

To better comprehend this, see Fig. 1-1. It demonstrates the dependence of the amount of energy of a certain microscopic object (e.g., an elementary particle and atom), shown in the figure as a little ball, on its distance from the object (again: elementary particle, atomic nucleus, atom) that interacts with (exerts a force on) it, and whose location is indicated by the place of intersection of the vertical y-axis with the horizontal x-axis (that is zero distance on the x-axis). We can see that energy is very high in the nearest neighborhood of the object exerting force; it quickly decreases as a function of the distance from it, reaches a minimum at a certain critical distance, and then increases from negative values tending asymptotically to zero, as the distance increases to infinity. A fundamental physical principle is that such a system tends toward the lowest energy state. In other words, this means that the object (a little ball) on which a force (influence) is exerted will be attracted more strongly when it is situated at a certain distance from the object exerting the force greater than the critical distance (corresponding to minimal energy), the closer to this object it is situated (at very large distances the attraction can be neglected, as it is very close to zero). However, when the distance between the objects is less than critical, a strong repulsion will appear in place of attraction. The system will reach a relatively stable state when the object experiencing the influence finds itself exactly at the critical distance from the object, constituting the source of the force. Additionally, it should be decidedly emphasized that the situation between the objects is fully symmetrical: both objects in any situation exert on each other the same, although oppositely directed force (influence). Therefore, the objects exerting force and experiencing force can be exchanged with each other, and this would change nothing. The apparent lack of symmetry results from the difference of masses between objects: it is simply more difficult to "move from a place" a heavier object using the same force, while it is easier to set in motion lighter objects. Because of this (and only because of this), it seems that, e.g., an atomic nucleus attracts an electron, while the earth attracts a spacecraft, and not inversely.

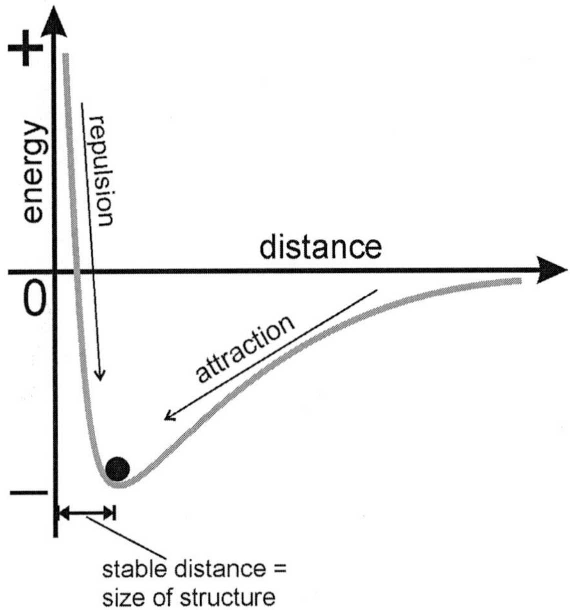

Fig. 1-1. Dependence (gray curve) of the energy of the system on the distance between an object experiencing influence (force) (little black ball) and an object exerting force (situated at distance 0, that is in the place of intersection of the vertical axis *y* and horizontal axis *x*). The system spontaneously tends toward minimization of energy, and therefore attraction will predominate at distances greater than the stable (critical) distance, repulsion will prevail at distances less than the stable distance, while the system will remain in equilibrium (or oscillate around it because of disturbances from the surroundings) when the distance between the objects will equal exactly the stable distance.

As a result, the reciprocal interaction between objects constitutes a balance between attractive and repulsive forces. Both forces become weaker with the distance from the objects generating them (in most cases); therefore, the closer the objects are located, the greater force they exert on each other. In order to reach an equilibrium (balance) between them at a certain distance, the strength of repulsion should be stronger at very small distances, but it should fall, with distance, quicker than the force of attraction so that attraction can prevail at greater distances. This property ensures not only the existence of a critical distance between objects, at which attraction and repulsion counter-balance each other, which predicates the existence of the structure itself, but also the dynamic stability and elasticity of this distance. It is expressed by the fact that every disturbance of the stable distance between objects (displacement of the system from the equilibrium state) results (as long as it is not too strong) in an automatic action in the opposite direction, causing the return to the equilibrium state. An increase in the distance will be compensated for by the prevalence of attraction over repulsion, while a decrease in the distance — by an intensified repulsion. Because of this, the distance between objects will oscillate around the stable equilibrium value in an environment full of disturbances (e.g., thermal movements and collisions, where the temperature is proportional to the average kinetic energy of molecules). All of the above statements can seem rather abstract and enigmatic. Therefore, I will try to explain, explicate, and exemplify.

Let us start with the structure of an atom. As it is commonly known, the atom is composed of a positively charged nucleus and negatively charged electrons. Why do electrons at all keep themselves in the proximity of the nucleus (they are bound within the atom) and not fly away into the surrounding space? Again, it is well-known that charges of the opposite sign attract each other. This asserts that (apart from conditions, when sufficient energy is supplied from the outside, see below), electrons remain in the immediate neighborhood of the nucleus. Does this suffice to explain the (frequently very complex) structure of the atom? Leaving aside the fact that electrical attraction by itself, which is, after all, identical for every electron, does not illuminate the fact that particular electrons occupy different electron shells and, within them, orbitals of various shapes, the attraction itself should cause electrons to fall on the nucleus and neutralize its charge, because of which the atom would shrink to an electrically-neutral point devoid of any dimensions and structure. However, physical studies indicate unequivocally that nothing like this happens. Moreover, it has been found that a certain repulsive force, which does not allow the excessive approach of electrons to the atomic nucleus, is responsible for this.

This force has more complicated properties than the "classical" repulsive force and is described by quantum mechanics. According to quantum mechanics, an electron bound in an atom does not find itself at any determined point but is as if "blurred" in space: it adopts the shape (form) of an "electron cloud" (called orbital) corresponding to the probability of finding the electron in a given point of space (Atkins and de Paula 2002 326-334; Chandrasekhar 1998, 35-62; Al-Khalili 2012, 52-77; Baggott 2013, 27-53). In the regions where this cloud is dense, the probability that the electron is situated there is large,[1] while in places where the orbital

[1] In reality, the electron is not located in any concrete place before the act of measurement (detection) or, strictly speaking, the act of interaction with the surroundings.

undergoes rarefication (dilution), the chance of finding the electron there diminishes. Therefore, even in the case of the simplest orbital (called orbital s), having generally speaking a spherical shape, the greatest density of the electron cloud takes place, not in its center (namely in the atomic nucleus) but on the surface of the sphere of a certain finite radius, determining the distance of this surface from the nucleus. In the direction of the surrounding space, inward and outward of this surface, toward the nucleus or outside, the density of the cloud falls to reach zero both in the atomic nucleus and, asymptotically, at the infinite distance in the emptiness of the universe. All this happens because, according to quantum mechanics, the energy of an electron bound in a given atom can adopt only strictly determined, discrete (discontinuous), quantized values of energy. An electron moves in space as the so-called matter wave. Hence, it can simultaneously pass through two slits in a septum and bend on them: the diffraction phenomenon, and then interfere with itself. A series of electrons passed singly, one by one, through such a pair of slits gives a characteristic pattern of stripes on a screen located behind the slits (Al-Khalili 2012; 78-105; Baggott 2013, 27-53).[2] Orbitals (their shapes, dimensions, and spatial arrangement) result from the fact that only for some energies (and therefore lengths of the matter wave, inversely proportional to energy) of an electron, the "hills" and "valleys" do not suppress or cancel each other ("hills" do not meet "valleys"), but enhance. The "hills" meet the "hills," and the "valleys" meet the "valleys," leading to the origination of stable "structures." Using this metaphor, an electron "goes around" (revolves) the atomic nucleus in such a way that the "hills" of the matter wave associated with this electron overlap with themselves in subsequent "rounds"—if the "hill" would meet the "valley," they would have canceled each other out. Therefore, the "circumference" of orbitals must, in a sense, correspond to the length of the matter wave of a given electron (the distance between the "hills" and the "valleys") or its multiplicity (Chandrasekhar 1998, 63-83; Baggott 2015, 105-113). The energy of an electron in an atom is proportional to its distance from the nucleus: the closer to the nucleus, the lower the energy. The lowest possible energy corresponds to the shortest distance of an electron (or, strictly, the smallest distance of the highest density of its electron cloud) from the nucleus. This specific repulsive force, forbidding the electron to come close to the nucleus excessively, and resulting from quantum mechanics (it would be difficult to find its equivalent in the classical world), enables the creation

[2] They result from the mutual enhancement of the matter waves being in the same phase (the "hills" of the wave meet "hills," while the "valleys" meet "valleys") and with the suppression (quenching) of waves being in the opposite phase (the "hills" overlap with the "valleys").

of the simplest structure (the nucleus plus electron cloud), determining at the same time the magnitude of this structure (the simplest atom) (Chandrasekhar 1998, 63-83). This constitutes the first exemplification of our general thesis that structure in the physical world is possible thanks to a certain equilibrium between attractive and repulsive forces. At the same time, the attractive force must prevail at a distance greater than some critical distance, while the repulsive force begins to predominate when the distance is lower than the critical distance. This maintains the distance between physical objects on a certain level, determining the size of the structure formed by these objects.

Why, then, do more complicated atoms exist in nature with such diverse structures determining the physicochemical properties of various chemical elements? Why don't all electrons simply gather on the orbital closest to the atomic nucleus, characterized by the lowest energy? Here, a second "repulsive force" enters the stage, namely the so-called Pauli exclusion (prohibition) principle (Atkins and de Paula 2002, 337-344; Chandrasekhar 1998, 63-83; Penrose 2002, 388-389, 428-430, 448-449; Baggott 2013, 34, 65), stating that a given energetic level (more strictly: orbital) can be occupied by only two electrons and only when they have oppositely-directed spins (that is, simplifying a lot, when one of them rotates clockwise, while the other, anti-clockwise; quantum spin is a quantum property that cannot be easily expressed in common language and that is to a certain extent an equivalent of revolving around an axis). Generally speaking, no two electrons can be in an identical quantum state, whose element is orbital, on which a given electron finds itself, and the orientation of its spin. When an atom possesses more than two electrons (this is determined by the positive charge of the nucleus and, therefore, by the number of protons contained in it), they have to look for other higher energy orbitals to occupy and, therefore, a greater distance of the maximum of the density of their electron cloud from the nucleus. Those can be spherical orbitals s, as described above, but with a greater radius, orbitals p having the shape of two tears laid opposite to each other and abutting each other with their sharp ends in the nucleus (the maximum density of the electron cloud is represented here by two points located in the center of the spherical parts of each of the "tears") (Al-Khalili 2012, 154-187; Baggott 2015; 105-107). Three such orbitals are on each energetic level oriented perpendicularly to two other orbitals p, along axes *x*, *y*, or *z*. There are also orbitals d and f with the yet more complicated spatial distribution of the electron cloud. The principles of quantum mechanics determine their shapes. Particular orbitals can be ranked according to the increasing energy associated with them, which determines the succession of the occupation of these orbitals by electrons when the number of electrons in atoms of subsequent elements on the Mendeleyev board increases. Generally, orbitals combine into groups of similar energies called electron shells. This is just the diversity of the shapes of orbitals, the location within various electron shells, and the degree of occupancy of particular orbitals and shells by electrons responsible for such a huge differentiation of (the properties of) the atoms of particular chemical elements. As an additional factor, there can be quoted the third (most classical) repulsive force, namely, the electrical repulsion between identically (negatively) charged electron orbitals, and also the attraction by the atomic nucleus, which causes "the same" orbitals to have different dimensions (and energy values associated with them) in atoms of different elements.

Generally, the combination of attractive and repulsive forces described above is responsible for the complex structure of the atom (as well as its diversity in particular elements) and its size. The greater the number of electrons the atom contains, the more high-energy, and thus distant from the nucleus orbitals they must occupy, and therefore the greater the atom (although this dependence is not so simple, as atomic nuclei containing more protons attract electrons more strongly, which causes the electron shells to shrink).

As has already been said, atoms are comprised of the nucleus and electrons. Electrons are elementary particles. Such particles can be defined as objects *devoid of structure*, that is, having the geometrical form of a point (although, e.g., string theory treats them as one-dimensional strings with a length comparable with the so-called Plank distance, vibrating in various ways) (Greene 2005, 327-375; Greene 2010, 3-22). However, this does not concern atomic nuclei that are not elementary particles and possess structure. The nuclei are composed of a certain number of nucleons: protons (positive charge) and neutrons (electrically neutral). What maintains the structure of the nucleus? Again, the answer is the same as in the case of the atom: the balance between attractive and repulsive forces. This time we will start with repulsion. Protons, like all objects endowed with an electric charge of the same sign (positive in this case), exert reciprocal repulsion on each other. It is counter-balanced by the so-called strong nuclear force, one of the four fundamental physical forces (we distinguish here strong and weak nuclear force, electric or, more generally, the electromagnetic force and gravity), to which nucleons (protons and neutrons) are subject. Strong nuclear influences are mediated by *gluons* (like photons mediate electrical influences) (Al-Khalili 2012, 199-204; Baggott 2013, 54-80; Baggott 2015, 35-60). Interestingly, the strong nuclear force between protons alone does not suffice to offset the electrical repulsion. Therefore, a certain number of neutrons, which deliver additional attraction through strong nuclear force while not contributing to increased repulsive force, is necessary for stable atomic nuclei. Indeed, only the nucleus of hydrogen (its basic isotope, protium),[3] formed by a single proton, can do without neutrons. The nuclei of all heavier elements contain a certain number of neutrons; proportionally, the higher the number, the heavier the element. Unstable (having unstable, radioactively decaying nuclei) isotopes of otherwise stable elements come, among others, from the fact that they lack one or more neutrons in their nuclei.

Does this exhaust the problem of the atom structure? No, because protons and neutrons are not elementary particles, contrary to earlier suppositions. We already know what this means—they possess structure (resembling, to some degree, "mini-atoms"). Namely, protons and neutrons are composed of already true (at least according to the present knowledge) elementary particles—quarks (Al-Khalili 2012, 154-185; Baggott 2013, 54-80). The proton is built of two quarks "up" (u) of a charge +2/3 and one quark "down" (d) of a charge -1/3, which gives the resultant charge of +1. The neutron is composed of one quark u and two quarks d; as it is easy to calculate, this is the reason for the neutron's electrical neutrality. What forces contribute to the existence of the structure of nucleons? The same force is responsible for attraction as in the case of forming the structure of atomic nuclei, namely a strong nuclear force. It acts between nucleons (protons and neutrons) and inside them—between the quarks. The equivalent of the repulsive force is constituted here by the property called asymptotic freedom. It consists in the fact that at very short distances, corresponding exactly to the dimensions of a proton and neutron, the attraction between the quarks decreases with distance, instead of increasing and eventually ceases to act as they approach each other because of which they can freely move within the limits of this distance, forming a chaotic clump (of course, this description is only an imperfect "classical" approximation of the quantum situation).

We see, therefore, that the reason for the existence of the quite complicated structure of the atom is to set itself at a certain distance equilibrium (balance) between the attractive and repulsive forces, and this phenomenon recurs at three levels of hierarchy: between the atomic nucleus and electrons, between nucleons (protons and neutrons in the nucleus), and between the quarks in the nucleons. Therefore, as we already generally understand the nature of the atom structure and the reason for its

---

[3] The protium (^{1}H) nucleus contains only a proton, the deuterium (^{2}H) nucleus: one proton and one neutron, while the nucleus of tritium (^{3}H): one proton and two neutrons.

dimensions, it is time to pass on to the structure at higher levels of the hierarchy of matter organization.

The best known structure on a microscopic level, and at the same time the most directly associated with the complexity of living organisms, is the structure of chemical molecules (Chandrasekhar 1998, 63-83). These molecules are composed of atoms of the same or different elements. As the reader presumably guesses, the structure (its size, the length of the bonds between different atoms) results from the balance (equilibrium) between the attractive and repulsive force. It has already been mentioned that electrons in atoms occupy still-not-filled orbitals with the lowest possible energy (at the same time, one orbital can be occupied maximally by two electrons with opposite spins). In atoms of many (most) elements, the most outer orbitals (or groups of orbitals of the same energy, e.g., three orbitals *p*), called "valence orbitals," are not filled—there are free places for additional electrons on them (Atkins and de Paula 2002, 320-344). Such atoms possess, depending on their electrical properties (determined by the charge of the nucleus and the amount and kind of electron shells), the tendency to either accept the "lacking" electrons on the unfilled orbitals or to get rid of "free" electrons from these orbitals. When electrons are completely taken over from or given back to the surroundings, there originate negative or positive ions (in a non-ionized atom, negative charges of electrons equalize the positive charge of the nucleus). This takes place in water solutions of salts or, e.g., in many crystals (including the crystals of sodium chloride, that is, common table salt). However, frequently, when two atoms meet, and one is slightly more disposed to give away electrons, while the other is somewhat apt to their acceptance, there does not follow a transfer of electrons, but sharing ("communizing"). They become bound to two atoms simultaneously, while they are more shifted toward (there is a greater probability of finding them at) the atom that is more disposed to accept electrons.[4] Instead of two atomic orbitals, one common molecular orbital is formed, completely filled with electrons (Atkins and de Paula 2002, 363-385; Patrick 2000, 17-18). At the same time, if an atom of some element lacks two electrons to fill its most outer (valence) electron shell, it can bind (by sharing electrons with them) two atoms of some other element, of which each has one "spare" electron. The number of electrons that an atom of a given element is disposed to give away or accept is called the valence of this element (atom) (Atkins and de Paula 2002, 363-367). For instance, in the water molecule H_2O, one oxygen atom partly takes over two electrons

[4] In the case of bonds between atoms of the same element, for instance carbon, electrons (their "probability clouds") are symmetrically distributed between atoms.

from two hydrogen atoms (as a result, the oxygen atom in the water molecules becomes slightly negatively charged, while the hydrogen atoms become positively charged; the whole molecule becomes *polar* and is called a *dipole*). At the same time, it should be added that the tendency to accept or give away electrons is a relative property: an atom of a certain element taking over electron(s) from atoms of another (more "electropositive") element can give away electron(s) to atoms of yet another (more "electronegative") element. Nevertheless, electrons can be shared (communized) between two atoms of the same element with the same tendency to accept/give away electrons, as it takes place in the case of the oxygen molecule O_2, which is commonly present in the earth's atmosphere. Such bonds, characterized by sharing electrons between (among) two (or more) atoms of the same or different elements, are called *covalent bonds* (March 1992, 3-93; Atkins and de Paula 2002, 363-368).

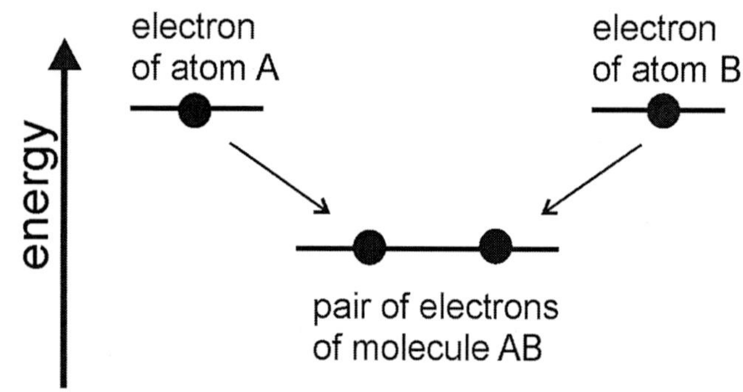

Fig. 1-2. The energy of unpaired electrons on valence orbitals in atoms A and B can be lowered after sharing electrons between atoms, equivalent to formation of a covalent bond and molecular orbital, and thus creation of a molecule of the chemical compound AB.

One could question why the sharing of electrons acts as an attractive force, which leads to bond formation. As mentioned, electrons tend to minimize their energy and occupy the atom orbitals with the lowest possible energy. It appears that forming the molecular orbital within a covalent bond from two atomic orbitals frequently causes the combined energy of electrons on the molecular orbital to be lower than the sum of energies of these electrons on atomic orbitals (the difference in energy is released to the surroundings in the form of heat during the occurrence of chemical reactions, e.g., in the process of combustion) (Atkins and de Paula 2002, 379-384). This happens in spontaneously occurring reactions and is illustrated in Fig. 1-2. One can see that electrons find themselves on relatively high-energy atomic orbitals of atoms A and B after forming a covalent bond between atoms, occupying a relatively low-energy molecular orbital AB. Besides, the above phenomenon concerns not only reactions between single atoms but also between chemical molecules; the only condition of the spontaneous occurrence of such a reaction is that the total energy of electrons on atomic/molecular orbitals of products is lower than their energy on atomic/molecular orbitals of substrates. For example, in the reaction involving the combustion of methane in oxygen, carbon dioxide and water are formed:

$$CH_4 + 2O_2 \rightarrow CO_2 + 2H_2O$$

only molecules participate, and not single atoms. Nevertheless, in many cases, more complex molecules can undergo decomposition into less complex molecules, while these then decompose into atoms if such a reaction is privileged energetically (strictly speaking: thermodynamically—see Chapter 2). A given molecule can also be transformed into another molecule of the same atomic composition (reaction of isomerization). Frequently, for an energetically "unfavorable" reaction that is associated with increased energy of products in relation to the energy of substrates to occur, it must be coupled with another reaction that is energetically (thermodynamically) "privileged," for example, the reaction of ATP (adenosine triphosphate) hydrolysis to ADP (adenosine diphosphate) and P_i (inorganic phosphate) (Marks et al. 1996, 275-290). Only the sum of these coupled reactions lowers the system's energy. Then, the newly formed chemical molecule is unstable and undergoes a spontaneous decomposition after delivery of sufficiently high energy from the surroundings (e.g., the kinetic energy of molecules proportional to temperature, higher than so-called activation energy).

Therefore, here, our attractive force that enables forming of the structure of molecules of the chemical compounds is a diminution of the system energy, where the excess of energy is released to the surroundings in the form of heat.[5] Similarly, a golf ball placed in a bowl tends to be in the most

---

[5] In principle, in spontaneously occurring reactions, there takes place a negative change of the so-called Gibbs free energy (Nicholls and Ferguson, 2002, 31-50), being a derivative of heat production and increase in the system entropy. For a given chemical reaction, this thermodynamic force can have a negative, positive, or zero value (in the last case we deal with the state of thermodynamic equilibrium), depending on the actual ratio of the concentration of products to the concentration of substrates related to the equilibrium ratio of these concentrations. Our

low-energy state, resting on the bottom of the bowl. On the other hand, what about the repulsive force, causing atoms not to approach each other too much? Here, the matter is simpler—this force is constituted by the aversion of electrons to occupy the same orbitals (to occupy the same places in space and generally the same quantum states). Therefore, when orbitals are fully filled with electrons, orbitals begin to overlap, which causes strong repulsion (the already mentioned Pauli principle). To express this in terms closer to classical mechanics, the negatively charged electron shells of different atoms repulse each other.

It should be stressed once more that covalent bonds are not stiff "sticks" joining the atoms but rather (although this is a broad analogy) springs, which can oscillate around the stable resting state (corresponding to the characteristic bond length). This is equivalent to the ball moving in Fig. 1-1 to the left and right in relation to the lowest location in the depression. These oscillations are caused by disturbances from the surroundings, particularly by the thermal movements of molecules, and are related to their collisions with other molecules endowed with kinetic energy; therefore, their intensity increases with temperature. They can also be excited due to the absorption of a quantum of electromagnetic radiation (photon) of an appropriate wavelength (energy). As long as the energy of such disturbances does not exceed a certain threshold value, the "springs" of the bonds shrink and expand around the resting location, and the amplitude of these oscillations is proportional to the energy of the disturbances. However, when, for example, temperature (that is, the kinetic energy of movements, vibrations, rotations, and collisions) is elevated sufficiently, or the matter is exposed to electromagnetic radiation of sufficiently high energy, the ball from Fig. 1-1 will fall out of the depression, and the spring will be broken. This will lead to the decomposition (or transformation) of the molecule (into) other (frequently simpler) molecules or atoms. In this way, a chemical compound undergoes degradation. The mechanism of melting solid bodies (their transformation into liquids) is quite similar, but here, covalent bonds are not most frequently broken, but rather ionic bonds (discussed earlier) or polar bonds (that are described in the later part of this chapter).

It should be clearly emphasized that Fig. 1-1 constitutes, at least in relation to interactions between atoms and molecules, a significant simplification. If atoms or molecules always attracted each other at a distance greater than the stable critical distance corresponding to minimum energy, all atoms/molecules that could potentially react with each other would instantly do this, forming

---

considerations concern physiological (appearing in living organisms) concentrations of reagents. More details are given in Chapter 2.

molecular bonds and sharing electrons, and no further chemical processes would occur. However, this does not happen; chemical reactions, even those advancing spontaneously that are thermodynamically privileged, occur with a finite, often very low speed. A kind of "hump" in the dependence of the system's energy on the distance between atoms/molecules is responsible for this. This is demonstrated in Fig. 1-3. This hump, the existence of which is responsible for the electrical interaction between negatively charged electron shells and orbitals, so that when atoms/molecules approach each other, a strong repulsion takes place between them until a certain critical limit (distance) is crossed, behind which the strong attraction between the atoms/molecules related to forming molecular orbitals and covalent bonds takes place. However, if so, why is the energetic barrier corresponding to the "hump" ever crossed, or, in other words, why do atoms and molecules react at all? The energetic barrier can be transgressed if energy greater than the so-called *activation energy*, corresponding to the height of the "hump," is supplied to the system (Marks et al. 1996, 104-108; Voet et al. 1999, 288-289; Murray et al. 2012, 57-69). Such energy can be the kinetic energy of atoms/molecules proportional to their mass and velocity and directly associated with the strength of their collisions with other atoms/molecules or of quanta of electromagnetic radiation. This is just the increase of the kinetic energy of atoms/molecules that causes the rise in temperature to speed up chemical reactions—more accelerated atoms/molecules can jump over the discussed "hump" more easily. Does this mean that when the temperature is lower than some critical temperature, the reaction does not proceed at all, and when it is higher, the reaction proceeds at the maximum possible velocity? Not at all, as particular atoms/molecules in a system with a determined temperature have very different speeds and kinetic energies (the distribution of kinetic energy in the population of molecules at a given temperature is described by the bell-shaped Gauss curve). This means that most atoms/molecules have energy close to the average kinetic energy of the system, while atoms/molecules with greater or less energy are much rarer. At lower temperatures, only very few molecules acquire energy greater than the activation energy due to accidental sequences of collisions with other molecules and therefore enter (participate in) chemical reactions. The proportion of such molecules rises very intensively with temperature, which causes the velocity of chemical processes to increase two-three-fold whenever the temperature is elevated by 10 degrees centigrade (or Kelvin degrees)[6] (Voet et al. 1999, 288-289; Murray et al. 2012, 57-69).

---

[6] This is the so-called absolute temperature, whose scale begins from zero 0 Kelvin degrees (K), at which there are no movements of atoms/molecules. In the Celsius scale, this corresponds to the temperature of about -273 degrees centigrade (°C).

One more matter remains yet to be explained. We stated previously that a high temperature can lead to the breaking of chemical bonds, while now we say that elevation of the temperature promotes forming of such bonds. Is this not a contradiction? Fortunately, it is not. Please pay attention to the fact that the "hump" (activation energy) in Fig. 1-3 is much lower than the depth of the depression corresponding to the stable distance related to the covalent bond length. Therefore, a much lower kinetic energy (temperature) is needed for the little black ball (atom/molecule) to jump over the "hump" and fall into the depression than the energy (temperature) necessary to cause (make probable) the ball to jump out from the depression. In other words, a moderate rise in temperature can accelerate forming a given chemical compound, while a very high temperature will bring about its decomposition (in extremely high temperatures, atoms lose their valence electrons, undergo permanent ionization, and the matter adopts the form of plasma; in yet higher temperatures, atoms can disintegrate to protons and neutrons, or even free quarks).

Nevertheless, this depends on the favored (spontaneous) direction of the reaction course. For many reactions, the depression left to the "hump" shown in Fig. 1-3 is much shallower than the broad valley to the right of this "hump." This means that the privileged direction of the reaction is, in this case, the jumping of the ball over the "hump" from the left to the right side and, therefore, the decomposition of a chemical molecule into simpler molecules or atoms. It also happens that the spontaneous direction of a given reaction is reversed when such a reaction (e.g., the synthesis of a complex chemical compound) is coupled with another spontaneously-occurring reaction (e.g., ATP hydrolysis), as mentioned above. This is the strategy ubiquitously applied by living organisms. This coupling of not-spontaneous reactions with spontaneous reactions takes place with the aid of enzymes, which otherwise serve to lower the discussed "hump" in the form of activation energy, that is, to speed up the chemical reactions (Marks et al., 1996, 104-108; Voet et al. 1999, 288-289; Murray et al. 2012, 57-9) (see below, Chapter 3).

Incidentally, the "hump" shown in Fig. 1-3 also appears in the case of the synthesis of the nuclei of heavier elements from nuclei of lighter elements, and the latter—from hydrogen nuclei (that is, protons; neutrons in the nuclei of heavier elements originate from the combination of a proton and an electron,[7] from where their electrical neutrality derives due to the

[7] In reality, this is somewhat more complicated. The weak nuclear forces cause the transformation within a proton of one quark u (charge +2/3) into one quark d (charge -1/3), because of which the proton is converted into a neutron. The charge conservation principle requires an emission of boson W^+ (one of particles carrying

mutual canceling out of the positive and negative charge). This synthesis (called thermonuclear synthesis and composed of many stages) occurs in the interior of stars, where gravitation compresses hydrogen to such a degree that enormous pressures and temperatures prevail (Baggott 2015, 121-145; Sparrow 2016, 72-75), which enable the repulsion between positively charged protons that is, in this case, responsible for the existence of the discussed "hump" to be overcome. The huge amount of energy released in the process of thermonuclear synthesis testifies that, generally, despite the presence of a huge "hump," the process of origination of heavier elements is, *per saldo*, very privileged energetically. This is confirmed by the fact that heavy elements, ejected into cold cosmic space during supernova explosions and forming so-called planetary nebulas (having, in fact, nothing to do with planets) in the terminal stages of the evolution of less massive stars, do not undergo spontaneous decay (decomposition), but participate in forming planets, moons, and (in some cases) living organisms.

The discussed "hump" does not appear in the case of the binding of the electrons in an atom—at the moment when (about 380,000 years after the Big Bang) matter in the expanding universe became sufficiently cool and rarified (diluted), free electrons bound very quickly to hydrogen and helium nuclei (nuclei of only these elements originated in the Big Bang). The remainder after this event is the so-called relic microwave background radiation—the binding of electrons in atoms (the *recombination* phenomenon) caused the universe to become transparent for electromagnetic radiation (earlier it remained in thermodynamic equilibrium with plasma containing free electrons), which since that time propagated freely in space (Baggott 2013, 102-128; Baggott 2015, 61-87).

While just after the recombination, electromagnetic radiation was very hot (high-energy, of a short wavelength), it cooled down greatly due to the expansion of the universe, and today its temperature equals only about 2.7°K.

Let us return to the most interesting in this context atoms and molecules. Already atoms or molecules of simple chemical compounds can organize themselves into relatively large, frequently macroscopic structures such as crystals and metals (Chandrasekhar 1998). Crystals are three-dimensional structures possessing one of several (strictly speaking: 14) distinguished

---

weak nuclear force) (charge +1), which in turn is decomposed to positron (anti-particle of electron) (charge +1) and electron neutrino (charge 0). The positron instantaneously annihilates with an electron (charge -1), due to which high-energy electromagnetic radiation is emitted. So, the aggregate reaction is: proton + electron → neutron + electron neutrino + hν (energy of the quantum of electromagnetic radiation).

kinds of symmetry. This symmetry results from the regular arrangement of atoms or molecules (or ions) in space, due to which the same structural unit is repeated periodically in determined spatial directions. The structure of a crystal is maintained either by ionic bonds (e.g., sodium chloride, that is, table salt) or by covalent bonds (e.g., graphite crystals made of carbon atoms) or else polar bonds (e.g., water ice). If these bonds are broken, the crystal breaks up, determining the fragility of crystals and their little resistance to deformation. Large crystals of table salt or quartz (SiO_2) are commonly known. However, microscopic crystals are ubiquitous in nature. In particular, most rocks forming the earth's crust are produced due to various geological processes—agglomerations of mixed-in and various combinations of crystals of different minerals.

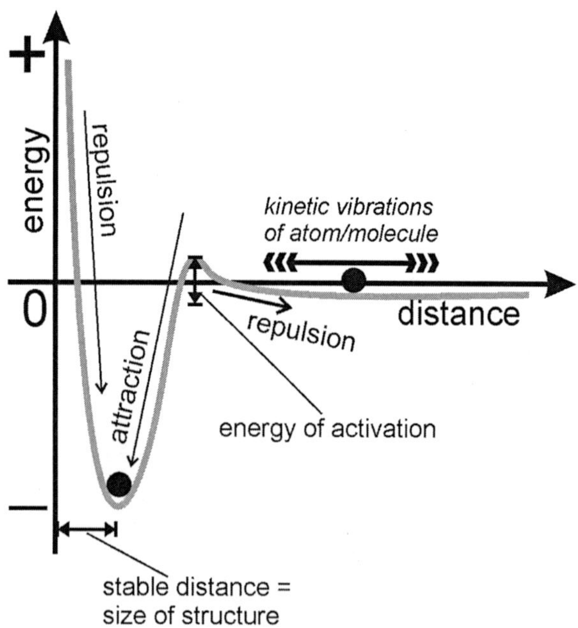

Fig. 1-3. Modified (made more realistic in relation to Fig. 1-1) dependence (gray curve) of the system energy on the distance between atom/molecule being the subject of force (little black ball) and atom/molecule being the source of the force. For the rapprochement (close up) between atoms/molecules and chemical reaction to occur, the "hump" corresponding to the activation energy must be jumped over by the ball (atom/molecule). The energy necessary for this can be delivered by thermal vibrations, movements, and collisions of atoms/molecules in the system. The rest as in Fig. 1-1.

As for metals (in a pure state), they form, in a sense, single huge molecules, as all unpaired (valence) electrons in a nugget of metal are shared among all atoms entering the composition of the nugget, which forms something like one enormous molecular orbital (Chandrasekhar 1998). This underlies certain features of metals as good electrical and thermal conducers and some of their magnetic properties. Hence, the great susceptibility to deformation: bending, shaping, and hammering (hard steel is produced by adding some other elements, for instance, carbon, to iron).

However, although crystals and metals possess a macroscopic structure, it is immeasurably monotonous. The same arrangement of one (metal, some crystals) or a few (other crystals) atoms is repeated in the whole bulk of metal or crystal many times. The rudimentary "structure" appears on only one level of the hierarchy. Therefore, this is certainly not the way that leads to the complexity of the structure of living organisms.

Structure in organic chemistry, biochemistry, and biology

In search of the roots of the diversity, complexity, and hierarchism of the structure of living organisms, we have to return to the molecules of chemical compounds, in which atoms are connected by covalent bonds. Most of these molecules are built of a few or a dozen atoms of two or more elements. As a matter of fact, only one element forms more complicated chemical compounds. This exceptional element is carbon. At the same time, this is about only one specific class of carbon compounds, called organic compounds. Organic compounds are significantly reduced,[8] i.e., containing hydrogen, carbon compounds, in which the C-H (carbon-hydrogen) bonds are present.[9] Apart from these two elements, organic compounds, chiefly those entering into the composition of living organisms, can also contain oxygen (O), nitrogen (N), phosphorus (P), sulfur (S), and various metals (iron, magnesium, copper, zinc, and others) bound non-covalently to them. It is commonly known that organic compounds, particularly large molecules of proteins, nucleic acids, fats, and complex sugars, constitute the basis of the structure of living organisms. Sometimes the chemical "formula of life" CH_2O is quoted, half-seriously, half-jokingly. It means, first of all, that organic compounds entering into the composition of living organisms are

---

[8] Reduced compounds (or metal ions) are, generally speaking, compounds with bound electrons or hydrogen atoms that easily give away electrons, while oxidized compounds are compounds without bound electrons or hydrogen ions.

[9] The C-H bond does not appear in urea, traditionally counted among organic compounds. However, for me personally this is a good reason not to classify urea as an organic compound.

moderately (but not maximally) reduced carbon compounds. Methane, CH_4, is a maximally reduced compound of carbon, while carbon dioxide, CO_2, is a maximally oxidized carbon compound. What is this special thing in organic compounds of carbon that allows them to form a complex, hierarchically-organized structure?

Carbon atoms exhibit three fundamental properties predisposing them to be the "carcass" or "structural basis" of complex, enormously diverse, and multi-functional organic compounds. These are the "spatiality" of valence electron orbitals, responsible for forming bonds in three dimensions, the ability to form chains composed of a few, dozen, or more carbon atoms (carbon "skeletons" of organic compounds), and moderately electro-positive/electro-negative properties, allowing forming bonds both with more electropositive (apt to give away electrons, such as H) elements and more electronegative (apt to take over electrons, such as O, N, or S) elements. Previously we were talking about spherical orbitals s and three, oriented perpendicularly to each other, and orbitals *p*, which have two tears that contact the sharp ends in the atomic nucleus. So, in the carbon atom, the so-called hybridization of one orbital s and three orbitals *p* is called sp^3 hybridization (Patrick 2000, 1-7). As a result, a complex orbital (sp^3) composed of four "tears" originates, whose sharp ends contact each other in the atomic nucleus, while the rounded ends (and the maxima of the density of the electron clouds situated within them) lie on the apexes of the solid geometrical figure called a regular tetrahedron. Equilateral triangles constitute the walls and base of this solid figure. Therefore, it resembles a pyramid with a triangle base. The hybridized orbital sp^3 is presented in Fig. 1-4. In the carbon atom, each of these "tears" corresponds to one unpaired electron and can form one bond by sharing this electron with another atom (of carbon or another element), having unpaired electron(s) on its valence orbital(s). Therefore, the carbon atom is tetravalent and thus can potentially bind four other atoms (the number of bound atoms can be fewer if a double or triple bond appears in the molecule). Importantly, there is not only the great number of potential bonds of the atom of carbon but also their spatial arrangement. The tetrahedral arrangement of the "tears" in the sp^3 orbital causes the "arms" of this orbital (and thus the covalent bonds formed by it) to be spaced out in all directions in the three-dimensional space. In fact, the tetrahedral arrangement implies the maximum possible spacing out and displacement from each other of the four "tears." This has principal importance for creating complex spatial structures by organic compounds.

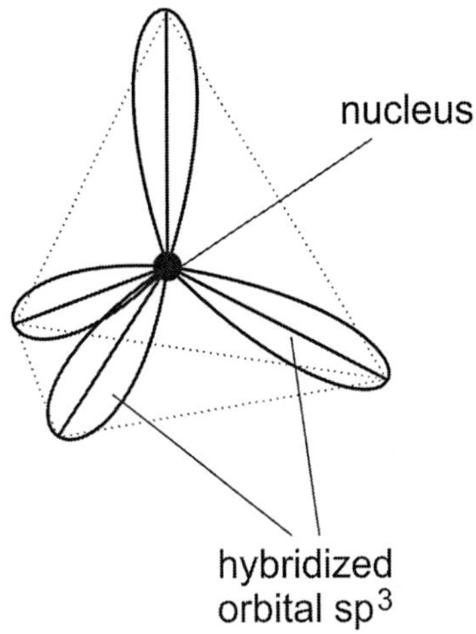

Fig. 1-4. Tetrahedral shape of the hybridized orbital sp^3 in the carbon atom.

The second of the aforementioned properties of carbon atoms is their ability to join into chains composed of many atoms of this element. This ability results mainly from the tetrahedral structure of the valence orbital of the carbon atom and the energy of carbon-carbon (C-C) bonds, ensuring the relative stability of such a bond (this energy results quantitatively from the structure of the entire carbon atom and can be derived from the laws of quantum mechanics).

The two above-mentioned properties (the shape of the sp^3 orbital and stability of the C-C bond) allow for the formation of a carbon chain (skeleton), in which two bonds in each carbon atom (one bond in terminal atoms) form bonds with other carbon atoms, while the remaining two (three) bonds can serve for joining of chemical groups of different properties. The carbon skeleton of a molecule can be linear, branched, or closed into a ring (cyclic) (in ramified or cyclic molecules, one carbon atom can bind three other carbon atoms). Atoms of other elements, such as nitrogen, oxygen, and sulfur, can enter into the composition of such linear chains or rings (in the case of ring-containing compounds, they are called heterocyclic compounds). In turn, the side "functional groups" can be hydrophobic (avoiding water), as bound hydrogen atoms (-H), hydrophilic (liking water, willingly finding themselves in its company because of the polarity of both these groups and water molecules), as, for example, the hydroxyl group (-OH) and thiol group (-SH), negatively charged: phosphate group ($-PO_4^-$) and the carboxyl group ($-COO^-$) or positively charged: amino group ($-NH_3^+$), metal ions (iron, Fe, magnesium, Mg, copper, Cu, zinc, Zn, and others) (they join to the carbon skeleton through other atoms, e.g., sulfur or nitrogen, as it takes place in chlorophyll, cytochromes, hemoglobin, and so-called iron-sulfur centers). As I mentioned, the rings can contain atoms other than C, namely N and O. The bonds between carbon atoms can be single and double or triple, considerably stiffening the bond structure. In so-called aromatic compounds, some electrons in the carbon ring are delocalized and distributed among many bonds (therefore, these bonds are neither single nor double, but "single with a fraction"). Not to go into detail, all this together has enormous potential for the diversity of structure and properties of chemical compounds. For instance, fats, having a long hydrophobic (carbohydrate, composed only of C and H) tail, are weakly dissolvable in water (on the other hand, the "tails" willingly contact with other "tails").[10] The dissolvability in the water by sugars, rich in hydrophilic hydroxyl groups (-OH), is much better, and the hem system binding metal (iron) ions, entering into the composition of, for example, cytochromes, hemoglobin (transports oxygen in the blood), and myoglobin (transfers oxygen in muscles), can easily accept and donate electrons (this is why cytochromes are components of electron transport chains, e.g., respiratory chain in mitochondria or photosynthetic chain in chloroplasts; see below, Subchapter 2.10).

Fig. 1-5 presents an example of a very simple molecule of amino acid glycine, in which the carbon chain is extremely short (two carbon atoms) and which contains three separate chemical groups: a carboxyl group (-COOH), an amino group (-NH_2), and a single hydrophobic hydrogen atom bound to carbon (-H) (the carboxyl and amino groups are shown here in the uncharged form, in opposition to this, what was presented above—in reality, both these forms co-exist in organic compounds).

Additionally, the diversity of organic compounds is enormously increased by the possibility of forming so-called *polymers*. Polymeric compounds are composed of many recurring units called *monomers*. These units can be identical (*homopolymers*), as it takes place in the case of many polysaccharides (multi-sugars) and many artificial polymers. Sometimes homopolymers are linear, possess great physical endurance, and therefore have an important structural significance (e.g., cellulose). In other cases, monomers are connected to form a network and can serve as nutritious reserve substances (e.g., starch deposited in plant cells or glycogen gathered in the liver and muscle cells in animals). The monosaccharide (simple sugar) glucose is the monomer in the enumerated *branched homopolymers*.

[10] However, this contact is in principle forced by ousting of hydrophobic "tails" from among water molecules.

Branched heteropolymers also exist, composed of different monomers (e.g., lignin composed of various aromatic alcohols). However, from a biological point of view, the most interesting (and key for the essence and identity of life and living individuals) are polymers composed of linearly-arranged monomers, which have the same "core," entering into the composition of the stem of the polymer chain, with very different side "functional groups" (these are *linear heteropolymers*). A very general scheme of such a linear polymer containing various monomers is presented in Fig. 1-6. The linearity of the discussed polymers creates the possibility of their easy synthesis based on some record, pattern, or one-dimensional matrix (see below), while the diversity of the side functional groups is responsible for the huge range of their properties and therefore executed in living organism functions. For these functions, the sequence of various monomers in the chain is extremely important; it should be added that the number of possible combinations of monomer sequences is truly astronomical.[11] Just linear heteropolymers lie at the base of the structure and function of living organisms (individuals), and the sequence of monomers they are composed of determines their identity.

The linear heteropolymers underlying on the molecular level the present life on the earth are *proteins* and *nucleic acids* (DNA, deoxynucleic acid, and RNA, ribonucleic acid). Proteins are composed of monomers called amino acids (Marks et al. 1996, 67-128; Voet et al. 1999, 77-160; Murray 2012, 17-47). There are 20 of them (in fact, there are many more of them, but only 20 form proteins). Glycine, shown in Fig. 1-5, is the simplest amino acid, whose functional group is a hydrogen atom. The carboxyl group of one amino acid binds with the amino group of another amino acid, forming the so-called peptide bond, which is responsible for joining particular amino acids into the protein (polypeptide) chain. Particular amino acids have side functional groups of different sizes (sometimes quite large), possessing various physicochemical properties (hydrophobic, polar, acidic, alkaline, and aromatic groups). Proteins constitute the basic functional and structural component of the living cell. They, as a whole, fulfill numerous functions: catalyze various chemical reactions (and thus form the entire metabolism), transport (as carriers) different substances and metal ions through cellular membranes, are receptors of signals from outside of the cell, participate in

[11] For example, the number of combinations of the sequence of 20 different amino acids in a relatively very short (in comparison with most proteins) polypeptide (multi-amino acid) chain containing 100 amino acids equals 20^{100}. This is a number greater by many orders of magnitude than the number of atoms in the observable universe. Most proteins contain many more than 100 amino acids and a living cell contains thousands of different protein types.

the intracellular signal transduction, regulate the reading of the genetic record (gene expression), constitute molecular motors (actin/myosin, kinesin, dynein, and rotor of bacterium flagellum), transfer electrons, transport oxygen, serve as hormones, constitute repair factors (blood coagulation cascade), function as morphogens steering embryogenesis (embryonal development), constitute the mechanical skeleton of the cell (cytoskeleton: microtubules, microfilaments, mediate filaments), join cells with each other (adhesive proteins), confer the shape and cohesiveness on tissues, create tendons and cords and confer elasticity on tendons (*collagen*, the most abundant protein in our bodies, corresponds for these functions), and finally—constitute the building material of the skin and its products such as hair, feathers, nails, hoofs, and antlers (e.g., keratin). Their structure and biological role are completely determined by the sequence of amino acids in the protein chain, which specifies the way the protein chain folds (and thus its ultimate shape), its connections with other molecules (proteins, sugars), the tendency toward incorporation into cellular membranes (whose frame is constituted by the lipid bilayer = double layer), the specificity of the catalytic niche, the binding of various effectors and co-factors. Assistant proteins called chaperones help to adopt functional native conformation by newly-synthesized proteins. Sometimes, just produced proteins are "processed" by other proteins, which, for instance, cut out fragments of protein chains or cause protein phosphorylation (attach the phosphate group to proteins) before final, functional, "ready to use" proteins are produced. Of course, the function of proteins results from their structure. One can speak about the "double nature" of proteins, which, on the one hand, as initially linear (one-dimensional) chains with a determined amino acid sequence, are carriers of information (constituting "recoded" information contained in the sequence of nucleotides in genes in DNA, and then in RNA, see below), while on the other hand, after spontaneous (frequently with the aid of other proteins) adoption of the ultimate three-dimensional native structure fulfill proper determined functions. Therefore, this double informative-structural/functional nature of proteins establishes a direct connection between the one-dimensional information expressed as the sequence of amino acids in the protein chain (corresponding to the sequence of nucleotides in the genes encoding proteins) and the three-dimensional structure leading to concrete function.

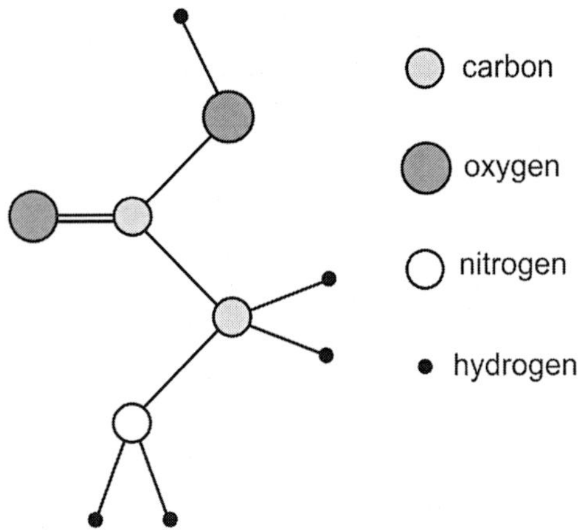

Fig. 1-5. Scheme of the structure of the molecule of a very simple organic compound—amino acid glycine. One can see here a short carbon skeleton (two C atoms) and atoms of oxygen (O), nitrogen (N), and hydrogen (H) that confer specific properties on the molecule. It is worth noticing that atoms and bonds in the glycine molecule do not lie in one plane, as the above scheme printed on a flat sheet of paper could suggest, but are arranged in three dimensions, and therefore the molecule itself has also a three-dimensional structure.

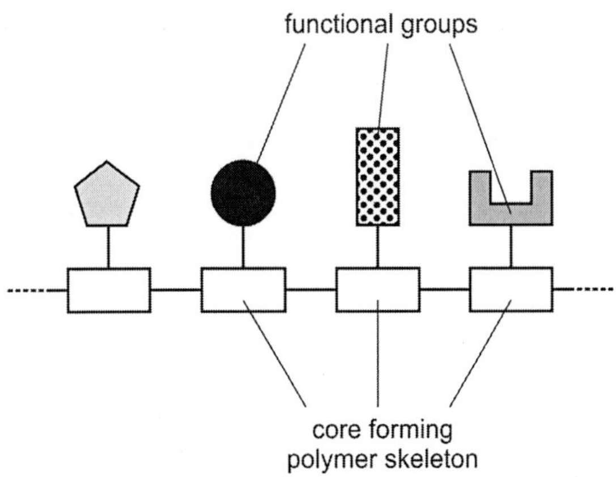

Fig. 1-6. Schematic diagram of an abstract linear polymer composed of monomers of various structure and physicochemical properties (linear heteropolymer). All its units—monomers—are composed of identical cores, forming the linear skeleton of the polymer, and side functional groups that decide about the properties of the monomer. The properties of the whole polymer depend on (are a derivative of) the sequence of particular monomers in the chain.

In turn, nucleic acids (DNA and RNA) (Marks et al. 1996, 149-166; Voet et al. 1999, 41-76; Murray et al. 2012, 343-353) contain monomers (in this case: nucleotides) called A, C, G, and T (in RNA, U appears in place of T).[12] They have (as in the case of amino acids) the same core participating in forming the stem of the nucleic acid strand (slightly different in RNA and DNA[13]) but differ by "functional groups."[14] The function of these groups mostly consists here in that the functional group C fits spatially and regarding electrical charge distribution to group G, while group A to group T (or U): the so-called *hydrogen bonds*, a variety of polar bonds, are formed between them; they are somewhat weaker than covalent bonds, but sufficient to form a stable spatial structure. As easily observed, a strand with the sequence of nucleotides -AGCAT- will fit spatially to the strand -TCGTA- (or -UCGUA-). This makes DNA form a double twisted ribbon composed of two complementary strands (the famous double helix). This enables RNA molecules to fold (curl up) into various spatial structures, e.g., the "clover leaf," whose form is adopted by the molecule of tRNA (transport RNA, see Chapter 3), in which there appear three "arms" and a "tail" built of a double RNA strand (in which single strands are complementary to each other—they possess complementary sequences of nucleotides lying opposite to each other in particular strands) and three loops ("leaves") and the end of the "tail" formed by a single RNA strand. This means one DNA strand can be copied onto the matrix of another DNA strand, which constitutes the basis for storing and copying genetic information. Finally, thanks to this, the information recorded in DNA can be transcribed to RNA (in particular–to mRNA, messenger RNA) and serve next to determine the sequence of amino acids in proteins (see Chapter 3).

The sequence of monomers in heteropolymeric linear molecules of nucleic acids and proteins is responsible for their diversity and multi-functionality, and because of this, it lies at the basis of the whole specificity, complexity, and identity of living organisms (individuals). Just this sequence determines the structure and function of biological systems,[15] constitutes the information (the genetic record) transferred from generation to generation, and underlies the structural and functional identity of living

---

[12] The tokens A, C, G, T, U are used both in relation to nucleotides and nitrogen bases entering into their composition.

[13] In RNA, a simple sugar ribose appears alternately with a phosphate group, while in DNA, ribose is replaced by deoxyribose.

[14] These are nitrogen bases: purines (A – adenosine and G – guanine) and pyrimidines (C – cytosine, T – thymine and U – uracil).

[15] On the molecular level, from which other levels emerge.

individuals based on it.[16] It is argued below that the elements from which biological evolution starts at the moment of the origination of life had to be sufficiently diverse and complex so that the living organisms could gradually (from generation to generation) self-organize and elevate their degree of complexity during this process. The entire structure and function of living organisms are created by a set of proteins whose amino acid sequence is recorded in DNA (in the form of the sequence of nucleotides in genes). As we see, the ability to form linear heteropolymers by organic compounds has key importance for the origin, existence, and very essence of the phenomenon of life in general and living individuals in particular (and also specific for these individuals' properties and identity, distinguishing them from other individuals).

Therefore, we can clearly and decidedly state that, ultimately, practically, the whole macroscopic structure and function of complex living organisms is a derivative of the sequence of monomers in linear heteropolymers—nucleic acids and proteins. As we will see, this has a crucial significance for the origin and evolution of life.

Before we move on to structures created by assemblies of organic macromolecules, we must mention the various bond types at the molecular level (March 1992, 3-93). Above, we discussed covalent bonds, in which unpaired electrons from valence atomic orbitals are shared between atoms forming the bond. Apart from this, as already mentioned, there are ionic bonds resulting from the electrical attraction between atoms or chemical groups endowed with opposite (positive and negative) charges (due to getting rid of electrons to or binding electrons from the surroundings) (that is, simply ions). Slightly subtler are the polar bonds, in which the end of one polar molecule (chemical group) with a predominance of the positive charge attracts (and is reciprocally attracted by) the more negatively charged end of another molecule (chemical group); here, the molecules remain electrically neutral as a whole, in contrast to ionized molecules participating in the ionic bonds. A certain type of polar bond, in which a hydrogen atom participates from one side, and an atom of oxygen or nitrogen from the other side, is called a hydrogen bond, was also mentioned. Very weak polar bonds resulting from quick oscillations of electrical charges distributed over molecules are van der Waals interactions. Most ephemeral are the so-called hydrophobic "bonds" (or interactions), which, in fact, are not bonds or

[16] Nevertheless, a continuity of some elements of the structure and function, in particular the whole genetic apparatus, composed of nucleic acids and proteins, is necessary for reproduction and inheriting of properties, for construction of the structure and function of the descendant individuals (see below).

interactions at all. I will explain this using the example of a fat drop suspended in water. As is commonly known that fat does not mix with water and does not dissolve in it; just the opposite, fat molecules group together, forming drops or "spots" on the water's surface. This results from the fact that dipolar water molecules willingly mutually attract each other (polar bonds are responsible for this; water molecules are dipoles, as more electropositive hydrogen molecules are situated in them on one side of a more electronegative oxygen atom) and remain in their own company, while they interact much more weakly with fat molecules, whose main part constitutes a non-polar, hydrophobic (hydrocarbon) "tail." As a result, fat molecules are ousted from mutually-attracting water molecules, become ostracized, and in the face of this, concentrate into drops to reach the least possible area of contact with water. For this reason, it is not the attraction between fat molecules that decides their grouping into "structures" (drops and spots) but the fact that they are "unwanted" by the water environment. In other words, a homogeneous mixture of water and fat is intrinsically unstable and spontaneously undergoes self-organization (a spontaneous breaking of the "isotropic symmetry" takes place) into a system having a certain macroscopic structure. The minimal contact area between water and fat molecules corresponds to a minimal energy (and maximal entropy) of the system.

The protein chain can fulfill its biological functions only after adopting an appropriate spatial conformation; that is, after folding into a native coil, within which sub-structures are present, for instance, so-called spring-shaped α-helices or β-structures resembling an accordion. Polar, ionic, covalent (so-called bi-sulfur bridges -S-S- between the "functional groups" of amino acid cysteine, containing the -SH group), and hydrophobic bonds participate in the spontaneous formation of a spatial structure by a protein chain (supporting proteins, chaperons, can help in this process). In adopting the ultimate spatial configuration of the protein chain, minimization of the system energy applies. Single proteins (polypeptides) can join into complexes composed of a few, dozen or so (or even some tens), identical or different protein sub-units (e.g., hemoglobin is a tetramer—a complex composed of four sub-units of two types, while cytochrome oxidase is composed of several different sub-units). Here, the main role is played by polar and covalent (bi-sulfur bridges) bonds. Hydrogen (polar) bonds bind two complementary DNA strands in the double helix. The lipid-protein membranes present in living cells (the cellular membrane surrounding the cell, nuclear membrane, outer and inner mitochondrial membrane, endoplasmic reticulum, and Golgi apparatus) owe their structure to hydrophobic and polar interactions. In such a membrane, the hydrophobic lipid "tails" from both layers of the lipid bi-layer, constituting the frame of the membrane, are turned inward to each other and make contact in the middle of the membrane, while the hydrophilic "heads" are turned outward toward polar water molecules (the transverse arrangement is: head-tail/tail-head). In addition, in protein molecules co-creating lipid-protein membranes, the fragments immersed into the membrane (lipid bi-layer) are hydrophobic, while the fragments that protrude from both sides of the membrane, and contact water molecules are hydrophilic (polar). Also, greater complexes of proteins and nucleic acids, such as chromosomes (in which proteins serve the possibly dense packing of DNA during the cellular nucleus division) or ribosomes (molecular machines serving the synthesis of proteins on the matrix of mRNA, that is translation, composed of proteins and three or four strands of ribosomal RNA—rRNA) are based on the enumerated bond/interaction types. This also concerns proteins of the cytoskeleton (microtubules, microfilaments, and medium filaments), constituting the cell's scaffolding, determining its shape, and participating in transporting various substances and structures within the cell. Finally, these influences enable the existence and functioning of the entire cell, merging its particular elements (lipid-protein membranes surrounding various compartments filled with cytoplasm, cytoskeleton, ribosomes, and so on) into one whole. Other proteins connect particular cells (adhesive proteins: integrins and cadherins). It should be noted that all these interactions are only varieties of the electrical force: attractive interactions occur between electrical charges of opposite signs, while the relative incompressibility of particular structures results from the repulsion between electron orbitals that find themselves too close to each other (in quantum formulation from Pauli's principle). In addition, all structures of higher order (such as tissues, organs, and whole multicellular organisms) are based on the electrical force that predominates in forming the spatial organization of living systems from the level of a single atom to entire complex organisms (individuals). Gravity "only" keeps living organisms at the earth's surface, imposing firm legs in animals and trunks in plants and efficient blood circulation system in animals and vascular system in plants—they transport blood and water, respectively, against the force of gravity.

Even more, all objects on the earth in the scale of magnitude comparable with living organisms (e.g., stones, snowflakes) are also based on the balance between electrical attraction, leading to lowering of the system energy, and repulsion responsible for the incompressibility property.

Structure in astronomy

And what about objects in a much greater, astronomical scale of magnitude: planets, stars, and galaxies? Where do their structure and size come from? Also, we have the same repeating motif: balance between attractive and repulsive forces. However, in the case of heavenly bodies, the only important attractive force is gravitation (Green 2005; Baggott 2013, 2015; Sparrow 2016). To be sure, the electrical force is greater by tens of orders of magnitude than the gravitational force, but astronomical objects contain a practically identical amount of positive and negative charges and therefore are perfectly electrically neutral; the attraction between charges of opposite signs and repulsion between charges of identical signs cancel each other, and there remains on the scene the much weaker gravitational force. On the other hand, the repulsive force has a different character in the case of different astronomical objects. Planets (including the earth) and moons were formed under the influence of gravity, leading to matter concentration in a relatively small volume. This matter was not compressed even more because of the electrical repulsion between negatively charged electron shells (orbitals) (alternatively, in the frame of quantum mechanics, the impossibility of overlapping orbitals, that is Pauli's principle). The condensation of matter due to gravity enables the formation of covalent bonds between atoms, thanks to which there are, for example, rocks. Under the influence of gravitation, there followed stratification (layering) of substances forming, for example, the earth. The heaviest iron underwent melting (as in a metallurgical furnace) and formed (with a slight admixture of nickel) the earth's core; the lighter compounds of silicon, oxygen, magnesium, aluminum, calcium, and other metals entered into the composition of the earth's mantle and crust; still lighter water formed the hydrosphere (oceans and seas), while the lightest gases (first of all nitrogen and oxygen) entered into the composition of the atmosphere.

The diameter of stars (and their spherical shape) is a result of the competition of two forces: gravitation that compresses the matter of a star (mainly hydrogen) to the center of gravity, and pressure of radiation emitted due to thermonuclear synthesis that pushes the star matter outward. Under certain conditions (in terminal stages of the evolution of massive stars), when gravitation is exceptionally strong, while the pressure of electromagnetic radiation is weakening due to the draining out of nuclear fuel, the matter reaches such a density that gravity becomes strong enough to overcome the incompressibility of matter resulting from Pauli's principle acting on the electrons. Consequently, electrons in the star's core become by force "pushed into" protons, and thus neutrons are created (the process of combinations of protons and electrons into neutrons was described above). The whole star's core of enormous density is now composed almost exclusively of neutrons and resembles a single monstrous atomic nucleus. This is how the so-called *neutron stars* originate. Astronomers observe neutron stars as so-called *pulsars*: objects revolving around their axis with a great speed and emitting at each revolution a pulse of radio radiation (as a matter of fact, a pulsar emits a beam of electromagnetic radiation continuously, but we can see it only when it is directed, due to rotation, toward the earth). The repulsive force, preventing the further collapse of a neutron star under the influence of enormous gravitation, is here the nuclear equivalent of Pauli's principle, related to the above-discussed asymptotic freedom of quarks. Can this force also be overcome by gravity? Yes, if the original (initial) star was sufficiently massive, already nothing can stop the further collapse of its core, and the whole collapses into a black hole, whose gravitation is so strong that nothing, even light, can escape from it. Therefore, black holes are remnants after explosions at the terminal stages of the evolution of massive stars known as *supernovae*. However, black holes of monstrous sizes can also be found in the centers of most (if not all) galaxies. Billions of years ago, when the universe was still young, they constituted the core of *quasars*, the most luminous objects in the universe (a huge amount of energy was emitted by them due to falling of matter into a black hole when the matter underwent a very strong heating within the so-called accretion disk). In the case of black holes, no force exists (known by contemporary science) anymore that could counter-balance gravitation, and therefore, there should not be possible, in agreement with the superior rule discussed in the present chapter, any structure. Moreover, this takes place: according to our present knowledge (or at least in agreement with general relativity), the matter in a black hole simply collapses to a geometrical point (the so-called *singularity*, an object of zero or infinitely small dimensions), where density, pressure, and temperature reach infinite values. Nevertheless, scientists hope that future scientific theories, which would unify general relativity and quantum mechanics, can eliminate these undesired infinities.

On the other hand, the external star layers are in both cases (e.g., during the origination of neutron stars and black holes) violently ejected into cosmic space, accompanied by releasing enormous amounts of energy in the form of electromagnetic radiation. In other words, such stars explode as supernovae. They enrich interstellar space with heavier (than hydrogen and helium) elements synthesized in their interiors and during the explosion, such as oxygen, carbon, silicon, sulfur, nitrogen, and iron. In the explosion, elements heavier than iron are also formed. This is the only source of these elements in the universe (apart from forming planetary nebulas in the late stages of the evolution of less massive stars, which also deliver some heavier elements to interstellar space), as only hydrogen and helium originated in the Big Bang. Therefore, essentially the whole matter of the earth, including the substance of our bodies, was synthesized in the interiors of massive stars or during their explosions as supernoave.

Going even further, the structure and size of planetary systems (composed of a central star and planets revolving around it) and galaxies (especially spiral ones), composed of hundreds of billions of stars, are a result of the balancing of the attractive gravity force (exerted by the "visible" matter[17] familiar to us and by the so-called "dark matter"[18] of unknown nature[19]) and the repulsive centrifugal force, which tends to eject the object constituting its subject—a planet or star—from the system into cosmic space (both planets in planetary systems and stars in galaxies revolve around the gravity center of the system).

Finally, the structure and current size of the entire universe constitutes the result of the competition between gravitation and the expansion of the universe.[20] The gravitational attraction wins on shorter distances (in the scale of stars, galaxies, and clusters of galaxies), while the expansion predominates in greater scales (than galaxy clusters); additionally, this expansion seems to accelerate with time under the influence of so-called "dark energy."[21] We do not know where this expansion comes from, although it is supposed to result primarily from a kind of "phase transition" of the vacuum, leading to the so-called inflation phase, when space underwent a truly enormous extension (Hawking 1988; Weinberg 1993; Barrow 1994; Baggott 2015). Without this, the universe would have collapsed to a single point long ago.

---

[17] Constituting about 4% of the content of the universe. These are first of all quarks building protons and neutrons, electrons, and particles carrying interactions between them.

[18] Constituting about one-fourth of the universe content.

[19] Perhaps weakly interacting massive particle (WIMPs), black holes, or so-called brown dwarfs—star-like objects sufficiently cool not to emit electromagnetic radiation in the visible range, but only infrared (heat) radiation.

[20] The initial impetus for this expansion was imposed by the Big Bang, frequently identified with the beginning of the universe. This expansion adopted monstrous dimensions during the inflation phase.

[21] Of unknown nature, constituting three-quarters of the content of the Universe, and exerting a repulsive force that increases with the distance, something like negative gravity. It is sometimes called quintessence or cosmological constant.

object / constituent elements	attractive force	repulsive force
nucleon (proton, neutron) / quarks	strong nuclear	asymptotic freedom
atomic mucleus / nucleons	strong nuclear	electrical (+/+)
atom / nucleus and electrons	electrical (+/-)	electrical (-/-) (atomic orbitals; Pauli principle)
molecule / atoms	energy decrease (molecular orbitals)	electrical (-/-), Pauli principle
astronomical object (planet, star) / building material	gravitation	electrical (-/-), (incompressibility, Pauli principle)
system of astronomical objects (planetary system, galaxy) / stars, planets	gravitation	centrifugal force

Fig. 1-7. Attractive and repulsive forces between building elements, responsible for the existence of structure (and its size) of various systems (objects) composed of these elements in different scales of magnitude.

Structure in various scales—summary

Having the form of a table, Fig. 1-7 summarizes the attractive and repulsive forces responsible for the existence of the structure (and its dimension) of various systems in the scale of magnitude from quantum to cosmic. It should be emphasized once more that, in principle, the structure of living organisms has the same source as the structure of chemical molecules: the attractive and repulsive electrical forces defined according to the quantum dynamics paradigm.

Structure of living individual vs. structure of literary work

However, let us leave this cosmic digression and return to our leading motif—the genesis of the structure of biological systems. Finally, we come to the point where we are already able to explain, at least in principle, how the structure of living individuals, with all its richness, complexity, and hierarchical organization, ultimately emerges from (is based on) various combinations of spatial locations of only three elementary particles: quark u, quark d, and electron (or, if somebody prefers, of proton, neutron, and electron). Is this not a paradox that a highly complicated structure, with the entire gamut of its properties, can be only based on an extremely little diverse set of only three elements? How can something so poor in the repertoire of properties yield something immeasurably complex and emanating with the splendor of various features and aspects? Of course, the discussed transition from simplicity to complexity does not occur immediately, in one step. Various combinations of elementary particles form atoms of several tens of elements naturally appearing in nature and important for the phenomenon of life. These elements already possess a broad range of physical and chemical properties: it is enough to compare hydrogen, carbon, oxygen, iron, sulfur, and phosphorus. The diversity of chemical elements significantly exceeds the diversity of elementary particles. However, this is only the beginning of climbing the ladder of complexity.

Next, we have simple chemical compounds whose molecules are built of atoms of a few elements. Again, the variety of their properties exceeds by several orders of magnitude the diversity of the properties of chemical elements. There come, in turn, more complex molecules, among which there have primacy organic compounds (e.g., monosaccharides, lipids, amino acids, nucleotides). Frequently, monomers enter into the composition of linear and branched polymers, homo- and hetero-polymers. For instance, cellulose is a linear homo-polymer, starch and glycogen are branched homo-polymers, and hemicellulose and lignin are branched hetero-polymers. All of them, apart from lignin, are polysaccharides. Linear hetero-polymers, most of all proteins and nucleic acids, have key importance for the essence and identity of living individuals, their structure and functioning, record and reading of genetic information, cellular metabolism, and so on. Their informational content is truly astronomical, while the potential range of properties is practically unlimited.

A somewhat different way toward greater complexity was taken by crystals, but this is a "dead end"—it leads, at best, to rocks and minerals, constituting in their essence conglomerates of macro- and microcrystals. Returning to more complex organic compounds, they can form yet greater structures composed of various elements, such as lipid-protein membranes, cytoskeleton, ribosomes, and chromosomes. Various cellular organelles, including the nucleus, mitochondria, chloroplasts, endoplasmic reticulum, lysosomes, and Golgi apparatus, create the next level in the hierarchy of complexity. Finally, we come to living cells, multicellular organisms, frequently built of hundreds of types of cells, tissues, organs, organ systems, individuals of various species, populations, ecosystems, and the whole boiling with life earth's biosphere. Nevertheless, this is only a complicated system of three types of elementary particles. Additionally, apart from the level of quarks, nucleons (protons and neutrons), and atomic nuclei, the electrical force is practically the only force involved in the structure and function of living systems.

Let us consider another very complex and hierarchically organized creation, namely a literary work. I mean here both the physical carrier of the work or simply the material object: the book—an assembly of printed pages between the covers, and its semantic layer, that is, content and meaning. How can the unquestionable complexity of the work help us understand the emergence of the complexity of living organisms? First, it should be noticed that the work's entire (material) structure is based only on two elements: black dye that serves for printing letters and the white background. Therefore, the starting point is even poorer here than in the case of living organisms, where we dealt with three elementary particles. One could say that the black dye and white background are accidental properties of the physical carrier of the work, having nothing to do with its informational content. Well, then, what are the most elementary units of the literary work? Letters. Depending on the language (I mean natural ethnical languages), the alphabet contains (roughly) between 20-something to 30-something letters.

Nevertheless, the number of letters is a matter of convention. Letters can be coded in the Morse alphabet, which contains only three elements: a dot, a dash, and a space. However, in the era of computers, it is easier to refer to the binary code, in which every text can be written down in a sequence of only two signs: zeros and ones. Therefore, these are our true "elementary particles" of the literary work. According to this view, letters would constitute an equivalent of atoms (please notice that in both cases, the number of elements on both levels of hierarchy is similar: two or three "elementary particles" and several tens of "atoms"). Similarly, like physical atoms, letters in literary work form more complex creations: words, sentences, paragraphs, subchapters, and chapters, a whole book. Finally, the whole "book ecosystem," as one could half-jokingly regard a library. On each of these levels, the diversity of the combinations of elements (and thus the richness of the expressed semantic content) increases greatly. On the highest level (corresponding to a living individual), we see the masterpiece of *The Castle* by Franz Kafka. Nevertheless, all this ultimately constitutes, after all, only a certain sequence of zeros and ones.

However, I would not like to prolong this analogy between a living organism and a literary work. Huge differences exist between them. The living organism is a dynamic ("happening in time") object with a three-dimensional spatial structure, while the literary work is a static linear string of symbols (although its meaning, its semantic and emotional layers, of course, are not such a linear string of symbols). However, just this fact is not the most important for our consideration. I intended to show that even the simplest set of elements of extremely poor assembly of properties (in the binary code, idealized signs differ by exactly one property) can, when *appropriately organized*, create a very complex and diverse structure. At the same time, it should be noted that in the above statement, the word "appropriately" is at least as important as "organized." In the case of living organisms, this is about the fact that their essence is not boiled down to being immeasurably complicated structures with a hierarchical organization but consists in the fact that these structures are purposefully directed on the realization of a certain very specific task: survival and copying themselves (or strictly speaking: their identity), they "do not care" about anything else. I will return to this topic when discussing the phenomenon of life and living individuals (Chapter 6). So far, the statement will suffice that the richness of the structure of a living organism and a literary work comes from the fact that their extremely simple building elements undergo such reciprocal connection that they form an appropriately organized SYSTEM.

However, there is one more important difference between the structure of a literary work and the structure of a living individual. Namely, the text of a literary work can be equally well written down using zeros and ones and red and green dots. The particular given form of the simplest element of the text structure is here without any significance—an appropriate cultural convention will be completely sufficient, as it can be observed in analogous texts written in different alphabets and ethnical languages. A diametrically different situation occurs in elementary particles—living organisms' most fundamental building bricks. These are their characteristic, very specific properties that enable the real existence of structure in the form known to us in general and of atoms, molecules of chemical compounds, and other higher-level structures such as living individuals. It must be like this. Someone arranges the symbols (letters or zeroes and ones) in sequences that are to carry some information (but this is conventional information because of the commonly accepted meaning of symbols, resulting from a social convention); elementary particles, atoms, and molecules must self-organize themselves into more complex systems. Most of the complex and purposeful structures of living organisms originated throughout the last billions of years of biological evolution in natural selection (although self-organization plays an important role until today).

Nevertheless, this evolution had to start from something, and to the present, it works on physical material, which exhibits at least the potential ability to create complex systems. This material, on the lowest level, which consists of elementary particles and atoms, possesses a rather unique set of properties, without which the origin of the higher-order structures would be impossible. We remember what a good coincidence it is that on at least several levels of hierarchy (atomic nucleus, atom, molecule, and so on), in still greater and greater scales of magnitude, that attractive and repulsive forces counterbalance each other, thus enabling the appearance of a structure. For this to happen, the "naked" condition that the objects forming the structure attract each other at a long distance and repulse at a short distance will not suffice (please imagine the consequences of the three remaining possibilities).[22] Indeed, a unique selection of properties and values of various physical parameters include the Planck constant, the magnitude of the elementary electric charge, masses of particular elementary particles, the strength of particular forces, and many others (Smolin 1999; Baggott 2013). One tiny change in one of them would cause that, for example, the entire universe would very quickly collapse to a single point or, inversely, would instantly explode into infinity, and no stars would be formed, or they would collapse into black holes just after their formation, the existence of stable atoms of heavier elements would be impossible, and so forth, not to mention the creation of living organisms, whose existence requires a just incredibly special set of properties of the physical world (e.g., the dielectric constant of water, transparency of the earth's atmosphere for only a narrow band of electromagnetic radiation, comprising visible radiation necessary for photosynthesis, but not containing high-energy radiation, for instance, most ultraviolet radiation, which could damage living organisms). If not for the completely improbable distribution of the energy of the electron shells of the carbon atom, the thermonuclear synthesis of heavier elements (such as carbon, oxygen, nitrogen, sulfur, and metals, indispensable for the origin and existence of life) would not be possible (Smolin 1999; Greene 2004; Baggott 2013, 2015). Examples of this type could be multiplied. Where do all the above properties of the physical reality that enable structure formation come from?

Many scientists believe (or would like to believe) that they result from some deep yet undiscovered laws of nature, most preferably the only possible and logically necessary. Do such laws exist, and if so, will we ever discover them and dress them in robes of mathematical equations? At the present stage of scientific development, we simply do not know the answers

---

[22] In the case of attraction at all distances, everything would very quickly collapse to single point, in the cases of repulsion at any distance, every system would undergo rapid dispersion in space, while in the case of repulsion at long distances and attraction at short distances, objects from some areas would collapse to points, which in turn would escape from each other with an accelerating speed. In fact, according to present knowledge, the situation is to some degree similar in the case of the universe, whose expansion as a whole accelerates due to the repulsive action of the dark energy (cosmological constant), while at small (in the global astronomical scale) distances matter concentrates under the influence of gravitational attraction into stars, galaxies, and clusters of galaxies.

to these questions. Others think that our universe is one of an infinite number of universes that continuously originate in the process of unceasing inflation. Each of these universes has slightly different properties (generated accidentally at their creation), and only a few are suited as the home for life (Smolin 1999). However, this is pure speculation that can never be verified and does not meet the basic requirements of scientific methodology (Baggott 2013).

Do elementary particles of matter exist at all?

We started our considerations concerning the structure of various objects and systems in the universe in general, and in living organisms in particular, from the lowest, fundamental level—the elementary particles of matter. We investigated how various configurations of elementary particles (electrons, quarks u, and quarks d) enter into the composition of all "common" material objects in the universe, from atoms and chemical molecules, through stones and living organisms, to planets, stars, and galaxies. However, the question remains concerning the "materiality" of elementary particles themselves. What does, in fact, constitute the substance, or "matter," of elementary particles, sometimes called after all elementary particles of matter? Do elementary particles and matter itself "really" exist at all, and if so, what does this, in fact, mean?

It is obvious to us that, for example, a stone held in our hand, possessing a shape, color, weight, a determined position in space, hardness, the texture of the surface, *etc.*, constitutes a material object. This happens as the stone is built of material substance. But what about the material substance of elementary particles? If we treat such particles as point objects or vibrating one-dimensional strings (of zero thickness) (Greene 2010), what constitutes the substance of these points or strings? The answer might seem surprising: such substance simply does not exist. In my opinion, the only thing that creates elementary particles, underlies their existence and constitutes their essence is the influences (forces) exerted by them on other elementary particles. An elementary particle is simply the center, the place of the greatest intensity of the forces associated with this particle. (I already leave aside the fact that, according to quantum mechanics, an elementary particle cannot be exactly localized in space. They are in a sense "blurred" and adopt the form of, e.g., a spherical probability wave that propagates in space as a wave of matter from the point of the emission of the particle or an atomic orbital, that is a wave of matter bound to the atomic nucleus and "revolving" this nucleus in a quantum way. In such a case, the probability of finding the particle in a given place in space is determined by the wave function. The discussion in this subchapter principally ignores the wave-particle duality of elementary particles, focusing on their corpuscular aspect). Furthermore, because other particles on which these forces are exerted are also exclusively “composed of” forces, then we have, as a result, only forces interacting with other forces. Physics describes these forces using mathematical models/formula as influences propagating with the speed of light in empty space. Therefore, in reality, the stones (and living organisms, planets, and stars) are ultimately built of… emptiness. Of course, this emptiness has, according to quantum mechanics, the form of a quantum vacuum that can possess various energetic states, within which there undergoes unceasing creation and annihilation of the so-called virtual particles, in which probability waves associated with elementary particles propagate, where electrical and gravitational potentials (fields) are present. However, physics cannot answer the most fundamental question: what is the elementary particle composed of? At best, physics can provide a more or less adequate mathematical description of various objects, phenomena, and processes, but it can say nothing about their essence. Physics provides descriptions and prescriptions and applies the mathematical formalism for the (approximate) representation of the world, objects present in it and processes occurring in it, but it does not investigate the essence of the physical world and the nature of the “substance” filling this world (Korzeniewski 2015, 2017a).

Furthermore, what about, as it is commonly assumed, the most inseparable attribute of matter—mass? It should be started from the statement that two completely different types of mass are always associated and equal to each other (that means that every object endowed with mass possesses both types of mass of equal value). These are gravitational mass and inertial mass (Faller and Cook 2019). The gravitational mass is simply the size of the “gravitational charge” and constitutes the measure of the force of attraction of other objects possessing mass. For this reason, it can be identified with the gravitational force; thus, all that was said above refers to it.

On the other hand, inertial mass constitutes the measure of how it is difficult to move a given object from a place or change its movement’s velocity and or direction (vector). The equality of both mass types implies that objects that strongly attract (set into movement) other objects, themselves, relatively unwillingly undergo (change their velocity and or direction of movement) the influence of gravitation (or, e.g., electrical forces) exerted by other objects. Can the inertial mass constitute the essence of matter?[23] Certainly not (at least of such matter, we formed our imagination based on a stone held in our hand). The inertial mass of an elementary particle only determines how strongly this particle (that is, the center of forces associated with it) is "anchored" in space. Therefore, in my opinion, matter is nothing more than a combination of a set of reciprocally-exerted-on-each-other forces (gravitational, electrical) and the centers of their greatest intensities, which are nevertheless "blurred" in agreement with the quantum indeterminacy and susceptibility to these forces. As mentioned when discussing the relations between the mutual balancing of attractive and repulsive forces and structure (Fig. 1-1 and the relevant text), interactions are strictly associated with energy. In this context, the famous Einsteinian equality of energy and mass ($E = mc^2$) is not surprising.

To sum up the above considerations in a slightly scandalizing (and simplifying) manner, it can be said that all matter in the universe (and in principle, the whole universe, together with matter, energy, and the space filling it) is composed of a structured, complex void. However, such a thesis is true only when we use the notions of matter and void shaped based on our everyday, common experience. Nevertheless, physics teaches us that actually nothing like "true" matter and void exists—instead, we have only curved space(-time) filled with a set of various interactions (forces, potentials, fields) that penetrate this space(-time) (but at the same time shape it), exert reciprocal influences on themselves, and as a result form all observed structures in the universe. In this sense, all objects and phenomena in the universe mutually co-condition each other (Korzeniewski 2013b, 2015, 2017a). Nonetheless, progress in physics is associated with advancing the subjectivization of such concepts as matter, space, and time. I will return to this problem when I discuss the problem of the objective/subjective existence of the phenomenon of (the special type of) complexity (Subchapter 7.3). In short, frequently, the apparently subjective and not truly real nature of the phenomenon of the complexity of structure and function is opposed to the apparently objective and real matter composed of atoms (and also energy, space, time). In an extreme case, this opinion is reduced to the statement that only atoms (or, more broadly, the physical level of reality) exist. I will argue below that a specific type of the complexity of structure and function characteristic for living systems has an equally objective (or, if somebody prefers, subjective) character as the matter these systems are built of, space, in which this matter is dislocated within, and time, in which various configurations of matter gradually

[23] One should notice that there is nothing like "electrical inertial mass."

undergo transformations one into another (see also Korzeniewski 2015, 2017a).

It should be remembered that such objects or phenomena as an electron, orbital, gravitational force or the collapse of the wave function constitute, to a similar degree, representations of "real" objects/phenomena/aspects of the physical world and constructs (to a great degree mathematical) of our mind. We cannot decide how much "things in themselves" (*noumena*) and our minds contribute to the creation and ultimate form of these concepts. Therefore, we cannot demarcate any principal limit between the external world and the content of our consciousness!

Reference to the next chapter

Do the very laws of "dynamic" physics (quantum mechanics, general relativity) suffice to explain the origination of structure in general and the structure related to living organisms in particular, always and under all conditions? After all, while galaxies, stars, and planets appear ubiquitously in the universe, according to our knowledge, life constitutes a rather (although most probably not very) rare phenomenon. As will be discussed in the next chapter, the structure of heavenly bodies (and their fragments, for instance, rocks) is completely different from the structure of living systems. Because we are interested mostly in the latter in this book, let us stop for a while at them. Life undoubtedly requires special conditions to come into being in the universe. The only place known to us where these conditions have been fulfilled is the surface of the earth. Even here, living systems seem fragile, and destroying them is easy.

Organic substances, such as proteins or sugars, are prone to spontaneous decomposition. If we put a cube of sugar on a table, it will convert to carbon dioxide and water after some (rather long) time. The decomposition of sugar can be accelerated by an elevated temperature or microorganisms (e.g., yeasts). Therefore, if the structure of organic compounds and living systems built on them have a strong tendency to decay spontaneously, where do they come from? How can life maintain itself on the terrestrial globe and even increase its degree of complexity in the process of biological evolution? We learned only from the above considerations how structure, in general, and the structure of living organisms in particular, is at all possible in our universe. Nevertheless, we still can say nothing about why nature at all "troubles itself" with the creation of such complicated structures, why they can originate spontaneously without the contribution of any intellect (let us notice that, although the laws of physics allow for the existence of, e.g., computers and robots, then computers and robots do not originate spontaneously in nature). In other words, the very presence of a chaotic mixture of LEGO play blocks (or balls and sticks, of which spatial models of protein or DNA molecules are built) is not enough for a splendid castle built of LEGO blocks (or DNA molecule model) to originate spontaneously. We have demonstrated that, in principle, the elements of physical reality can potentially form complex, hierarchically-organized structures. However, we still do not know why, at least under certain circumstances, they really do this and what conditions must be fulfilled for the structure and function based on it, particularly the structure and function of living organisms, to originate spontaneously. The possibility of something coming into being is by no means equivalent to the fact that this something will anywhere and ever come into being. These problems will be dealt with in the next chapter.

CHAPTER 2

WHY DOES STRUCTURE ORIGINATE?

Introduction

There are even more fundamental differences between the structure of living organisms and the structure of constructions made of LEGO blocks or models of chemical macromolecules built of balls and sticks (and also the structure of computers, cars, robots, *etc.*), apart from the spontaneous genesis of the former, as opposed to the "artificial" creation of the latter by intelligent beings. First, while the aforementioned inanimate creations are static structures, "congealed," and unchangeable in time (I mean their internal structure and not, for instance, the movement of car wheels, robot arms, or work of a computer), biological systems represent dynamic structures, "happening" in time. Second, the basic level of the structure and functioning of living organisms is the molecular level, based on proteins, nucleic acids, saccharides (sugars), lipids (fats), and other compounds (first of all, organic compounds). They comprise tens, hundreds or thousands of atoms arranged in determined complex configurations. On the other hand, the basic, "lowest" elements (units) of structure and "carriers" of function in computers, cars, and robots (not to mention LEGO blocks and balls and sticks in the plastic models of macromolecules) are much bigger, supra-molecular, macroscopic creations composed of billions of atoms and molecules entering into the composition of metals, semiconductors, and various plastics, creating on the molecular level a disorderly, chaotic, homogeneous mixture, and shaped in the technological process by people, such as elements of microprocessors, chips, diodes, capacitors, and cables. The structure and function of a protein result univocally from the sequence of amino acids in its chain, while the structure and function of the elements of a chip do not result at all from their building materials but from the shape and connections of these elements imposed on them by engineers and realized during the technological process. Of course, this relates to their genesis: while molecular structures can originate spontaneously, supra-molecular structures must be shaped due to some external, purposeful, intelligent action. Third, the building materials and elements of inanimate structures are not replaced by new materials/elements and do not change in time (leaving apart a very slow decomposition process related to entropy increase), while the building material and subparts of living individuals are in unceasing, frequently cyclic motions and undergo continuous replacement. Fourth, in order for this to happen, a continuous flux of matter, energy, and information must flow through living organisms. This is just the dissipation of energy, loss of information, and production of entropy that can ensure the growth, development, and ceaseless persistence in time of the biological structures, despite various disturbances coming from the environment. Fifth, living individuals are equipped with purposeful, shaped-by-evolution functions and regulatory mechanisms to sustain the integrity of their structure and internal homeostasis during individual life. Sixth, these functions and mechanisms that cause biological systems are capable of self-repair, an example of which is the healing of wounds or fighting off parasite infections. Seventh, a living individual, with its very complex structure and function, can self-develop from some relatively simple germ, for instance, from the fertilized egg cell (zygote).[1] Eighth, finally, the structure and function of living organisms can undergo modification and sometimes an increased complexity, resulting from biological evolution, which is based on the already existing structure and function (Korzeniewski 2011). On the other hand, new creations of people must be, in principle, designed each time from the very beginning. In the process of biological evolution, all transient forms must be fully functional and constantly win the competition with other co-existing living forms (organisms). The matter is different with the devices and mechanisms people create—no such limitations are present here. Let us, for example, imagine an engineer that has, so far, produced cars and now wants to produce a plane and must gradually, step by step, rebuild a car into a plane in such a way that all intermediate forms are completely functional! Even more, this car would have to be constantly in motion! This seems completely unworkable. And yet, living organisms must cope with just such a modification and a possible increase in the complexity of structure and function during biological evolution.

A living individual as a whole is ultimately directed toward reproduction, toward self-copying (its identity), in a possibly large number

---

[1] Admittedly, for instance, in mammals the female carries the fetus in the uterus during pregnancy, feeds and protects it, and after birth feeds the neonate with milk, teaches it, and prepares it for independent life. However, the structure and function of the neonate, and later of the adult individual develop independently through self-steering by a set of control mechanisms in the process of morphogenesis during embryonic (embryogenesis) and individual (ontogenesis) development, principally without contribution of the mother.

of copies, possibly similar to the original. Living individuals (their identity) can evolve from generation to generation and, as a result, undergo drastic (although principally gradual) changes, sometimes associated with a huge increase in complexity over long periods. At the dawn of life, it emerged spontaneously from inanimate matter, together with germs of structure and function. All these properties do not concern artificial complicated static structures. Castles built of LEGO blocks, models of the DNA molecule made of balls and sticks, and cars or robots will never self-organize spontaneously. Those created by people will eventually decay/degrade (which is related to the entropy increase) due to various influences from the environment, for instance, chemical reactions with its components, chaotic collisions with various molecules, or more dramatic damages caused by the winter, water currents, rolling stones, electrical discharges, temperature variations, or (seldom) meteorite falls.

On the other hand, living organisms can maintain the continuity and unchangeability of their structure through an unceasing exchange of used and damaged elements, and their complex functioning as dynamic structures is perpetually sustained by the flow and dissipation of energy. Apart from this, living organisms reproduce so that even when some of them undergo destruction due to environmental influences or simply old age, they will be replaced by other organisms, identical or similar to them. Therefore, their identity, or an identity very similar to it, will endure. In a sense, living organisms are unceasingly "boiling" like water in a pot, with the matter and energy flowing through them driving several cyclic, hierarchically-organized processes. As we will see, this analogy is deeper than it might seem at first glance.

Not only the physical interactions between various objects (elementary particles, atoms, chemical (macro)molecules, and their complexes, cells, and so on), are based primarily on electrical forces, but also changes over time occurring in complex systems composed of these objects and interactions between them are crucial for the essence of the structure of living organisms. At the same time, it is important that, generally speaking, these changes are unidirectional and irreversible—this will be easy to comprehend for everybody that has observed developing an oak tree from the stage of seed and seedling to the adult tree or pondered over the course of biological evolution.[2] Let us see what contemporary physics says about the dynamics, unidirectionality, and irreversibility of changes in time.

[2] This does not change the fact that a huge amount of cyclic processes that keep them alive are present in biological systems.

All four fundamental forces underlying the existence of any structure in the universe mentioned in the previous chapter are described mathematically within the two greatest and most famous theories of contemporary physics: quantum mechanics and general gravity (Penrose 2002; Baggott 2013; Greene 2005; Baker 2007). Quantum mechanics concerns the world on a microscale: elementary particles, atoms, and molecules, and its domain comprises, among others, electrical (electromagnetic), strong nuclear, and weak nuclear forces (the last mediate, for instance, in the transformation of protons into neutrons and inversely). On the other hand, general relativity formulated by Einstein is, in essence, a theory of gravitation, identified with the curvature of space-time, and as such, it refers, first of all, to astronomical objects: heavenly bodies and the entire universe. As mentioned above, in the scale of magnitude appearing on the earth's surface and therefore relevant for living organisms, the only important force is electric interaction, while gravity fulfills only this trivial role that it keeps living objects on the earth's surface and, for example, impedes the transport of water to great heights by trees and flying by birds. Generally, quantum mechanics and general relativity complement each other in the description of the fundamental, elementary level of reality. The problem is that these theories are mutually contradictory—for deeply fundamental reasons, it is impossible to reconcile their mathematical formalisms and conceptual constructs (physical interpretations). Furthermore, at least one of them, and most probably both, are incomplete and will have to be replaced by some unified theory, reaching a deeper, more elementary level of reality. However, it is an entirely different problem whether such a theory will be developed at all and whether it is within reach of people's mental abilities.

However, an even bigger paradox, which is more related to the topic of this book, is the fact that the mentioned theories, being, after all, the greatest pride of physicists and the entirety of science, cannot shed light on such fundamental properties of physical reality as the amount and degree of the complication of the structure, complexity, degree of order, or amount of information, and the arrow of time related to its unidirectional flow from the past to the future, conditioning the real existence of the present (Coveney and Highfield 1991; Penrose 2002; Baggott 2013).[3] Simply put, the

[3] The arrow of time alone does not imply yet the real unidirectional flow of time, as it can appear in its absence and only point from past to future in the "congealed" block of space-time, in which the past does not principally differ from the future, and the present does not really exist. I discuss the topic of the real unidirectional flow of time in Supplement 1 at the end of this book, and in much more detail in my article (Korzeniewski 2016).

mathematical equations describing movement in the discussed theories are symmetrical in relation to time: an exchange of the past with the future changes nothing within them. Both the movement of bodies in the gravitational field and the evolution of the spatial distribution of wave function in time are, in principle, reversible (and determined). The only exception seems to be the so-called *reduction* of the wave function, its *collapse* or *decoherence* (Al-Khalili 2012; Al-Khalili and McFadden 2014; Korzeniewski 2016), which constitute not only truly irreversible but also indeterministic (probabilistic) processes/phenomena (see Supplement 1). Information, structure, complexity, ordering, and also the irreversible and *real* flow of time are key concepts for the characteristics of the essence of life (or psychical phenomena). Nevertheless, the greatest physical theories of our times have practically nothing to say about them. Therefore, the incompleteness of our physics has a deep and multi-dimensional character (Coveney and Highfield 1991; Penrose 2002; Baggott 2013).

Overview of the thermodynamic basis of life

Nevertheless, a branch of science is traditionally classified as a part of physics (although many physicists treat it as something worse, less elegant than "dynamic" theories: quantum mechanics and general relativity) that deals with the phenomena enumerated above. This is *thermodynamics* (Penrose 1990, 391-410; Coveney and Highfield 1991; Coveney and Highfield 1996, 147-181; Atkins and de Paula 2002, 3-116).

The appearance in physical theories of the arrow of time (practical irreversibility), amount of information, degree of order, and, as a consequence, the spatially and temporarily organized structure was possible because of a transition from single objects to systems composed of many (a huge number of) objects, which forced the replacement of the deterministic approach (movement in the gravitational field in general relativity, the evolution of the wave function in time in quantum mechanics) or the probabilistic approach (the collapse/decoherence of the wave function), prevailing in dynamic theories—with the statistical approach, averaging the behavior of the whole system over a huge number of elements. Such a transition is enabled just by thermodynamics, which is, in its essence, a statistical theory in which a macroscopic state (macrostate) appears due to averaging over microscopic states (microstates) of a huge number of atoms/molecules (Penrose 1990, 400-406) (I will discuss macro- and microscopic states in a while). In principle, despite the existence within it of the arrow of time, thermodynamics, most probably, is not contradictory with dynamics, including quantum mechanics and general relativity. The two last theories are simply incomplete in describing reality on the so-called physical level of reality. Something completely new (called emergent properties) appears when we pass from the behavior of single objects to their sets. This simply escapes the mathematical structure and conceptual apparatus of dynamic theories (again: quantum mechanics and general relativity) (Coveney and Highfield 1991). Similarly, the essence of the biological level completely escapes the system of concepts offered by physics (Korzeniewski 2001, 2005, 2013a) (and the nature of the psychical level escapes the conceptual apparatus of biology; Korzeniewski 2013a, 2015, 2017a). Thermodynamics says what life is from the point of view of physics and what conditions must be fulfilled for it to exist and function, but it cannot reach the very essence of the phenomenon of life. It seems clear from the above discussion that physics has great problems with a coherent description of reality, even on its "own" level. Indeed, the so-called uniform physical level (as a level, I understand here a certain layer of reality that can be described within one coherent system of notions and mathematical formalism) appears to be an illusion, existing mostly because of historical reasons.[4] There are at least two levels in physics: the dynamic and the thermodynamic levels. They are not contradictory to each other—the thermodynamic level constitutes a kind of "superstructure" over the dynamic level. The key point is that the emergence of the former from the latter cannot be described within the mathematical and conceptual formalism taken from the dynamic level (and remains largely a conundrum until the present day). The emergence (in whatever way it would occur) of the thermodynamic level from the dynamic level constitutes a very important step toward the biological level and the evolution of the living world occurring on it. First, I mean a spontaneous origination of dynamic macroscopic structures, organized spatially and temporarily.

Living organisms are dynamic structures undoubtedly highly organized in space and time. The origination of the organized structure of living organisms during biological evolution (phylogenesis) and individual development (ontogenesis) seems to contradict the universal rush to create chaos and increase entropy, which thermodynamics alludes to.

Two laws of thermodynamics are commonly known (Penrose 1990, 391-410; Coveney and Highfield 1991, 147-181; Atkins and de Paula 2002). According to the first law, energy is conserved in an isolated system: energy cannot be created from nothing or disappear without a trace. On the other hand, some forms of energy can be transformed into other forms, for

---

[4] Thermodynamics is no more similar to the "classical" dynamic physics than chemistry, and certainly less than quantum chemistry is.

instance, potential energy into kinetic energy and the inverse. In contrast, the famous second law of thermodynamics states the preferable direction of these changes. Namely, it (one of its formulations) ascertains that more ordered, capable of performing useful work, associated with low entropy forms of energy, have an immanent tendency to pass into less ordered, less able to perform useful work, associated with high entropy forms of energy. In the conditions prevailing on the terrestrial globe's surface, heat energy (equivalent to the energy of infrared radiation) is the energy form associated with greatest entropy. In other words, entropy (a measure of disorder or "chaos") must increase or, when it reaches the maximal value, remains constant in every isolated system. This is just the inevitable increase of entropy that determines the thermodynamic arrow of time and imposes practical irreversibility to most processes occurring in the universe. However, at the same time, which seems most surprising and interesting, this global rush toward disorder can give birth to order locally (Prigogine 1980; Prigogine and Stengers 1984; Coveney and Highfield 1991). This phenomenon appears not only to be not contradictory to the principles of thermodynamics but results from these principles as their inevitable consequence. It appears that, paradoxically, local "islands" of order speed up the global disorder increase.

The second law of thermodynamics refers only to isolated systems. However, living organisms are open systems that exchange matter, energy, and information with their surroundings. Only the entire system: an organism and its surroundings (ultimately—the rest of the universe, or at least the earth's surface and the sun—see below) constitutes an isolated system. This is just within this whole where the law of increasing entropy in time applies (Prigogine 1980; Prigogine and Stengers 1984; Coveney and Highfield 1991; Schrödinger 1992). If the building of the structure of an organism (e.g., during individual development) and sustaining of its existence (e.g., unceasing recreation of its elements) is associated with an increased degree of organization, and thus with a decrease of entropy, the entropy of the surroundings must increase even more, so that the total change of the entropy of the system: organism + surroundings is positive. Likewise, for someone to grow and keep themselves alive, for example, they must oxidize various nutritious substances (e.g., sugars and or fats) with the participation of oxygen, with the products of this process being carbon dioxide and water. It appears that the increase of entropy in the process of oxidation of, for example, glucose (see below) is greater than its decrease associated with the building/recreating of the body, the gathering of memory (information) in the brain, and so on. In this way, the second law of thermodynamics is saved, and the local origination of complex structures appears not to be contradictory to this law.

However, this lack of contradiction alone insufficiently explains where the organized dynamic structures in nature, such as convective currents, whirls in water or gas (e.g., twisters), or living organisms come from. Why does entropy production— the eventual dissipation of energy in the form of heat—lead to the origination, or even force the existence, of some forms macroscopically organized in space and time? Theoretical (mathematical) analysis of the laws of thermodynamics has led to the conclusion that if a given system is near thermodynamic equilibrium (its entropy is, in given conditions, close to maximal), the speed of entropy production (dissipation of energy) will reach the lowest possible value and in such a situation no structures can appear. However, this is not the case for living organisms, which "feed on" the negentropy (inverse of entropy) of the surroundings far displaced from thermodynamic equilibrium and are thus characterized by entropy much lower than maximal.

At greater thermodynamic gradients (e.g., temperature differences), the dissipation of energy proceeds in a turbulent way instead of proceeding mildly and gradually, as in states close to equilibrium. If we slightly warm up the bottom of a vessel with water, heat will be transduced between the bottom and top layer of water through the transfer of kinetic energy (velocity of movement) by molecules located lower to molecules located higher. However, when the temperature difference between the bottom and the top layer of water exceeds a certain critical value, this energy dissipation mechanism (transduction) becomes insufficiently efficient. Instead of simple transduction of heat, there will appear, discussed in more detail below, ascending (upward) and descending (downward) convective currents. An analogical situation will occur in the flow of water through a pipe—in a slow stream of water, the flow has a monotonous (uniform) character, while after exceeding a certain threshold value, the flow becomes turbulent—macroscopic structures in the form of whirls appear (Coveney and Highfield 1991, 203). A microscopic fluctuation (disturbance) forms a convective current or whirl, which grows huge very quickly (in relation to molecular), that is, to macroscopic dimensions. Once such a fluctuation appears, it appropriately organizes its surroundings (creating a properly oriented, up or down, clockwise or anti-clockwise, convective current or whirl, respectively). In this way, the system retains to some extent the memory of its history; that is, in this, and not another place, there appeared such-and-not-another microscopic fluctuation. Interestingly, the dependence of the large-scale history on microscopic, unpredictable disturbances and preserving the memory of this history in even much longer time scales also characterizes the process of biological evolution.

The phenomenon of the appearance of the dynamic structures described above, called *dissipative structures*, when the thermodynamic gradient (the amount, or rather "density" of the energy "demanding dissipation") in the system exceeds a certain limit value, was predicted theoretically by Ilia Prigogine (Prigogine 1980, 103-104, 112-119; Prigogine and Stengers 1984, 13-14, 148-150, 171; Coveney and Highfield 1996, 168, 187, 193-195, 219-225), based on the properties derived by him of non-equilibrium thermodynamics. Therefore, while thermodynamics states that every isolated system tends to the state of the highest entropy that is, the lowest ordering, at the same time, it results that on the way to reach this state, there can appear highly ordered structures, whose "role" is, surprisingly, the acceleration of the rise of global disorder. In this way, a bridge was built between physics (thermodynamics) and biology. This is because, from the point of view of thermodynamics, life is nothing else but one of many processes on the earth's surface participating in the dissipation of the energy of visual electromagnetic radiation emitted by the sun (this energy is ultimately dissipated in the form of infrared = heat radiation sent into the abyss of cosmos). Because the earth dissipates in a unit of time sufficiently great amounts of energy (producing sufficient entropy), various dissipative structures such as living organisms or atmospheric phenomena (e.g., cyclones) originate on its surface. Whatever else it is, life is *also* a dissipative structure or a hierarchically-organized complex of such structures, which can come into being only in a system intensively dissipating energy (producing entropy), such as the earth's surface. However, being a dissipative structure does not constitute an *exhaustive* characterization of life (after all, neither convective currents nor whirls are alive) and barely touches the heart of its essence (Korzeniewski 2005). Therefore, while thermodynamics quotes a condition (one of many) necessary for the coming into existence of living organisms as spatially and temporarily organized dynamic structures, it does not formulate criteria sufficient for the univocal defining of biological systems. Such criteria must be supplied by laws ruling a higher than the dynamic, or even thermodynamic, level of reality—namely, the biological level (Korzeniewski 2005, 2013a, 2015, 2017a). In Chapter 6, I will argue that cybernetics is the only formal science able to shed light on the essence of the phenomenon of life and living individuals.

Essentials of thermodynamics as a statistical science

The above considerations allowed us to look at the thermodynamic fundaments of dynamic ("happening in time") structure in general and the structure of living organisms in particular, albeit in a general way, from a certain distance. In other words, we tried to understand why matter not only has the potential to organize itself into complex systems but also actually "troubles itself" to create complicated structures. Using the paradigm of talking about structure in the previous chapter, the point is not only to demonstrate why atoms and molecules are virtually able to interact and mutually join together in such or another way, which we did previously, but first of all, to investigate why, and in which conditions, this can lead to forming more and more complex and organized *configurations* of atoms and molecules. From LEGO blocks, a bar composed of just three blocks put one over another or a splendid castle with a beautiful and complicated architecture composed of tens of thousands of blocks can be arranged. Therefore, one can pose a question of why nature sometimes (in most cases) builds from the elements accessible for it (elementary particles, atoms, and molecules plus interactions) rather simple constructions (bars) while rarely (in specific, determined conditions) it brings into existence the splendor of functional, purposeful, and hierarchically-organized structures of living organisms ("dynamic castles"). Below, we will analyze this problem in more detail.

We will start from the arrow of time which is the unidirectionality (and irreversibility) of its flow (Coveney and Highfield 1991; Greene 2004, 127-176). We all experience the flow of time in everyday life. People are born, grow up, grow old, and die; rain falls from the sky, heat flows from places with a higher temperature to places with a lower temperature, an egg dropped on the floor breaks into tiny fragments, and the universe undergoes expansion. In each quoted case, the sequence of events is determined and cannot be reversed. We would instantly recognize a movie running backward from the end. A broken cup that spontaneously joins itself from pieces into an intact whole and flies onto the table would shock us with its unnaturalness. Unthinkable then is the universe, in which galaxies, stars, and planets break down and disperse, against the gravity force, into a homogeneous mixture of elementary particles that existed just after the Big Bang. Our psychical introspection, where the future always follows the present, preceded in turn by the past, makes us sure about the unidirectional flow of (the arrow of) time.

In the face of this, it can seem strange that, as discussed above, the two greatest contemporary physical theories (their mathematical formalism), general relativity and quantum mechanics, ignore the existence of the arrow of time (Penrose 1990, 391-410; Coveney and Highfield 1991). This means that if to stop all particles/objects in the universe and run them at the same speed in the opposite direction, the course of all events would be perfectly reversed. We would observe apples "falling" up from the ground to a branch of a tree, hear thunder preceding lightning, and light rays would fall out of our eyes, reflect from various objects, and ultimately be absorbed by the sun.

The above contradiction, an obvious absurdity, has its roots among others in the fact that both relativity theory (together with its predecessor—Newtonian dynamics) and quantum mechanics (and the earlier Maxwell's electromagnetism) were created to describe single objects, such as heavenly bodies or elementary particles, or relatively very simple systems composed of at most a few elementary objects. If we consider the earth revolving around the sun or the system (hydrogen atom) composed of a proton and electron, reversing the direction of the flow of time would change completely nothing. The earth revolving around the sun in the opposite direction would not be unnatural.

On the other hand, the arrow of time appears in thermodynamics, describing the transformation of various forms of energy (e.g., mechanical, electrical, kinetic, potential, heat) from one another (Coveney and Highfield 1991, 169-178; Greene 2004, 143-176). The previously mentioned second law of thermodynamics states that an isolated system's entropy (degree of disorder) grows in time (or remains constant after reaching the maximal value). The major fact that differs thermodynamics from relativity theory and quantum mechanics is that it deals not with single objects (or pairs of objects) but with complex systems containing billions (this is a metaphor—in reality, these are much larger numbers, say 10^{25}) of elements (e.g., atoms, chemical molecules, elementary particles). For obvious reasons, thermodynamics cannot analyze the fates of all such elements in detail. Instead, it represents a statistical approach, describing the average behavior of elements in a system (and its distribution). For instance, it is possible to calculate that at a given temperature, the mean speed (or kinetic energy) of molecules in gas equals x, while the percentage of molecules within a certain determined range of speed amounts to y. Thermodynamics also states, for example, that in the system of two chambers with gas connected by a pipe, gas will flow from the chamber with a higher pressure to the chamber with a lower pressure, regardless of the behavior of particular molecules (Penrose, 1990, 400-406). Coveney and Highfield 1991, 159-162). To be sure, it is theoretically possible to imagine such a set of velocities and locations of molecules that would result in the opposite behavior, but the probability of the origin of such a set is so astronomically small that it does not pay at all to consider it.

Similarly, there is a finite (non-zero) probability that water will freeze in a pot put on a gas flame, and the egg smashed on the floor will spontaneously assemble itself into an intact whole and jump up to a hand catching it, but billions of years in the existence of the universe would not suffice for such an event to happen at least once. Let us notice that the flow of gas upward the gradient of pressure in time is identical to the normal flow of gas (from the place of higher pressure to the place of lower pressure) back in time, up the stream of its flow. Therefore, thermodynamics states no more, no less, that the occurrence of certain phenomena (equalization of the pressure in chambers connected by a pipe, the smashing of an egg) is much more probable in the direction of the flow of time than in the opposite direction. Thus, a well-expressed arrow of time appears here. Effects of this type are possible thanks to the statistical approach, adequate in systems containing (very) many elements. This is how the arrow of time emerges during the transition from simplicity to complexity, from single elements to their complicated systems, which constitutes undoubtedly a compelling puzzle of modern physics. Let us only notice that the (thermodynamic) arrow of time appears gradually when we pass from systems containing one, two, three, five, ten, a hundred, thousand, million, billion objects to real macroscopic systems, in which the number of objects is described by such numbers as 10^{25}. In a system containing only one object, for instance, an elementary particle, the thermodynamic arrow of time cannot come into existence.[5] Generally, the greater the number of elements the system is composed of, the stronger the arrow of time in this system is expressed. According to the so-called ergodic theory (Penrose 1990, 182; Coveney and Highfield 1991, 269-286), every system fulfilling certain conditions (not necessarily concerning our universe, which, e.g., expands, in which the quantum and chaotic indeterminacy take place) will eventually return to the starting point, and it will repeat its history circularly. The problem is that an essentially infinite period would be necessary for this to happen in real systems, certainly by many orders of magnitude exceeding the time of the existence of our universe.

[5] This differentiates the thermodynamic arrow of time from, for example, the cosmological arrow of time, associated with the expansion and evolution of the universe from the moment of the Big Bang into the future.

The thermodynamics of physical systems

In a system composed of many elements (say, 10^{25}) in a container with gas, it is impossible to follow the behavior of every single element in separation. The number of measurements necessary for this would be so enormous that it does not pay even to take it into account, leaving aside the fact that exact measurements and calculations would not be possible for principal reasons because of the indeterminism (and indeterminacy principle) of quantum mechanics governing the behavior of single elementary particles, atoms, and molecules. Fortunately, a detailed knowledge of the state (position and vector of velocity) of every particle/atom/molecule entering into the composition of a considered system appears, to a large extent, needless, at least from the macroscopic point of view, dealt with by *statistical mechanics* (Coveney and Highfield 1991, 169-173).

This happens because they are canceled mutually after averaging the various opposite behaviors and therefore do not manifest themselves on the macroscopic level. If we take a closed container filled with gas, particular gas molecules move in various directions with very different, frequently very large, speeds (Penrose 1990, 400-411). However, on average, the same number of molecules move in a given direction with a given speed as in the opposite direction with the same speed. For this reason, the resultant (mean) speed vector will be exactly zero, which seems obvious, as the gas as a whole does not move because of the limitations of the container walls. We can also be interested not in the resultant direction (vector) of the velocity of molecules but in the mean value of this velocity and the kinetic energy of molecules proportional to it (that is, in turn, proportional to temperature). Here, we can also determine the mean value of this energy, which will be non-zero (unless at absolute zero, in Kelvin degrees, but then the gas would freeze). Therefore, the essence of thermodynamics is the statistical approach, consisting of determining an average value of some parameter in the investigated system, expressed, for instance, in the fact that a given velocity (strictly speaking: a certain narrow range of velocities) is possessed by a certain percentage of gas molecules, while another velocity—by another percentage of these molecules.

Based on such microscopic, single-molecule parameters, we can determine that the value of some macroscopic whole system parameters has no sense when referring to particular system elements. Such a macroscopic, statistical, phenomenological parameter is, for instance, temperature, proportional to the average kinetic energy of molecules (microscopic parameter). It seems obvious that in a physical system, a determined temperature can be possessed by the gas but not by a single molecule.

Similarly, in language, the meaning can be attributed to a sentence or word but not to a single letter. Apart from temperature, the gas can be characterized by other macroscopic parameters, such as pressure (the average pressure of molecules on the wall of the container and other molecules caused by incessant collisions with the gas molecules) and volume (Atkins and de Paula 2002, 3-14). These three parameters describe in a complete (in some important sense) way the present state of the gas (assuming that this gas remains in thermodynamic equilibrium). The thermodynamic description of a system composed of many elements using macroscopic parameters seems very economical in comparison with the dynamic description—in the latter, it would be necessary to determine the three coordinates of position and three coordinates of the velocity vector for each of, say, 10^{25} molecules entering into the composition of a given gas sample (Penrose 1990, 400-406). This brief description, of course, results from the statistical approach—ignoring the individual behaviors of particular system elements and limited to only determining the mean values of the microscopic parameters and their frequency distribution.

Reaching the very essence of thermodynamics directing attention on systems composed of many elements enables the formulation within its frame of such concepts as "ordering" or "entropy" (the measure of disorder) and even "information" and "structure" (Penrose, 391-410; Coveney and Highfield 1991; Greene 2004, 127-176) (nevertheless, this is information and structure in the thermodynamic sense, and have little to do with the information and structure essential for the nature of living organisms; Korzeniewski 2015, 2017a).[6] It is obvious that when we deal with a single object (element), it cannot be ordered. On the other hand, when we consider a system composed of many objects, their very distribution (arrangement) in space can be uniform and chaotic (and thus not ordered) or form a certain structure. A spatial structure can be accompanied by a temporal (functional) structure, comprising a determined pattern of changes in the distribution of elements in time. Thermodynamics, therefore, enables the introduction of the concepts of structure, information, and (dis)order, absent on the dynamic level, or rather it makes these concepts sensible.

From the point of view of dynamics, in particular quantum mechanics, an atom of carbon entering into the composition (as one building element of a molecule of some organic compound) of a highly-organized structure (let it be the human brain) is exactly the same sort of atom as an atom of carbon in CO_2 molecule being an element of homogeneous (uniform) gas,

---

[6] A large amount of thermodynamic information and structure is necessary for living organisms to come into existence, but it does not allow for an explanation of the heart of their essence.

carbon dioxide, in a container that reached maximal entropy (degree of disorder) under given conditions. Dynamics states that the movement of this atom and its possible formation of chemical bonds is subject (in both cases) to determined physical laws and nothing more. It cannot enter on a higher level and predicate anything about the rules governing the behavior of systems composed of many atoms. This impossibility by no means results from the temporary (transient) imperfections of present dynamic theories, which will be replaced in the future by better theories, allowing these limitations to be overcome. This is because these limitations are principal and result from the very essence of the general dynamic approach (paradigm). In order to pass from single objects to systems composed of many elements, it would be necessary to supplement dynamics with a new complex of concepts and way of describing the considered systems, absent within it and not derivable from it (similarly, as it is not possible to describe the essence of life using only terminology drawn from the physical level of reality).

To be sure, as discussed in the previous chapter, even a single object, such as an atom, can possess a certain kind of structure, expressing itself among others in the fact that an atom is composed of the nucleus and electron shells that contain particular orbitals occupied to a different extent by electrons. However, this is not the sort of structure thermodynamics deals with. The structure of an atom is necessary, "the only one possible," as it results univocally (in given conditions) from the laws of physics (although, e.g., in very high temperatures, this structure can undergo destruction). On the other hand, in a thermodynamic system, any potential structure is by no means necessary—it may exist, or equally so, it may not. It is possible to attribute to every thermodynamic structure a probability of its spontaneous formation, and the lower this probability, the more ordered (in the thermodynamic sense) this structure is. Structures with the greatest probability of spontaneous appearances, such as gases homogeneously distributed in a container, are characterized by the lowest ordering, and their entropy adopts maximal value.

Because the concept of structure can seem inadequate in the case of the homogeneous distribution of gas in a container, which in fact means a complete lack of structure, a much more general notion of *macroscopic state (macrostate)* was introduced (Difference Between Microstate and Macrostate 2020; LabXchange Microstates, Macrostates, and Entropy 2019).

It denotes a somehow distinguished macroscopically (phenomenologically) spatial (in the case of purely physical systems; in the case of chemical systems, this is more complicated—see the next Subchapter) ordering of system elements. For instance, a chaotic, homogeneous mixture or a human organism can form from the same atoms. Both cases correspond to different macroscopic states of the same system (set of elements). Nevertheless, potentially an unlimited amount of various macroscopic states can be imagined here, for instance, atoms of all chemical elements segregated in separate vials. An analogous situation occurs in a container with gas—apart from a uniform distribution of the gas, it is theoretically possible that all gas molecules are concentrated only in one half of the container; there is also a huge amount of yet different potential macroscopic states.

The key element for the "later" (resultant) features of thermodynamics that emerge from its departure points is the fact that various macroscopic states have different (indeed, very different) probabilities of spontaneous formation. This means that the chance of a spontaneous origin of various macroscopic states due to accidental movements of its elements is very (as a rule, truly astronomically) different. There is no need to convince anybody that due to random translocations of a certain set of atoms, a homogeneous mixture will originate much faster (with a much greater probability) than the human organism. The same concerns the container with gas—due to the accidental movement of its molecules, they tend toward a uniform distribution within the container. Such a selection of the velocities and movement directions of molecules that will undergo concentration in one half of the container is attributed with a similar relative probability as to tossing only tails in, say, 10^{25} coin tosses (assuming that the number of molecules in the container equals 10^{25}).

How does thermodynamics cope with determining the (at least relative) probability of various macroscopic states? The number of *microscopic states (microstates)* corresponding to given macroscopic states is of fundamental significance. The microscopic state is a such-and-not-another unique state of all elements of the considered system (Difference Between Microstate and Macrostate 2020; LabXchange Microstates, Macrostates, and Entropy 2019). In the case of gas, this means an exact determination of the position and vector of velocity (that is, the value of speed and the direction of movement) of all molecules entering into the composition of the gas. By definition, all microscopic states are bestowed with the same probability of spontaneously coming into existence. Various macroscopic states represent various classes of microscopic states—many (usually enormous amounts of) microscopic states correspond to one macroscopic state. It seems obvious that the very different chaotic (disorderly) distributions of gas molecules will roughly look identical, and in the face of this, they will correspond to the macroscopic state called "uniform distribution." A much lower, but still enormous (astronomical), number of microscopic states can be attributed to the macroscopic state “all gas molecules in the left half of the container”—many combinations of positions and velocities of particular molecules having just such state are thinkable.

However, what does it have to do with the probability of macroscopic states? A lot. Because the probability of coming into existence of a given macroscopic state depends strictly on the number of microscopic states corresponding to this macroscopic state—the more microscopic states contained in the latter, the greater the probability. All microscopic states have, in principle, the same probability (such that the sum of the probabilities of all possible microscopic states gives the result 1). Moreover, because of the incessant movement of molecules, transitions continuously occur from one microscopic state to another. This next microscopic state can belong to the same class of microscopic states (that is, macroscopic state) or to another class than the previous microscopic state (Penrose 1990, 407-411). During the random wandering from one microscopic state to another, the probability that the system will pass from one class of microscopic states (macroscopic state) to another will depend on the relative numerosity (number of contained elements) of these classes. It seems obvious that a much more probable scenario will be a transition from a macroscopic state containing a relatively small number of microscopic states to a macroscopic state with much more numerous microscopic states than the opposite transition. In such a case, we say the second macroscopic state is more probable than the first. Thus, we can see how the number of microscopic states entering a given macroscopic state's composition determines the latter's probability.

Let us consider as an example a set of four natural numbers in the range from 1 to 100. Let us assume that each of these numbers can change completely randomly at any moment into any number in this range. The microscopic state of this system will be represented by the current set of four concrete numbers. The system will, every once in a while, pass randomly from one microscopic state into another. Let us now distinguish macroscopic state X, in which all numbers are identical, and macroscopic state Y, corresponding to any other situation (when it is not so that all four numbers are the same). The transition probability from state X to state Y is much higher than the probability of the opposite transition. Combinations of four identical numbers will therefore appear relatively seldom. This idealized example differs from real situations mostly by the fact that the probability of the spontaneous appearance of the distinguished macroscopic state X will be small but not very small.

Every system has an immanent tendency toward a transition from a macroscopic state of low probability to a more probable macroscopic state. This means that if the initial state of the system is at low probability, the system will inevitably evolve toward a state of higher probability. On the other hand, when we start from the most probable state, nothing will happen—the system will remain in this state. The evolution from a less probable (more ordered) to a more probable (less ordered) state is, therefore, (in practice) *irreversible*.

The idea of macroscopic and microscopic states, increase of entropy and spontaneous evolution of a system is illustrated in Fig. 2-1 using the example of a container with gas. Here, we have two distinguished macroscopic states—state 1: all gas molecules in the left half of the container, and state 2: gas molecules uniformly (randomly) distributed throughout the container volume. Of course, the number of gas molecules is extremely low here compared with real systems for the purpose of clear presentation on the diagram. State 1 has a minor chance of spontaneously coming into being. For it to originate, external work (associated with increased entropy of the surroundings) would have to be done, consisting of, for instance, gas compression within one-half of the container. Although this state contains (in real systems) a huge number of microscopic states, it is still, which is crucial for our considerations, by an astronomical factor lower than the number of microscopic states entering into the composition of state 2. Therefore, state 1 will possess an immanent tendency to pass into state 2. If we start from state 2, nothing will happen—the distribution of molecules will remain homogeneous—this is the state of maximal entropy in given conditions. The probability of the opposite transition is lower by many (very many) orders of magnitude and will most probably never happen in our universe. State 2 is characterized by a much higher entropy than state 1, and therefore the state 1 $\rightarrow$ state 2 transition is associated with increased entropy. In the face of this, the ordering of the system falls. State 1 is also characterized by a greater "complexity of structure" (understood thermodynamically) and related to it information. Even intuitively, we receive the distribution of molecules "left half of the container filled with gas, right half empty" as possessing some structure, while the arrangement of "uniform distribution" seems to us completely devoid of structure. As for information, if our task is to find a gas molecule in the container, it will be easier when we know that it is somewhere in the left half than when we are informed that it is anywhere in the whole volume of the container. Therefore, in state 1, we have greater information about the localization of particular gas molecules at our disposal. This information is twice as great in state 1 than in state 2 for a single molecule. However, if we take, say, 10^{25}

molecules, the difference in the informational content between both states will become gigantic. Summing up, state 1 is characterized by a lower entropy, greater ordering, higher complexity of structure, and a greater amount of information (all in the thermodynamic sense), and containing less (in real systems—by many orders of magnitude) microscopic states than state 2, and therefore state 1 will spontaneously pass into state 2, characterized by greater entropy, less ordering, lower complexity of structure, and a smaller amount of (thermodynamic) information and a greater number of microscopic states (Fig. 2-1).

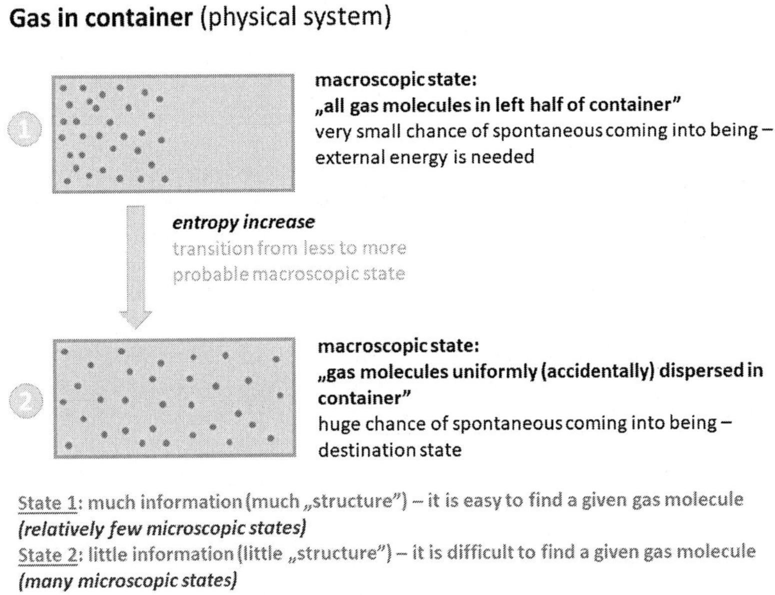

Fig. 2-1. The idea of macroscopic states presented using the example of a container with gas. Macroscopic state 1 with a smaller number of microscopic states, greater ordering, higher complexity of structure, and a larger amount of thermodynamic information passes spontaneously into macroscopic state 2 with a smaller number of microscopic states, less ordering, a lower degree of structure complexity, and a smaller amount of thermodynamic information, which is associated with entropy increase.

The thermodynamics of chemical systems

However, this purely physical example may seem distant from the case of living organisms (individuals), which in their very nature are, first of all, chemical systems (although physics also plays a significant role in them). Therefore, analogously, let us now consider a simple chemical system: the reaction of glucose oxidation with the participation of oxygen. In this reaction, a 6-carbon (whose molecule contains six carbon atoms) sugar glucose ($C_6H_{12}O_6$) reacts with oxygen (O_2), due to which carbon dioxide (CO_2) and water (H_2O) are produced. This is shown in Fig. 2-2. It is always possible to set reagent concentrations so that the reaction will be in thermodynamic equilibrium, which means that the rate of glucose oxidation (forward reaction) will be equal to the rate of its synthesis (backward reaction). However, in such a condition, the glucose concentration will be marginally low compared to the concentrations of other substances participating in this reaction.

On the other hand, at physiological concentration ratios of reagents, this reaction is very far displaced from thermodynamic equilibrium: the forward reaction (from left to right) proceeds millions of times faster than the backward (from right to left) reaction. It is possible to speak about the rate of a reaction if we deal with many molecules so that the statistical approach can be applied. At the level of molecules, we can talk only about probabilities. In the discussed case, the probability of the forward process is greater by several orders of magnitude than the probability of the backward process. This is why in the equilibrium state, the concentrations of CO_2 and H_2O exceed millions of times the concentrations of glucose (and oxygen): in order to cause the backward reaction to be as equally fast as the forward reaction, the low probability of the reaction must be compensated by a huge number of reacting molecules (Nicholls and Ferguson 2002).

Generally speaking, glucose has a tendency (once again: at physiological concentrations of reagents) to a spontaneous decomposition, which can be greatly accelerated by enzymes (biocatalysts). This is associated with releasing huge amounts of energy, a part of which can be stored in some useful form (the rest undergoes dissipation as heat). In living organisms, energy is conserved most frequently in the form of ATP (more strictly—in the form of a high, far from thermodynamic equilibrium, ratio of the concentration of ATP to the [product of] concentrations of the products of its hydrolysis: ADP and inorganic phosphate—P_i) (Marks et al. 1996, 275-290; Voet et al. 1999, 13-22; Nicholls and Ferguson 2002). This compound is regarded as a "universal energy carrier," as it is synthesized by a few processes specially designed for this purpose, while its decomposition (hydrolysis) is coupled with, and thus drives, numerous processes in the cell that need energy for their occurrence, including various chemical syntheses (of proteins, nucleic acids, and other cell components), mechanical work (muscle contraction, transport of various substances along microtubules by kinesins and dyneins) or osmotic and electrical work (transport of metal ions: Na^+, K^+, Ca^{2+}, and various organic substances through lipid-protein membranes against the gradient of concentration and or electric charge). Because glucose tends to spontaneously decompose, its synthesis, for instance, through the assimilation of CO_2 from the atmosphere by photosynthesis (its light-independent phase), requires coupling with the spontaneously-proceeding reaction of the hydrolysis of ATP (produced in the light-dependent phase of photosynthesis).[7]

However, let us return to the glucose oxidation reaction and Fig. 2-2. Also, in this process, as in the container with gas, two macroscopic states can be distinguished, although they are not so easy to visualize in this case. In state 1, carbon, oxygen, and hydrogen atoms are arranged in such a way that they are organized into $C_6H_{12}O_6$ and O_2 molecules, while in state 2, they form CO_2 and H_2O molecules. This system is also more complicated than the container with gas because covalent bonds (based on the electrical force) play a crucial role—they are not as easy to illustrate as the spatial distribution and collisions of gas molecules. Nevertheless, we can state that state 1 is characterized by lower entropy, higher ordering, greater structure

---

[7] The spontaneous direction of the course of a reaction and its energetic balance at a given ratio of its products and substrates concentrations is determined by a thermodynamic potential called the Gibbs free energy change (free enthalpy change) ΔG (Nicholls and Ferguson, 2002, 39-64). Its value determines how far and in which direction the reaction is displaced from thermodynamic equilibrium. When ΔG is negative, the reaction proceeds spontaneously and can perform the greater work, the greater (more negative) the value of ΔG is (the more the reaction is displaced from thermodynamic equilibrium in the direction of converting of substrates into products, the more the ratio of substrate concentrations to product concentrations exceeds the equilibrium ratio). And inversely, reactions with positive ΔG, such as glucose or protein synthesis, cannot proceed spontaneously. For them to proceed, they must be coupled with (one or more) reactions (processes) with negative ΔG, so that the resultant ΔG is also negative. Just such a reaction driving various processes that cannot proceed spontaneously is the hydrolysis of ATP to ADP and P_i. Finally, in the state of thermodynamic equilibrium (in principle absent in living organisms) ΔG equals zero.

For readers interested in more detail: the Gibbs free energy change in a given process is described by the equation: $\Delta G = \Delta H - T\Delta S$, where ΔH is the change in the amount of heat (it is negative when heat is released to the surroundings), T is absolute temperatures (in Kelvin degrees, K) and ΔS is the entropy change (when entropy increases, then the expression "$-\Delta S$" has of course a negative sign). Therefore, in processes proceeding spontaneously (negative ΔG), a heat release and rise of system entropy, or at least one of these processes, takes place (Nicholls and Ferguson, 2002, 39-64).

complexity, and a greater amount of thermodynamic information than state 2.[8] The structure of the glucose molecule itself looks quite complex. When we consider this information, if our task is to find a certain atom of carbon, if we know the position of one atom, we know that the remaining five atoms are located nearby (in a determined configuration) in the glucose molecule—the locations of carbon molecules in the glucose molecule are correlated with each other. On the other hand, in state 2, knowing the position of one atom of carbon, we only have information that the remaining five atoms are anywhere in the considered system, "inbuilt" into a CO_2 molecule.

At the same time, it should be added that this intuitive and simplified description cannot always be directly applied to real systems. For instance, because of strong covalent bonds formed as a result of this process, atoms (or simple molecules) frequently spontaneously combine (join) themselves into more complex molecules, which is associated with an increase in entropy, heat release and a decrease in the thermodynamically understood (this understanding seems counterintuitive for us) complexity of structure and amount of information. Generally, when physical forces, for instance, an electrical force, enters the stage, which is practically absent in the container with gas (thermal collisions between molecules predominate there), what we intuitively receive as a structure of a macroscopically-organized object associated with information, is not such a structure/ordering/information at all from the point of view of thermodynamics. Spontaneous creation of quite complex chemical molecules (although more complex organic compounds cannot originate spontaneously) or a purely physical self-organization taking place in the case of formation of lipid bi-layers, when hydrophobic tails of lipids become oriented toward each other, while the hydrophilic heads—outside, toward the water phase, or the simple formation of oil spots on the surface of water, although "for our eyes" lead to forming spatially-organized structures, occur in agreement with the laws of thermodynamics and are associated with increased entropy and release of heat to the surroundings, and decrease of thermodynamically understood structure complexity and amount of information (Coveney and Highfield 1991). This also refers to the spatial distribution of matter in the gravitational field, which is discussed below. In other words, matter self-organization is frequently associated with entropy increase.

[8] Strictly speaking, state 1 should be represented here by the whole reaction (with its substrates and products) with reagent concentrations far displaced from thermodynamic equilibrium, and state 2 by the reaction with the equilibrium ratio of reagent concentrations.

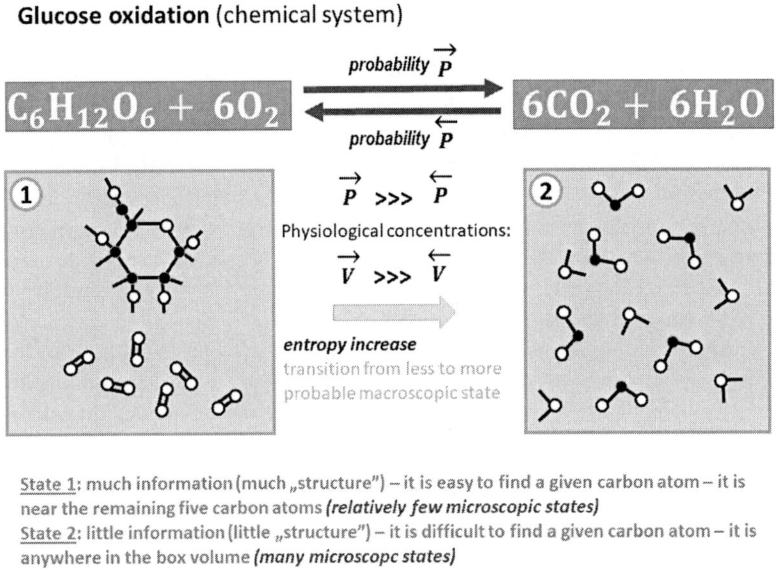

Fig. 2-2. The idea of macroscopic state using the example of the chemical reaction of glucose oxidation. The macroscopic state 1 with a lower number of microscopic states, higher ordering, greater degree of complexity, and higher amount of thermodynamic information passes spontaneously into the macroscopic state 2 with a higher number of microscopic states, lower ordering, lower degree of complexity, and lower amount of thermodynamic information, which is associated with entropy increase. In order to equalize the rate of the reaction in both directions (and thus reach the state of thermodynamic equilibrium), it is necessary to counter-balance the much higher probability of the state 1 → state 2 transition ($\vec{P}$) than the probability of the state 2 → state 1 transition ($\overleftarrow{P}$), by a large increase of the concentration of products (CO_2 and H_2O) in relation to the concentration of substrates ($C_6H_{12}O_6$ and O_2). At physiological concentration of reagents, the rate of the forward reaction ($\vec{V}$) is millions of times greater than the rate of the backward reaction ($\overleftarrow{V}$). Black balls represent carbon atoms; white balls, oxygen atoms; and ends of short lines, hydrogen atoms.

One more example

Let us consider once more the problem of the irreversibility of the evolution of dynamic systems using another figurative example. Let us consider a rectangular vessel with a divider and two compartments: a small compartment A and a large compartment B. Such a vessel is shown in Fig.

2-3. A single little ball is located in the vessel, and the whole vessel is shaken. The ball can either jump from compartment A to compartment B or, inversely, from compartment B to compartment A. Each of these transitions is endowed with a determined probability, while the probability of the A → B transition (pAB) significantly exceeds the probability of the B → A transition (pBA). This is quite obvious: the ball falling into compartment A requires rare luck, consisting of a specific selection of the velocity and the direction of the ball, while practically any significant movement will cause the ball to jump from compartment A to compartment B. Therefore, the system possesses an immanent tendency of evolution from state A toward state B.

At the same time, the single ball (its location) represents the state of the whole system, not only the state of one element (e.g., a single gas molecule in a container). On the other hand, the bottom of the vessel designates the set of possible ball locations, and so of microscopic states of the system. Of course, there are many microscopic states (that is, simply, the possible ball locations on the bottom of the vessel). On the other hand, we deal here only with two classes of microscopic states: macroscopic states. As we remember, all microscopic states can be regarded as equivalent, that is, associated with the same probability. In turn, the probability of macroscopic states corresponds to the relative number of microscopic states comprised by them. It can be clearly seen in Fig. 2-3 that macroscopic state B is much more spacious (contains many more possible microscopic states) than the distinguished macroscopic state A (comprising only relatively few microscopic states). Therefore, in accordance with what was said above, state B is much more probable. State A is, in this sense, much more ordered (possesses low entropy); that is, it supplies much more information about the ball location than state B—if we already know that the ball is located in the smaller compartment A, it is much easier to find it than when we have only information that it is somewhere in the vast compartment B. Taking into account the general probability, the ball almost certainly resides in compartment B; therefore, the information that it actually is there carries very little information. Macroscopic state B is therefore characterized by a large disorder of the ball location (it can reside in one of the very many microscopic states), and so by high entropy.

As stated earlier, the described system exhibits an irreversible tendency toward evolution from state A to state B or toward remaining in state B once it is reached. It differs from real systems among others because the transition B → A in this state is only little, but not dramatically little. We previously saw that state B, being more probable, is also characterized by less information, a lower degree of ordering, and higher entropy than state A.

Therefore, we learned step by step, no more, no less, that in every (isolated) system, there must take place a fall of ordering and amount of information, and an entropy rise, until the moment when the values of these parameters stabilize on a constant minimal (in the case of entropy—maximal) possible level under given conditions.

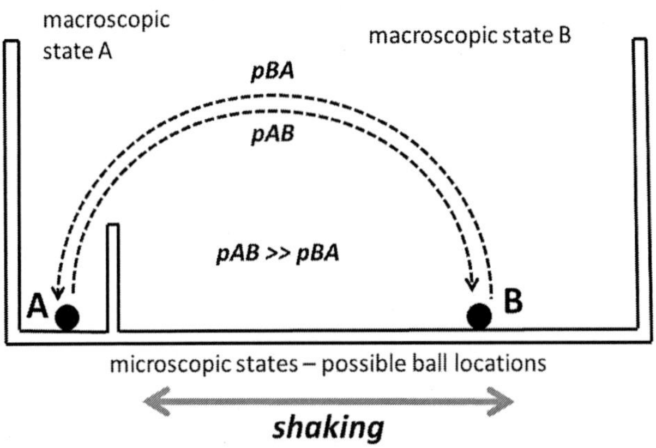

Fig. 2-3. A vessel with a divider and two compartments: A and B. A little ball located in one of these compartments represents a separate macroscopic state. Microscopic states are represented by all possible locations of the ball on the bottom of the vessel. Because of shaking of the vessel, the ball can transit from state A to state B, or inversely. The probability of the first transition (pAB) is much greater than the probability of the opposite transition (pBA). Therefore, the system will exhibit a strong tendency toward the evolution from macroscopic state A to macroscopic state B.

Thermodynamics, arrow and flow of time, irreversibility, entropy, orderliness, information, the complexity of structure

In this way, starting from the initial assumptions of thermodynamics (its "departure points"), we were successful in deriving its most fundamental, macroscopic properties, which appeared to be inevitable consequences of a description of systems composed of *many* elements. First, we arrived at the famous *second law of thermodynamics*, announcing that in isolated systems, entropy rises as a function of time until it reaches the maximal value. This means a universal tendency of physical systems to increase the degree of disordering. In particular, during the transformation of one form of energy into another on the surface of the earth, energy tends to be dissipated in the most disordered (in this system) form, namely heat energy (Prigogine 1980; Prigogine and Stengers 1984; Penrose 1990; Coveney and Highfield 1991; Voet et al. 1999,13-21; Atkins and de Paula 2002; Green 2004).

The second very important consequence resulting from the statistical nature of thermodynamics, following from only the averaged behavior of elements in a system containing many such elements, is the emergence of the (thermodynamic) *arrow of time*.[9] It constitutes, in a sense, an alternative face of the second law of thermodynamics—the arrow of time corresponding simply to this direction in time, in which entropy increases. In other words, the universal tendency to decrease the degree of ordering not only imposes the unidirectional arrow of time but, in a sense, is even identical to this (thermodynamic) arrow of time. If a particular system reached the state of thermodynamic equilibrium, the (thermodynamic) arrow of time would disappear. Time is "oriented" in this direction in the space-time of the universe, in which entropy increases. The decrease of order with time means that at the beginning of the universe, the degree of ordering reached the maximal value. I will come back to this matter later on.

To digress, it seems very interesting how the real flow of time relates to the arrow of time and whether the former takes place objectively. Because this topic exceeds the main topic of the present chapter, my opinion on this matter is shortly presented in Supplement 1 at the end of this book and in more detail elsewhere (Korzeniewski 2016).

The second law of thermodynamics also implicates a practical *irreversibility* of all processes in which a significant change of entropy takes place. This means that for such processes associated with entropy increase, we will not observe opposite processes in nature. A cap can be broken into pieces but will not re-assemble itself spontaneously from such pieces. Gas in a container will tend to a possibly homogeneous distribution of molecules in space. To be sure, these processes are reversible *in principle*, at least according to the essence of dynamic theories, but are irreversible *in practice* because of the extremely small probability of the occurrence of opposite processes. Generally, every system aims for the state of *thermodynamic equilibrium*, in which entropy adopts a maximal value, and after reaching it, nothing macroscopic happens. In such a state, atoms and molecules (between which invariably the same kinds of interactions occur) are arranged into configurations possibly devoid of structure and organization (in the thermodynamic sense).

---

[9] Which, as mentioned, does not yet imply the real *flow* of time.

Finally, thermodynamics attributes a specific and, at the same time, very (sometimes too overly) general meaning to such phenomena as "ordering," "structure," and "information" (Prigogine 1980; Prigogine and Stengers 1984; Penrose 1990; Coveney and Highfield 1991). The macroscopic state A in Fig. 2-1 is endowed with *information* responsible for the spatial ordering of gas molecules and, therefore, just for the type of *structure* in the thermodynamic meaning. Gas molecules in system A are ordered in the most common sense. Similarly, there is order in a flat when clothes are in a wardrobe, food is in a fridge, and waste is in a wastebasket. Therefore, the spatial segregation of particular objects takes place. Of course, the concentration of all gas molecules in one half of a container serves nothing practical (in opposition to the ordering of things in a flat), but distinctions of this kind are completely unimportant from the point of view of thermodynamics.

It may be more difficult to accept the statement expressed intuitively that state A already possesses a well-determined structure. Let us consider a rectangle filled with small squares arranged one by one in, say, 1,000 rows and 500 columns. Therefore, we have 500,000 squares. Each square can be white or black. The squares filling the rectangle form a certain pattern or picture, like grains on a black-and-white photograph. Let us assume that an equal number of white and black squares is distributed randomly on the surface of the rectangle. A uniform gray picture will arise ("gray noise"), which does not possess, of course, any structure.

On the other hand, if white and black squares are arranged to form a reproduction of the painting *Mona Lisa*, we will deal with a very complicated two-dimensional structure (e.g., the human body represents a three-dimensional structure). This will result from the relative arrangement of white and black squares in a special, meaningful, determined way that can be easily shortly characterized by some statement, for instance: "this is a human face." Another, albeit much less complicated picture of the structure, would be represented by a rectangle, in the upper half of which there would be only black squares, while in the lower half—only white squares. Such a structure could be specified as "a rectangle divided into a black and white part." It represents, in a sense, an intermediate case between the *Mona Lisa* and a chaotic mixture of white and black squares. If some abstractionist painter created it, it could bear with it even some germs of meaning. Nevertheless, such a structure would certainly be a little less monotonous than the uniform gray background, which deserves the name structure. Similarly, a certain kind of structure is formed by the gas grouped in one half of the container, in opposition to gas uniformly dispersed over the whole container.

However, the matter is not as simple as it might seem based on the above superficial analysis. In the discussed example, the statement that the coarse (because of the resolution of little squares) reproduction of the *Mona Lisa* is more complex, it contains more structure than the rectangle divided into a white and black part seems intuitively obvious. However, a serious problem appears when purely thermodynamic (mathematical) criteria are applied to both cases. Although this could sound paradoxical, from the point of view of thermodynamics, both pictures mentioned are characterized by the same complexity of structure (although both have more structure than the chaotic mixture of squares). As in both cases, black squares would have to find themselves in the same number of precisely determined locations (assuming that the reproduction of the *Mona Lisa* contains an equal number of white and black squares), the probability of a spontaneous formation of both pictures will be equally small.

Both the notion "degree of system complexity" and the term "degree of structure complication" can be defined, expressed by the amount of information in the macroscopic state corresponding to a given structure (Coveney and Highfield 1991; Gell-Mann 1995). It was stated above that the more information a given macroscopic state possesses, the fewer microscopic states correspond to it; that is, the smaller the probability of a spontaneous coming into being of a given macroscopic state due to an accidental arrangement of the elements of the considered system. In the above analogy, it is obvious that white and black squares will probably arrange themselves into a (macroscopically) gray mixture, while significantly less probability will be attributed to the spontaneous segregation of the squares into two equal parts: black and white. However, a spontaneous "painting itself" of the coarse-grain reproduction of the *Mona Lisa* is equally probable as this last event. Hence, the conclusion is that, from the thermodynamic point of view, the reproduction of the da Vinci masterpiece contains an identical amount of information as the rectangle containing the black and the white part, insofar as the resolution of the picture (size of squares) is the same in both cases, while the chaotic mixture of squares is endowed with a lower amount of information.

Therefore, this what constitutes a distinguished structure from our subjective point of view, does not have to be such a structure for soulless thermodynamics. Fortunately, there is an alternative and, in my opinion, a much more adequate way of "measuring" the information associated with a given macroscopic state. According to this approach, *the amount of information contained in a given macroscopic state is proportional to the length of the possible shortest description/distinction of this state* (alternatively, to the smallest possible number of instructions/procedures necessary for its generation, for instance, by some computer program). In other words, the more information a macroscopic state contains, the longer the description is necessary to show how this state is distinguished and what it consists of (Gell-Mann 1995, 30-58, 80-101).

Let us apply this procedure to our rectangle. Let us introduce coordinates for each little square determining which column (*x* coordinates) and row (*y* coordinates) it is located. Therefore, the location of a given square will be fixed univocally by a pair of numbers having the form (*x*, *y*), and *x* can now adopt a value from 1 to 500, while *y*—from 1 to 1,000; the square coordinates can therefore amount, for example: (123, 698). The complete description of every possible microscopic state of the considered system will look in the following way: "squares with coordinates (here enumeration takes place) are black" (implicitly: the rest of squares are white). This is the longest possible description of any macroscopic state, especially when there is an equal number of white and black squares. At the same time, this describes a concrete microscopic state. However, many macroscopic states can be more economically characterized. In particular, in the shortest way, namely by the words "random distribution," the macroscopic state can be described as the "gray background," which is the case when white and black squares are scattered chaotically, and entropy reaches the maximal value. A little more information must be contained in the description of the rectangle divided into two halves: black and white. It can look as follows: "all squares with coordinate *y* between 1 and 500 are black" (implicitly: squares with coordinate *y* between 501 and 1,000 are white) or "all black squares are located in the upper half" (however, it would be necessary to define first the upper half). The description of this macroscopic state is, therefore, longer than the description of the "chaotic" macroscopic state, and therefore the state "half black, half white" is more ordered and possesses a higher degree of structure complication.

On the other hand, to fully characterize the macroscopic state of the "reproduction of the *Mona Lisa*," it would be necessary to specify the coordinates of all the black squares, and therefore the length of the description of this macroscopic state would approach the maximal length. Therefore, this macroscopic state would contain the greatest possible amount of information, the most complex structure, and the highest organization degree.[10] For the same reason, we can describe a human being

---

[10] This statement is a certain simplification, as the position of black squares (black pixels on the reproduction) is not completely independent of each other. Some squares force to some extent the position of other squares, in order to preserve, for example, the continuity of an oval face. For this and many other reasons, finding the possible shortest description of a complicated macroscopic state would be in practice

as a much more complex structure (with lower entropy) than a chaotic mixture of atoms. Therefore, we can see that it is possible to state sensibly that the system "coarse *Mona Lisa* reproduction" contains more information and structure than the system "a rectangle divided into black and white parts." The black-and-white rectangle is much more "compressible" (a technical term used in the case of electronic picture recording), meaning the possibility of shortening such a record in relation to a record specifying the coordinates of each pixel—the point of the picture). However, the problem is that the criterion of the shortest possible description cannot be easily inbuilt into statistical thermodynamics.

Fig. 2-4 illustrates the above example using, for the sake of presentation clarity, fewer little squares (for the same reason, black squares were replaced with gray ones so that the borders of the squares can be seen). The upper rectangle represents the macroscopic state "random distribution," the middle rectangle—the macroscopic state, "all squares with coordinate *x* from *a* to *j* are gray," while the lower rectangle—the macroscopic state corresponding to some complex structure (because of the resolution of the picture this is not the *Mona Lisa*, but something relatively simpler). The specification and distinction of this macroscopic state require the longest description decidedly.

Therefore, we can see that the criterion of the shortest possible description offers us a tool allowing at least an intersubjective estimation of the degree of structural complexity and ordering (and therefore informational content) of an analyzed macroscopic state. However, this is by no means a fully objective criterion. For an extraterrestrial, the arrangement of squares representing a reproduction of the *Mona Lisa* would not have to differ significantly from a purely random (chaotic) distribution of squares. In addition, biological structures (including the organism of humans) can be regarded as specially distinguished mainly from the point of view of functionality imposed by the "purposes" of biological evolution (and recognized as such by our brain). After all, food plus oxygen[11] that are taken up by the human body for forming its structure during growth and its

very difficult, as estimation of the mutual correlation of various points seems very troublesome. On the other hand, this is only a technical problem, strictly related to the problem of optimal picture compression, which does not discredit in any way the general rule.

[11] Strictly speaking, this concerns here the system: food + oxygen → carbon dioxide + water + excreted products of metabolism + undigested food remains. The entropy increase during the conversion of food and oxygen into the products of metabolism is greater than its decrease related to the formation and maintenance of the structure and function of the human body.

maintenance (re-synthesis of decomposed body components) during adulthood, possess, in accordance with the second law of thermodynamics, more thermodynamic structure and information than an adult, although we are inclined to attribute to the latter a greater amount of traditionally understood, biologically purposeful structure and information. The huge majority of life on the earth's surface is sustained and driven, which is discussed in detail below, by the dissipation of visible radiation into infrared radiation. Therefore, the set of absorbed quanta of visible radiation is characterized by lower entropy, higher structure organization, and a greater amount of information than the earth's biosphere (strictly speaking, the transformation of visible radiation into infrared radiation is associated with a greater increase of entropy than its decrease associated with sustaining biosphere functioning). Generally, matter was endowed with minimal entropy and maximal "structure complexity," degree of ordering, and amount of information at the beginning of the universe's existence when its spatial distribution was homogeneous (Penrose 1990; Coveney and Highfield 1991). As discussed above, the force of gravity is responsible for this apparent paradox. However, of course, this is not the kind of structure and information we mean when we want to characterize the specificity of the phenomenon of life. Therefore, thermodynamics cannot shed light on the value, purposefulness, and specificity of information, order, and structure of biological (and psychical and cultural) systems (Korzeniewski 2013a, 2015, 2017a).

As it results from everything said above, thermodynamics allowed, through describing sets of many elements, the definition of such properties of the physical world as the tendency to increase entropy, arrow of time, and irreversibility of all physical processes. It gave the possibility of speaking about properties of physical systems evading the dynamic paradigm: structure complexity, degree of order, and amount of possessed information. Unfortunately, as I indicated previously, it was necessary to pay the price for this in the form of some subjectivism of thermodynamics (at least statistical thermodynamics), as an inseparable element of the conceptual construction of this theory is the macroscopic state. As we remember, this is just the tendency to pass less probable macroscopic states into more probable ones that underlaid the second law of thermodynamics, the arrow of time, and the property of irreversibility. This macroscopic state is characterized by determined structure, ordering, and informational content. However, the very distinction of some macroscopic state, that is, the grouping of some set of microscopic state into a distinguished class, seems a principally subjective (or, at best, inter-subjective) procedure. The best accessible measure of the amount of information deciding about the essence of the macroscopic state seems the shortest description necessary for full characterization and distinction of this state. However, this length is relativized in relation to the state of our mind, its knowledge, and the categories in which it perceives and conceives the world. The contamination with subjectivism of the very idea of macroscopic states is a deeply fundamental property. Only microscopic states endowed with equal probability and equivalent to each other in any other aspect exist completely objectively.[12] A given, concrete microscopic state, represented by all atoms coming into the composition at some moment of a given human organism, is equally probable as another concrete microscopic state of identical atoms forming a chaotic mixture. In both cases, a specification of a given microscopic state requires information about positions and vectors of velocity (and what results from this, interactions with other atoms) of all atoms constituting a given system. A human is a much more ordered macroscopic state than a chaotic mixture of atoms, as the amount of possible microscopic states giving a homogeneous mixture outnumbers by a really astronomical factor the number of microscopic states that can be classified as the macroscopic state "human." However, as we cannot specify any objective criteria of such classification, in particular strictly define the macroscopic state "human" or "John Smith" (Korzeniewski 2013b), we must agree for the immanent subjectivism of thermodynamics[13] or change our concept of objectivity, accepting the fact that on the thermodynamic (and next: biological and psychical) level there are features or aspects of the world absent on the purely dynamic level. This would mean granting a more objective status to such aspects of the world as ordering, intrinsic biological purposes, and psyche (or accepting that complexity constitutes an equally real aspect of the world, as space, time, matter, and energy) and in principle—negation of any principal difference, opposition, the demarcation line between objectivism and subjectivism (Korzeniewski 2015, 2017a). I am decidedly an advocate of the latter alternative, as I think that our association of objectivity with only the (sub)atomic level of reality

---

[12] Even this statement is not so clear, taking into account the indeterminacy of quantum mechanics and potentially infinite number of positions in space, which is composed of an infinite number of points of zero dimensions.

[13] At least in relation to biological, psychical, and cultural phenomena. Of course, gas dispersed over the entire container has more entropy in an objective manner than gas gathered in one half of the container. However, as mentioned, for example, sugars, from which the body of a growing animal is being built, is also characterized by a greater amount of information than an adult animal. However, just this animal is regarded to be more complex as it is endowed with a complicated, hierarchically-organized structure and intrinsic purposefulness.

has already become slightly too archaic, especially that just on this level, quantum mechanics meets huge problems with objectivity (Korzeniewski 2015, 2017a).

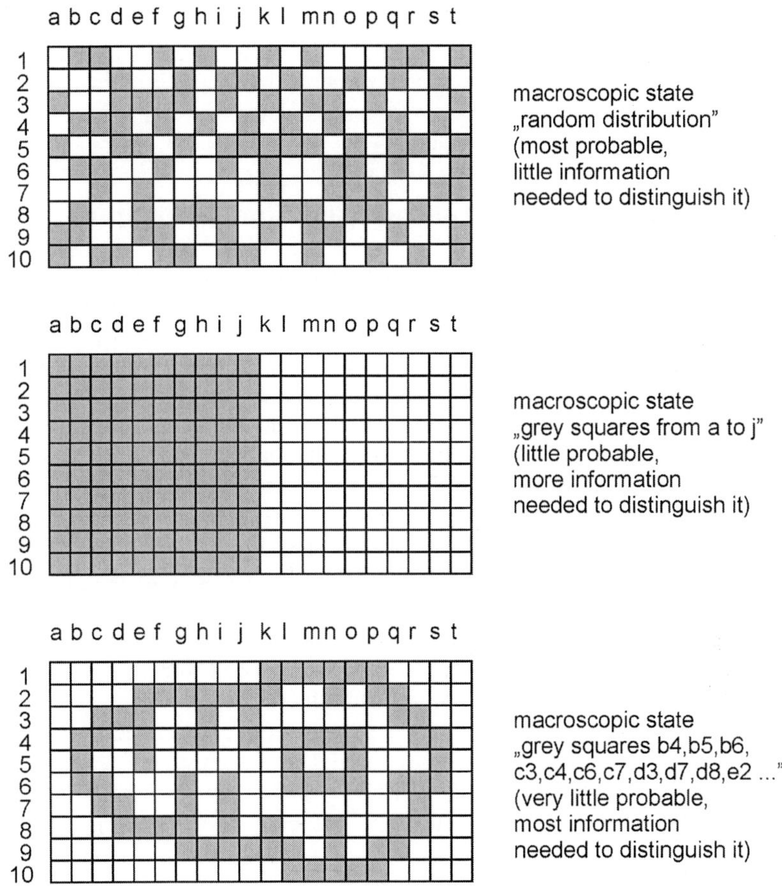

Fig. 2-4. Example of three macroscopic states with different degrees of ordering and amounts of information. The presented system is a rectangle composed of 200 little squares with horizontal coordinates from *a* to *t* and vertical coordinates from 1 to 10. Each of squares can be either white or gray.

We can conclude that in the case of biological systems, thermodynamics offers us a measure of quantity but not quality (specificity, value, purposefulness) of information, distinguishing living systems from physical objects. The same set of atoms that forms the human body can be arranged in such a way that the atoms will form the macroscopic state having the same or greater amount of thermodynamic information than the organism of humans, but devoid of any sense, purposefulness, and value from the point of view of biology. Therefore, a measure (and definition) of purposeful and "valuable" biological information is needed (Korzeniewski 2001, 2005, 2015). In Chapter 6, I propose a cybernetic paradigm of characterizing the phenomenon of life, particularly of a living individual and biologically-relevant information.

Thermodynamics in cosmology and astronomy

Before I explain the thermodynamic genesis of dynamic structures organized in space and time, particularly of living organisms, I would like to make a small digression and, for a while, deal with the structure of astronomical objects. Where do they come from? After all, we argued that according to the second law of thermodynamics (tendency to increase entropy), the gas in a container evolves from an uneven to even distribution. On the other hand, something apparently quite the opposite happens in the universe on an astronomic scale. Just after the Big Bang, the matter distribution was unusually uniform (which is expressed, for instance, in the almost perfect homogeneity of the microwave background radiation, a relic after the early stages of the evolution of the universe[14]); afterward, matter concentration started to form in different scales of magnitude, which gave rise to the beginning of stars, galaxies, galaxy clusters, and yet bigger structures (strings and walls composed of galaxies and surrounding empty "bubbles" devoid of galaxies), while according to our knowledge, the final stages of the evolution of matter are still greater and greater black holes, in which the maximum entropy in the scale of the universe is achieved (Hawking 1988; Penrose 1990; Weinberg 1993; Barrow 1994; Baggot 2015).[15] Therefore, does the second law of thermodynamics become invalid or even reversed in the scale of heavenly bodies and the entire universe?

---

[14] The deviations from the average temperature of the microwave background radiation are of the order of 1 per 10,000 to one per 100,000, and anyway this radiation was emitted only 380,000 years after the Big Bang, when the perfect homogeneity of matter distribution was already somewhat disturbed.

[15] Black holes can "evaporate" due to emission of the so-called "Hawking radiation" (Hawking, 1988, Penrose, 1990, 468-469) . This consists in creation of particle-anti-particle pairs (for instance electron-positron pair or two photons, of which each is the anti-particle for the other) on the event horizon (a sphere surrounding the area from which nothing, even light, can escape). One of the particles in the pair, having

Of course not. The force of gravity is the key to understanding the thermodynamics of the universe as a whole and the distribution of matter directly associated with it (Hawking 1988; Penrose 1990; Weinberg 1993; Barrow 1994; Baggott 2013, 2015). It is exerted by every object endowed with mass (or having energy equivalent to mass), and at the same time, every such object is a subject. As is commonly known, the gravitational force exerts an attractive effect that causes the mutual approaching of bodies to each other. Because of this property, matter naturally tends to concentrate in one place. The fact that all the matter in the universe (together with its space, related to gravitation) did not undergo condensation long ago into one point, forming something like an enormous black hole or terminal singularity, is due to the expansion of the universe. A ceaseless competition takes place between the moving away of material objects from each other due to the expansion of the universe (resulting initially from the impetus related to so-called inflation and presently also from the repulsive force exerted by the so-called dark energy) and their approaching of each other caused by gravitation. Up until the present day, gravitation won over shorter distances, which caused planets, stars, galaxies, and galaxy clusters to form. On the other hand, over longer distances, the impetus imposed on matter (and space) during the Big Bang still predominates, thanks to which galaxy clusters still move away from each other. Additionally, the expansion of the universe undergoes acceleration because of the repulsive force of dark energy (quintessence, cosmological constant) mentioned previously. However, generally, the force of gravity promotes a heterogeneous distribution of matter in space. Such distribution in the gravitational field is associated with high entropy, low ordering, little structure complexity, and little information. Homogeneous distribution of matter in the gravitational field implies very low entropy and, paradoxically, high structure complexity, a high degree of ordering, and a large amount of information (of course, all in the thermodynamic sense). Indeed, as I proposed previously (Korzeniewski 2013b, 2015, 2016), the genesis of the universe from nonentity was equivalent to the origin of *information* about the mutual distribution of matter and space (compare with Subchapter 7.4.).

Conversely, the reciprocal gravitational influence between particular gas molecules in the container (and also within biological systems of the earth's

---

positive energy, is radiated into cosmic space, while the other, endowed with negative energy, falls into the black hole, diminishing its total energy, and thus mass. However, the efficiency of this process is negligible, and additionally the larger the black hole, the smaller the efficiency of the process. Therefore, most probably, the growth of black holes due to absorption of surrounding matter or merging of black holes will be faster than their evaporation through Hawking radiation.

surface) is so negligible that it can be ignored. In gas, the dominant role is played by the chaotic thermal movements and collisions of molecules resulting from kinetic energy possessed by them. Just because of the accidental direction of the movement of particular molecules, gas tends towards the state of random (and thus homogeneous) distribution of those molecules in space. If the molecules, because of an action of some external force, were compressed into a ball in the middle of the container (forming a "planet"), after removing the compressing forces, the gas molecules would disperse themselves over the whole volume of the container. On the other hand, in the case of objects on the astronomic scale of magnitude, their movement is by no means random—its direction is mainly forced by gravity (Penrose 1990). As a result, astronomical objects tend to approach the common gravity center. This explains why the primarily (almost) homogeneous gas and dust clouds underwent concentration into galaxies, stars, and planets, and the reverse process does not occur, as in the container with gas, that is, evolution in the direction of the homogeneous matter distribution. The difference is a derivative of the difference in the size of the discussed systems. On the cosmic scale, because of the enormous masses we deal with, the gravitational force (proportional to mass) predominates over the random thermal movements of molecules. After all, even the container with gas (or the terrestrial biosphere) does not fly away into cosmic space but remains in direct contact with the surface of our planet because of its gravitational attraction. However, on a small scale, gravity allows gas molecules to move "as they want," as the mutual gravitational attraction between them is negligibly weak because of their small mass.

According to the general relativity theory, material bodies exert a gravitational attraction and curve ("condense") space(-time) around them in the shape of a three-dimensional funnel (we can imagine it as a ball, the closer to the center of which, the curvature, the "density" of space is greater). For electromagnetic, including visible, rays for seeing the world, because of their enormous speed (speed of light) and the zero rest mass of their "carriers" (photons), space seems much flatter than for much slower bodies endowed with mass and thus tending to fall into the gravitational funnel. If we were able to "see with the eyes" of gravitating bodies, we would see a much more homogeneous distribution of matter in the universe and, what is most important, the evolution of the matter distribution from the homogeneous state to still denser and more massive concentrations would look to us as a transition from a highly inhomogeneous (with holes of the rarified matter distribution in places corresponding to funnels) to homogeneous matter distribution.

A similar situation occurs in other systems where we deal with strong interactions between system elements, for instance, in chemical systems where electrical forces predominate. In addition, frequently (although decidedly not always) an entropy increase accompanies the forming of a more inhomogeneous (spatially ordered) distribution of atoms due to the joining of atoms into molecules of chemical compounds and of simple molecules into slightly more complex molecules. For instance, two oxygen atoms, O, spontaneously merge into the oxygen molecule O_2, while carbon undergoes spontaneous oxidation with the participation of oxygen to carbon dioxide (CO_2). All these processes are associated with entropy increase (and heat release) and fall in the thermodynamically-understood structure complexity and amount of information. However, this concerns chemical reactions leading to synthesizing relatively simple chemical compounds. Formation of more complex compounds, first of all, more complicated organic compounds, is, as a rule, associated with entropy decrease (the positive Gibbs free energy change) and therefore, their synthesis must be coupled with processes (e.g., ATP hydrolysis), in which entropy significantly increases (strictly speaking—the Gibbs free energy change is strongly negative), so that the net balance of entropy change is positive and the net Gibbs free energy change is negative (compare with Fig. 2-2 and the accompanying discussion).

Dissipative structures

Let us deal now with a specific kind of systems far displaced from thermodynamic equilibrium, namely dynamic structures associated with energy dissipation and entropy production—*dissipative structures* (Prigogine 1980; Prigogine and Stengers 1984; Penrose 1990; Coveney and Highfield 1991, 168, 193-195, 219-225). Examples of such structures are *convective currents* in a fluid. Let us place some fluid, for example, water, between two metal plates—one at the bottom and one at the top (Coveney and Highfield 1991, 185-187). If the temperature of both plates (and the fluid between them) is identical, nothing will happen—the system will find itself in a state of thermodynamic equilibrium with maximal entropy (under the given conditions). When we start to heat the bottom plate, a difference in temperatures will appear between the plates and across the liquid phase—the bottom limit of the water layer will be simply warmer than the top limit, and the difference of temperatures between the limits will be the greater, the more intensively the bottom plate is heated up. A thermodynamic gradient will originate that will "drive" the transfer of heat (kinetic energy) from the bottom metal plate, through the water layer, to the top metal plate. In other words, the layer of liquid will "strive" to transfer heat[16] from its lower (warmer) to its upper (cooler) limit. However, how this will occur depends on the gradient (difference) of temperatures between both layer limits.

At a small temperature gradient, heat is transferred through *conduction*. Liquid molecules located at the bottom of the layer take over kinetic energy from vibrating metal atoms in the lower plate, thus increasing the velocity of their movement. Due to collisions between molecules, the energy is transferred to molecules located slightly above; they, in turn, transfer it yet further and further, and finally to the upper liquid limit, where heat is conveyed to the top plate, increasing the intensity of vibration of its metal atoms (their kinetic energy). This is illustrated in the left upper panel of Fig. 2-5. The movement of molecules in the liquid layer is completely chaotic (the movement of a given molecule is not correlated in any way with the movement of other molecules), but the lower the molecules are located, the faster they move as they possess higher kinetic energy.

However, at the moment when the difference of temperatures between the lower and upper layer limit exceeds a certain critical value, dependent, for instance, on the density, viscosity, weight proper of the liquid, and gravity strength, conduction becomes too little an efficient mechanism leveling the temperature gradient and a new type of behavior appears. The movement of molecules ceases to be chaotic and undergoes specific organization (self-organization) into macroscopic dynamic structures ordered in space and time, called *convective currents* (and the whole phenomenon is called *convection*). Fluid molecules that have "warmed up" by taking over kinetic energy from the lower layer limit, instead of transferring this energy to the molecules located above, travel upward in an organized stream, forming an ascending current. This journey against the force of gravity occurs as a group of liquid molecules with higher kinetic energy is lighter (as these molecules "jostle" themselves more and exert higher pressure, so the volume of the whole group increases) than the surrounding areas of "cooler" liquid. Therefore, the movements of particular molecules become mutually correlated. Reaching the upper layer boundary, the molecules remove the obtained excess kinetic energy, transferring it to the upper layer limit (the metal plate). As a result, the whole group of molecules "cools down," becomes heavier, and then moves collectively downward due to the force of gravity as a descending current. Due to contact with the warmer lower layer boundary (the metal plate), their

[16] Heat is the total amount of kinetic energy in the system, while temperature is proportional to the average kinetic energy of molecules. In other words, temperature is proportional to the kinetic energy per a given amount of molecules. It can be said that temperature corresponds to the "density" of heat.

kinetic energy rises again, and the entire cycle closes. The mechanism of convection is illustrated in the lower left panel of Fig. 2-5. Convection causes a much faster heat transfer than conduction and therefore appears at higher thermodynamic gradients, where the significant displacement of the system from the equilibrium state forces the tendency toward a quick return to this state (that is intense entropy production). When the temperature gradient falls below the critical value, the convective currents disappear instantaneously. Therefore, sufficiently large and ceaselessly sustained thermodynamic gradients are necessary for the origin and persistence in time of this sort of dynamic macroscopic structure organized in space and time.

The tendency toward decay of thermodynamic gradients (in this case: temperature gradient), that is, an entropy increase (entropy is maximal when a system has a uniform temperature), or in other words—to a dissipation of energy (forming a temperature gradient requires a supply of an organized form of energy, and therefore its decay is equivalent to energy dissipation) causes an origination of macroscopic structures ordered in space and time when the magnitude of the gradient exceeds a certain critical value. These structures are called dissipative structures because they are directly associated with the dissipation of energy. As mentioned above, at the moment of a decrease of the temperature gradient (or of a thermodynamic gradient in general) below the critical value, when the entropy of the system reaches a high value (maximal in the case of complete disappearance of the gradient), dissipative structures (in this case—convective currents) will vanish completely, as they are dynamic phenomena driven by continuous sufficiently intense energy dissipation (entropy increase). Therefore, although it may sound paradoxical, we can see that the common impetus toward the decay of order can give birth to local (in space and time) order. In other words, the decay of order proceeds in an ordered way. Ilya Prigogine contributed fundamentally to explaining the genesis of dissipative structures (Prigogine 1980, 90-119; Prigogine and Stengers 1984, 13-14, 138-150).

Convective currents similar to those described above occur in a pot of boiling water. What we see as water boiling is nothing more than ascending convective currents "jumping up" over the water's surface. In this case, the upper border of the water phase is simply air, to which heat is lost by steam. However, it is difficult to see these currents, even if we boil water in a glass vessel, among others, because of water's small density and viscosity. They can be seen much better in silicon oil placed between two glass plates (in the so-called Petri dish). When the lower plate is not heated up, we observe a smooth surface of silicon oil. When the heating is sufficiently intense, "out of nowhere," hexagonal cells, called Bénard cells, appear in the oil layer (Coveney and Highfield 1991, 185-187). They are shown in the right panel in Fig. 2-5. On first look, they can even resemble the cells of living organisms (or honeycomb), although they are, of course, much simpler. The point in the middle of the hexagonal cell is the place where an ascending current reaches the surface (limited by a glass plate) of silicon oil; the "rays" running off from it diverge out as horizontal currents of silicon oil while the borders of the cell are the places where descending currents sink into the oil layer. When the heating of the lower plate is halted, the Bénard cells instantaneously disappear. Therefore, in the presence of sufficiently large thermodynamic gradients, clear macroscopic dynamic structures ordered in space and time, namely dissipative structures, can spontaneously self-organize, driven by energy dissipation and entropy increase.

It is worth mentioning that convective currents also occur in the earth's mantle, which is heated up by the disintegration of radioactive isotopes in the interior of our planet; they lead to mutual movement of the plates of the earth's crust, some of which carry continents with them (Macdougall 1996; Ozima et al. 2012; Baggott 2015). As a result, in some places, the plates diverge, which leads to the creation of new oceanic crust from lava flowing out from the slits formed (*spreading zone*, for instance, the ridge in the middle of the Atlantic Ocean), while in other places one plate enters below another plate and ultimately melts in the earth's mantle (*subduction zone*, for instance, east of Japan; generally: Pacific "Ring of fire"). Both phenomena are accompanied by intensive volcanism and earthquakes, and sometimes also orogenesis (mountain range formation, as takes place in the case of the Himalayas, formed and still growing due to the collision of the Indian subcontinent with Asia). As discussed below, convective currents in the earth's mantle drive the drift of the earth's crust plates and lead to spewing out of the mantle substances on the crust surface, which enables life on earth. The gases and minerals washed out from the mantle material sustain (drive) energetically (thermodynamically) ecosystems existing in the neighborhood of the so-called hydrothermal vents and presumably played an important (maybe fundamental) role in the origination of life on earth. Convective currents also appear in the interior of the sun heated up by thermonuclear reactions and participated in dissipation of energy released in this process.

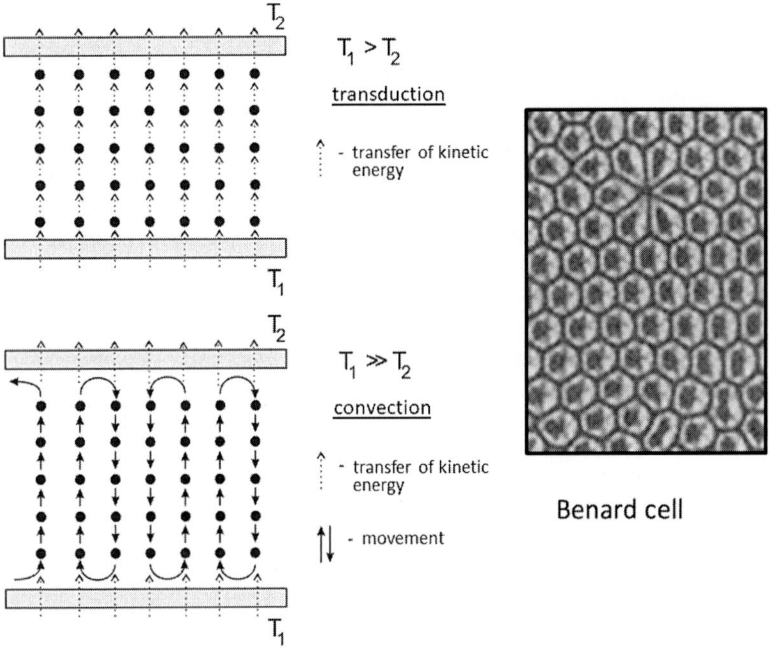

Fig. 2-5. Convective currents in fluid. Left: Flow of heat through a liquid layer in the way of conduction and convection. Heat flows from a warmer (continuously warmed up) bottom layer limit (e.g., metal plate) with temperature T_1 to a cooler top limit (metal plate) with temperature T_2. If the difference of temperatures between the bottom and top layer limit is small, there takes place conduction of heat consisting in transfer of kinetic energy by moving faster liquid molecules (black balls) located below to moving more slowly molecules located above. In the case when the temperature gradient exceeds a certain critical value, there appear convective currents—macroscopic dynamic structures organized in space and time. They consist of a group of liquid molecules that have taken over kinetic energy from the bottom layer limit and move together as a warm (relatively lighter) ascending current, in order to, after giving away the energy to the top layer limit, fall as a cold (relatively heavier) descending current. This process is driven and sustained by intensive dissipation of energy (production of entropy), and therefore such structures are called dissipative structures. Right: Hexagonal Bénard cells—seen from above convective currents in silicon oil "driven" by the dissipation of temperature gradient.

Dissipative structures on the earth's surface and thermodynamic-energetic fundaments of life

Let us pass now to thermodynamic fundaments of the genesis of various dynamic dissipative structures on the earth's surface, particularly of living organisms. As we remember, the first law of thermodynamics states that the total amount of energy remains constant in an isolated system (not exchanging matter, energy, and information with the surroundings). This means energy cannot originate from nothing or disappear without a trace. On the other hand, it can transform from one form into another. As we know, energy can appear in many forms: kinetic, potential, radiant, gravitational, and thermal. One should remember, however, that matter constitutes one of the forms of energy, according to the famous Einstein's formula for the equivalence of energy and mass ($E = mc^2$). The first law of thermodynamics allows, therefore, for converting one energy form into another under the condition that the overall amount of energy remains unchanged (Penrose 1990; 391-410; Coveney and Highfield 1991, 150-155).

On the other hand, the first law of thermodynamics does not indicate the direction of these conversions, that is, it says nothing about which forms of energy more willingly transform into other forms of energy or which kind of energy is preferred as the terminal stage of this entire sequence of transformations (evolution of forms) of energy. This problem concerns the second law of thermodynamics, which talks about the inevitable increase of entropy in time (Penrose 1990; 391-410; Coveney and Highfield 1991, 153-155). After a slight re-formulation, it proclaims that energy tends to transform from a form more ordered (more able to perform some useful work) to a less ordered form (possessing the lower ability to perform useful work). On the scale of magnitude in which life processes occur (so small from an astronomic scale that the force of gravity loses its key role), the most disordered form of energy is heat energy. Therefore, according to the second law of thermodynamics, energy transformed by living organisms ultimately dissipates in heat (in particular—infrared = heat radiation).

Important implications result from both laws of thermodynamics for the physical aspects of the functioning of living organisms. The first law proclaims that life by no means produces energy, as it is frequently stated in relation to the "production" of energy in the form of ATP. On the contrary, life only transforms energy from one form into another. The energy stored in the form of ATP comes from the energy released during the oxidation of nutrients (in animals) or from the quanta of electromagnetic radiation in the visible range (in plants), and its advantage is its common accessibility for various processes in the cells needing energy for their occurrence. More generally, life constitutes only one of the processes participating in energy dissipation by the earth's surface, a kind of a stop in the chain of transformation of some energy forms into others. This process is only a microscopic contribution to the totality of the energy transformation and entropy production in the cosmos.

The key fact for the existence of various dissipative structures on the earth's surface, including living systems, is the fact that the earth gives away energy in a different form than it receives it from cosmic space. The earth's surface receives significant amounts of energy of only one kind from only one source (if we ignore little amounts of energy still being released from the interior of our planet due to the disintegration of radioactive isotopes).[17] This is, of course, electromagnetic radiation, particularly visible radiation (light) emitted by the sun. On receiving energy, the earth must also get rid of it somehow so that the amount of energy accumulated at any given moment on the earth's surface remains unchanged—if only an inflow of energy to the earth's surface existed, its temperature would grow to infinity. Practically, the only way of getting rid of energy by our planet is the emission of infrared (heat) radiation into cosmic space. This is illustrated in Fig. 2-6. As I mentioned, the energetic balance of the earth (averaged over longer periods of time) is exactly zero—the same amount of energy is absorbed as is emitted. However, of key importance is the fact that visible radiation is absorbed while infrared (heat) radiation is emitted. Such a state of affairs is possible because the sky above our heads is characterized by a high degree of heterogeneity: a small area is focused on by the hot spot of the sun, while the rest falls on the cold emptiness of the cosmos (Penrose 2002, 413-414). If the entire sky possessed the temperature of the sun, the earth's temperature would also quickly acquire the same temperature, reaching a state of thermodynamic equilibrium with the sky, the state with the maximal possible entropy, in which all processes, like life, would become impossible (the same would happen if the entire sky were cold—the earth's surface would cool down to very low temperatures and all thermodynamic gradients would vanish). Fortunately, this does not happen. Cosmic space is much colder than the surface of stars (in fact, its temperature is the temperature of the relic microwave background radiation—about 2.7 K, that is, about minus 270°C), and the earth constitutes only a stop for electromagnetic radiation emitted by our star into space, a stop peculiar in this sense, that it changes one kind of radiation into another.

---

[17] The amount of energy released in a unit of time from the interior of the earth constitutes only 1/10,000 (one-ten thousandth) of the energy received from the Sun (Ozima et al., 2012).

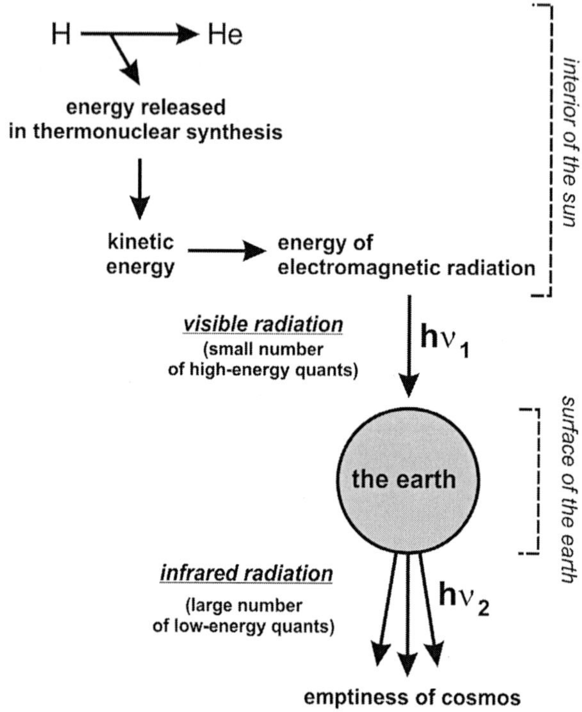

Fig. 2-6. Transformation of energy on the earth's surface. The earth receives from the sun visible radiation energy coming indirectly from thermonuclear synthesis in the interior of our star. Very high-energy electromagnetic radiation is produced in this process, whose energy remains in thermodynamic equilibrium with the kinetic energy of elementary particles. The temperature in the sun's interior amounts to about 15 million K. As we get closer to the sun's surface, the value of both kinds of energy decreases, and at the surface of our star (with a temperature of about 6,000 K), energy is emitted into cosmic space in the form of electromagnetic radiation with the maximum of intensity in the visible range, which reaches the earth's surface, by which point it is partly absorbed, and partly reflected. The earth emits the energy of previously absorbed radiation into the void of the cosmos in the form of infrared (heat) radiation. The overall energetic balance of the terrestrial globe is zero.

Visible radiation (that is, electromagnetic radiation of a certain range of energy of photons inversely proportional to the wavelength) that reaches us from the sun originates from another form of energy, namely from the kinetic energy of the plasma in the sun's interior composed of electrons, protons, and helium nuclei. Within the sun's core, the energy of electromagnetic radiation and kinetic energy remain in thermodynamic equilibrium with each other: elementary particles endowed with huge kinetic energy unceasingly emit high-energy quanta of electromagnetic radiation that are almost instantaneously absorbed by other elementary particles, which increases their kinetic energy, again and again. The temperature in the center of our star amounts to about 15,000,000 K. The temperature, pressure and energy of electromagnetic radiation decrease going from the core to the surface of our star. On the sun's surface, the radiation emitted "outside" is not absorbed but sent into the emptiness of the cosmos (in other words, the sun shines; the temperature of the sun's surface equals about 6,000 K, which corresponds to the energy of electromagnetic radiation, or, more strictly, to the maximum of emission in the visible range). Therefore, to preserve the dynamic equilibrium (constant temperature, pressure, the radius of the sun and so on resulting from the competition between the pressure of radiation and gravity force), the loss of radiant energy must be compensated by a supply of kinetic energy. The source of the latter is, in turn, the energy released during thermonuclear synthesis, consisting of the transformation of four hydrogen nuclei (protons) into a helium nucleus through proton cycle or CNO cycle (Baggott 2015, 93-96; Sparrow 2016, 90-94). Eventually, two protons merge with two electrons forming neutrons (in fact, this process is more complex, as discussed previously), and as a result, the helium nucleus is composed of two protons and two neutrons. Nucleons (protons and neutrons) in an atomic nucleus are bound by a strong nuclear force. Nucleons moved away from each other (similarly like masses exerting gravity force) possess potential energy with a negative sign. On the other hand, the potential energy of nucleons gathered in one place amounts to zero. The joining of nucleons into a helium nucleus (associated with the transformation of two protons into neutrons) means, therefore, a transition from negative energy to zero energy, that is, the system energy has simply increased. It is just this energy that is released during thermonuclear synthesis and transformed into the kinetic energy of elementary particles. Formulating this differently, energy is equivalent to mass according to Einstein's famous formula $E = mc^2$. The helium nucleus has slightly less mass (by 0.7%) than the four protons and two electrons from which it is being formed, and this tiny difference transforms into a huge amount of energy, considering the enormous re-calculation factor (square of light speed). The transformations of energy in the interior of our star are shown in the upper part of Fig. 2-6.

Thermonuclear reactions are possible thanks to gravity, which compresses protons and electrons in the interiors of stars so much that strong nuclear forces, acting at very short distances, can manifest themselves.

Therefore, the primary trigger of the entire described sequence of conversions of some forms of energy into other forms appears to be just gravitation.

Let us be reminded that this sequence comprises the transformation of energy released in the process of binding of nucleons in the nucleus of helium into the kinetic energy of protons, electrons, and helium nuclei (of which the sun is mostly composed), and into high-energy electromagnetic radiation (eventually—mostly into visible radiation), which proceeds in the interior and on the surface of our star, and the transformation of visible radiation into infrared (heat) radiation, occurring on the earth's surface (heat radiation next undergoes dispersion into cosmic space). Of course, we will first be interested in the last process as the driving force of life processes. The transformation of visible radiation into infrared radiation means that the latter has a higher entropy. According to quantum mechanics, waves of electromagnetic radiation propagate in space in packages (quants) called photons (the so-called wave-particle duality) (Penrose 1990, 221, 280; Al-Khalili 2012; Baggott 2013, 254-260). The wavelength of visible radiation is less than the wavelength of infrared (heat) radiation, and therefore (again, in agreement with quantum mechanics), the energy of its quants (photons) is greater. According to the first law of thermodynamics (energy conservation principle), the total energy of the radiation absorbed and emitted by the earth's surface must be the same over long periods if the average temperature on the surface of our planet does not change. Therefore, the number of high-energy quants of visible radiation (photons) absorbed by our planet is smaller than the number of low-energy quants (photons) of infrared radiation that is emitted by the earth's surface. For this reason, the energy of infrared radiation is characterized by a higher degree of dispersion in space (is scattered over a greater area, less ordered, it is more difficult to "localize" its energy quanta), which of course implies higher entropy and a smaller amount of (thermodynamically understood) information. It seems astonishing that such an elementary physical process as the dissipation of visible radiation into infrared radiation drives the huge majority of the processes forming the enormously complex and hierarchically-organized terrestrial biosphere (some organisms draw chemical energy from hydrothermal vents located on ridges on the oceanic floor that are described in more detail in Chapter 3 devoted to the origin of life; nowadays, however, this process has a minute contribution to the energetic balance of the whole biosphere).

Considering everything said above, from the point of view of the thermodynamics of the cosmos, the main "role" of the earth's surface constitutes the dissipation of the energy of visible radiation into the energy of infrared radiation and, therefore, entropy production. Many various processes are responsible for the "realization" of this "task." The simplest is the absorption and emission of radiation by naked rock or, e.g., sand in a desert. The rock surface is heated up during the day through the absorption of electromagnetic radiation in the visible range of the spectrum. Afterwards, during the day, and especially at night, the rock gives away heat (and itself cools down) by sending away infrared radiation. This is presented in Fig. 2-7 A. The transformation of visible radiation into heat radiation occurs here principally in two stages. First, the energy of visible radiation is transformed into the kinetic energy of molecules building the rock, which in turn undergoes conversion into the energy of infrared radiation.

A somewhat more complicated situation occurs when the ocean surfaces are heated by the sun. Transforming one kind of radiation into another launches a sequence of hydrological and atmospheric processes associated with cyclic water circulation. This is illustrated in Fig. 2-7 B. The surface of the oceans (of course, this also concerns smaller water reservoirs such as seas and lakes) gives away the sun's heat from the absorption of visible radiation through the evaporation of water in the form of water vapor. Light rays transfer their energy to water molecules in the ocean, increasing their kinetic energy, and enabling water vapor to transport to higher layers of the atmosphere against gravity, thus gaining potential energy. There, the water vapor undergoes cooling, getting rid of a part of its energy in the form of infrared radiation emitted into the cosmos. As a result, the water vapor forms clouds, within which it undergoes condensation and falls on the earth's surface in the form of rain (or snow), converting its potential energy into the kinetic energy of the raindrops. If the rain falls into the ocean, the cycle closes. This part of precipitation (rain, snow, hail) that falls on land (especially on mountains) forms streams and rivers flowing back to the oceans. The kinetic energy of the water flowing in them, driving, for instance, the turbines producing electricity in water power stations, is nothing but the transformed energy of the sun. When the water finally returns to the ocean, it can again evaporate, and the whole cycle also closes. Therefore, transforming visible radiation into heat radiation drives a certain cyclic process, forming a certain dynamic structure organized in space and time. The heating up and cooling down of air masses causes additional atmospheric phenomena, such as atmospheric pressure fronts, winds (including hurricanes), storms, and cyclones. The transformation (dissipation) of energy in the hydrosphere and atmosphere does not lead, therefore, to the completely chaotic behavior of water and air molecules (although the atmospheric and hydrological phenomena should be considered as chaotic in some important sense—compare the famous butterfly effect)—just the opposite, some regularities, and therefore order, can be noticed in the behavior of atmospheric currents. It has nothing in common with the structures created by gravitation, such as astronomical objects—gravitational interactions between various parts of the atmosphere are negligible. Nevertheless, gravitation plays an important role in the circulation of water in the atmosphere and hydrosphere: the "heated up" water, being lighter, is transported as a vapor to the upper atmosphere layers against the force of gravity, while the atmospheric precipitation (cooled down, heavier water) proceeds in line with the force of gravity.

Yet another system dissipating energy on the earth's surface is the biosphere, that is, the assembly of all living organisms. The phenomenon of conversion of some energy forms into others is here, as in the case of atmosphere and hydrosphere, a multi-stage process and is closed into a cyclic sequence of transformations. This already becomes characteristic of the systems participating in entropy production (energy dissipation)—that is, dissipative structures. But, first things first. The visible radiation is absorbed by plants, mostly unicellular, including algae and photosynthetic bacteria, and its energy transforms in the process of photosynthesis into the chemical energy of bonds between atoms in organic compounds. Sugar glucose (strictly speaking, three-carbon glyceraldehyde 3-phosphate, from which six-carbon glucose is synthesized), produced from carbon dioxide and water, constitutes the primary product of photosynthesis, while oxygen is released (in the light-dependent phase of photosynthesis). In other words, electrons (together with protons, that is, generally—hydrogen atoms) are transferred under the influence of the energy of quants of visible radiation from oxygen to carbon. Due to this, the "reduced form of oxygen," that is, water (H_2O), undergoes oxidation to molecular oxygen (O_2), while the extremely oxidized form of carbon, that is, carbon dioxide (CO_2), undergoes reduction to organic compounds (containing C-H and C-C bonds). The organic compounds produced by plants and other photosynthetic organisms serve, in turn, as the building material and energy source for animals and other, most frequently unicellular, heterotrophs (organisms needing an external source of organic compounds) feeding on autotrophs (not needing an external source of organic compounds) (photosynthetic, drawing energy from light or chemosynthetic, drawing energy from inorganic chemical reactions), or on other heterotrophic organisms, and for decomposers consuming dead organic matter.[18] They oxidize organic nutrients with the participation of oxygen, converting them to carbon dioxide and water, and

[18] The most abundant organisms on our planet are bacteria from genus *Pelagibacter* feeding on organic matter dissolved in oceans.

the whole cycle closes. The gained energy (stored mostly in the form of ATP) is used by animals and other heterotrophs for driving various types of work: chemical (synthesis of various compounds), mechanical (movement, e.g., muscle contraction), osmotic (transport of substances across cellular membranes against the concentration gradient), electrical (transport of ions across cellular membranes against the gradient of electrical potential), and so on. Each process (which also concerns plants' metabolic processes) is associated with some energetic losses. As a result, a part of the energy is dissipated in the form of heat and ultimately radiates into cosmic space in the form of infrared radiation. This is presented in Fig. 2-7 C.

Additionally, decomposers disintegrate dead bodies of living organisms (or their fragments), producing heat. Therefore, life as a whole does exactly the same as what naked rock, the atmosphere, and hydrosphere do—it transforms the visible radiation emitted by the sun that reaches the earth's surface into infrared radiation dispersed in turn in the abyss of the universe. Similarly, as in the case of atmospheric phenomena, this is associated with forming ordered dynamic structures based on cyclic processes—dissipative structures. There is perhaps no need to convince anybody that living organisms are characterized by highly-organized design. Nevertheless, the entire biosphere is a complicated dynamic structure (organized not only in space but, first of all, in time). The origination of such structures is undoubtedly associated with a local entropy decrease.

In addition, every living individual is an unmeasurably complex, hierarchically-organized dissipative structure, or, if somebody prefers, an assembly of such structures. Within the living individual, a huge number of cyclic processes are intensively dissipating energy and producing entropy. An example of such a dissipative structure is the cyclic circulation of protons (hydrogen ions H^+) across lipid-protein membranes associated with energy transduction (ATP production), presented schematically in Fig. 2-7 D (Nicholls and Ferguson 2002). This process will be discussed in more detail below.

Summing up, a simple, elementary process, namely the dissipation of visible radiation into infrared radiation, drives most living processes, enabling the existence of the biosphere. From the point of view of thermodynamics, life constitutes only one more (although undoubtedly unprecedentedly complicated) *dissipative structure* on the surface of our planet. Life is, in its nature, a circular process associated with a cyclic conversion of matter and energy, sustained by the thermodynamic gradient between the hot surface of the sun and the cold space of the cosmos. If the sun faded away for some reason, life (or at least most of it) would not be possible.

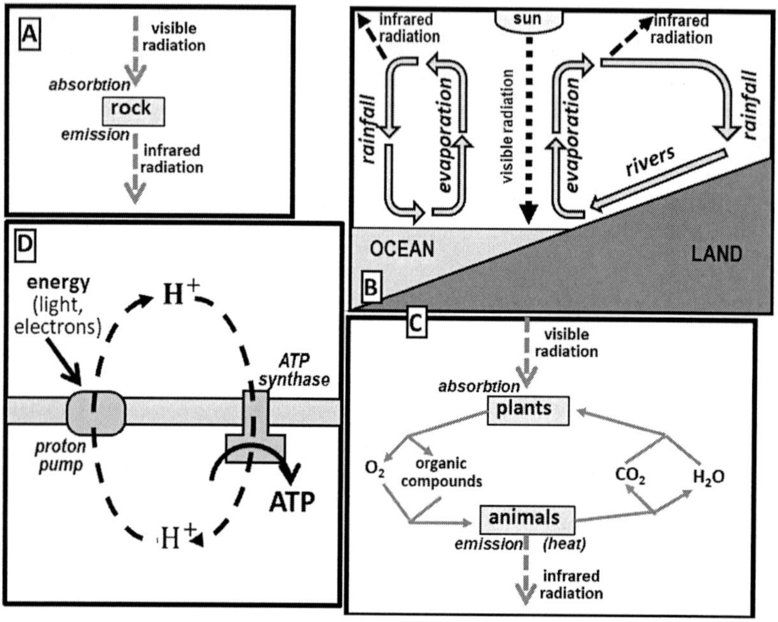

Fig. 2-7. Various systems dissipating energy and producing entropy on the earth's surface. A. Naked rock (or desert sand) absorbs visible radiation and emits infrared radiation; this does not lead to formation of any dynamic structures. B. Cyclic circulation of water in terrestrial atmosphere and hydrosphere driven by the dissipation of visible radiation into infrared radiation. C. Life as one of ways of energy dissipation by the earth's surface consisting of the transformation of the energy of visible radiation into energy of infrared radiation. D. Circulation of protons across lipid-protein membranes associated with energy transduction (inner mitochondrial membrane, membrane of thylakoids in chloroplasts, cellular membrane of bacteria), driven by a conversion of the energy of electrons in redox (reduction/oxidation) reactions (coming from the energy contained in respiratory substrates or energy of visible radiation) into the energy stored in ATP. This energy is used to carry out various kinds of work in the cell (chemical, mechanical, osmotic, electric, *etc*., work), and becomes ultimately dissipated in the form of heat.

Some minor part of the present biosphere does not draw energy from the sun and is not driven by the thermodynamic gradient between visible and infrared radiation. Whole ecosystems exist that do not depend directly or are completely isolated energetically (thermodynamically) from the visible radiation emitted by our star. They exist in the neighborhood of hydrothermal vents occurring in the region of spreading zones, that is, places where plates of the terrestrial crust diverge (move away from each other). The drift of the crustal plates (discussed in more detail in Chapter 3) is driven by convective currents in the plastic mantle of the earth. In the places where the ascending convective currents reach the mantle-crust border and afterward spread out on the sides, they draw (transport) crust fragments with them. Hot lava pours into these splits (rifts) between the tectonic plates of the earth's crust, which are moving away from one another, and solidifies due to cooling after contact with water, thus forming a new oceanic crust. In this way, mid-ocean ridges, with rifts running along their spines, are formed, for instance, the Mid-Atlantic Ridge that runs from north to south across the Atlantic Ocean (America moves away from Europe and Africa at a speed of about three centimeters a year, and the Atlantic Ocean widens by this length). In another place, the crust is being destroyed. This happens in the so-called *subduction zones*, where one crustal plate submerges into the mantle below another crustal plate and undergoes melting. Its material, being lighter, moves up in the mantle, punctures itself through the crustal plate to its surface, and flows out as lava. This is accompanied by intense volcanism and violent earthquakes. The subduction zone occurs, for instance, east of Japan or west of the Andes (entering into the composition of the so-called "Ring of fire" around the Pacific, characterized by a high intensity of tectonic phenomena, in particular, volcanism and earthquakes) (Macdougall 1996; Baggott 2015; Smith and Morowitz 2016, 115-140).

Let us, however, come back to the spreading zones. In the actual zones or their direct vicinity, water heated to the temperature of some hundred (200–> 400)[19] degrees centigrade spews out from a certain type of hydrothermal vents called acidic vents or "black smokers," carrying with it substances washed out from the newly formed crust, including first of all hydrogen sulfide H_2S and metal (Fe, Cu, Zn, Pb, Ba, and Ca) sulfides (which cause the black color of the released solution) (Lane 2015; Baggott 2015, 215-219; Smith and Morowitz 2016, 121, 368-369). The walls of vents are formed due to metal sulfide precipitation, and the solution that spews out from them has, as their name suggests, a pH significantly below 7.0 (as low as 2.8). They are hundreds of years old, and their height is several meters. Chemoautotrophic (autotrophic, carrying out chemosynthesis) bacteria accept hydrogen from H_2S (elementary sulfur S and sulfates SO_4^{2-} are produced), which is used for the reduction (assimilation) of CO_2, due to which organic compounds are formed (this type of chemosynthesis is called *sulfurization*) (in turn, other bacteria oxidize organic compounds in the

[19] Water can occur in the liquid state in these temperatures, as it experiences a huge pressure (200–400 atm) present on the depth of about 2,000 m (1,500–4,000 m).

process of sulfur respiration, or *desulfurization*, transferring electrons/ protons from organic carbon compounds to sulfates, reducing them to S, and finally to H_2S). In turn, whole organism assemblies feed on or live in symbiosis with these bacteria (Lane 2015; Baggott 2015, 215-219). In this way, differentiated, astonishingly rich local ecosystems originate from tube warms (reaching 2.5 m in length), polychaete worms, bivalves, starfish, shrimps, fish, snails, crabs, octopi, and many others. However, contrary to frequently repeated false statements in popular texts, these ecosystems are not fully autonomic energetically/thermodynamically because they depend on the sun's radiation. Namely, the reduction of CO_2 with the participation of H_2S is unfavorable energetically (thermo-dynamically) and, therefore, cannot proceed spontaneously. The energy for the assimilation of carbon dioxide is supplied by the oxidation of H_2S through a transfer of electrons (hydrogens) from this compound to oxygen O_2, due to which water H_2O is produced. Moreover, the only source of free oxygen on the earth is photosynthesis occurring at the surface of oceans (and on the land), driven by the absorption of quanta of electromagnetic radiation in the visible range and "transformation" of it into quanta of infrared radiation, ultimately emitted into cosmic space. Therefore, the ecosystems associated with the acidic hydrothermal vents (black smokers) are not energetically (thermodynamically) independent.

However, it should be noted that certain green sulfur bacteria (Chlorobiaceae) are thermodynamically independent and carry out photosynthesis in complete darkness at a depth of about 2.5 km, drawing both energy and "reductive force" from hydrothermal vents (Beatty et al. 2005). Doesn't this seem an obvious paradox: photosynthesis in the absence of light? However, the photosynthetic pigment of these bacteria is bacteriochlorophyll, a relative of chlorophyll, whose maximum absorption of electromagnetic radiation lies in the infrared range, invisible to the human eye. Such radiation (also called heat radiation) is emitted by hot objects, including acidic hydrothermal vents. Its energy is too low to knock out electrons (hydrogens) from water (H_2O) and release molecular oxygen (O_2), as plants or cyanobacteria do it. Instead, the discussed green sulfur bacteria use hydrogen sulfide (H_2S) as the donor of electrons (hydrogen). It so happens that acidic black smokers deliver not only infrared radiation, but also H_2S. These bacteria use it for the reduction (assimilation) of CO_2 and the production of organic compounds, which is driven by the absorption of infrared radiation. However, does this not contradict the repeated statement that infrared radiation constitutes the terminal stage of the transformation of energy on the terrestrial globe, its form of highest entropy, and in the face of this, is unable to perform any useful work? No, because infrared radiation "has many faces;" it comprises a very broad range of wavelength, and therefore also energy, of electromagnetic radiation: from 780 nm (an nm is a one billionth part of a meter, that is, 10^{-9} m) to 1 mm (millimeter) (that is 1,000,000 nm), while the visible radiation is situated only between 380 nm and 780 nm. Therefore, while the wavelength (and energy of photons inversely proportional to it) varies a little in the case of visible radiation, with the longest wave being around twice the length of the smallest, for infrared radiation, the longest wave can be over one thousand times the length of the smallest. Now, infrared radiation, with relatively little wavelength and high energy, emitted by bodies with a temperature of some hundred Celsius degrees,[20] for instance, black smokers, drives the photosynthesis of green sulfur bacteria, while infrared radiation with a relatively large wavelength and low energy, emitted by bodies with temperatures of a few to some tens of Celsius degrees, is released into the cosmic space. Therefore, one infrared radiation is not equal to another infrared radiation (the classification of electromagnetic radiation based on wavelength is to a large extent arbitrary, in the "natural" way, only visible radiation can be distinguished, received by our eyes). As the acidic hydrothermal vents deliver to green sulfur bacteria both energy (high-energy heat radiation) and reductive force (in the form of H_2S), while CO_2 is commonly accessible (and was even much more abundant billions of years ago, during the early stages of the evolution of the terrestrial atmosphere, when the oxygen-based photosynthesis was still absent), these bacteria are self-sufficient both energetically and materially. As will be discussed in Chapter 3, this could have great importance for the origin and early stages of the evolution of life on earth.

However, it is improbable that just photosynthesis was the first energetic process in the earth's history because of its complexity. The components of the photosynthetic chain, in particular, quinone (one of a few types) and complex bf (analog of complex III or bc_1 in the respiratory chain), are principally similar to elements of the chemosynthetic or respiratory chain (Nicholls and Ferguson 2002). The main novelty during the origination of photosynthesis was the "invention" of chlorophyll. It seems that it evolved only in bacteria (including ancestors of chloroplasts), but not in archaea or eukaryotes (see below). Nevertheless, this was not such a great novelty: porphyrin, the "core" of the chlorophyll molecule binding a magnesium ion, possesses a very similar structure to heme, constituting the "core" of the cytochrome molecule binding an iron ion. Furthermore, there are organisms

[20] As we remember, the visible radiation is emitted by the surface of the sun with a temperature of 6,000 °C.

(e.g., bacterium *Rhodobacter sphaeroides*) in which the respiratory and photosynthetic chains are "crossed" (Nicholls and Ferguson 2002, 158-163); they possess a large common part comprising NADH/NAD^+, NADH dehydrogenase (complex I), ubiquinone (UQ), complex bc_1 (complex III), and cytochrome c. The exclusively respiratory elements are three different terminal cytochrome oxidases transferring electrons to oxygen, including aa_3 oxidase (complex IV), while the exclusively photosynthetic element is the reaction center containing chlorophyll. Depending on the conditions, electrons flow through the common part either in one or the opposite direction, and thus either respiration or photosynthesis occurs. Probably this is how photosynthesis originated—through adding the reaction center (and entire light-collecting antennae, composed of bacteriochlorophyll) absorbing quanta of electromagnetic radiation and knocking out electrons first from H_2S and later from H_2O. We do not know, however, whether this initially happened near the ocean surface (visible radiation) or in the vicinity of black smokers (near-infrared radiation).

We finally come to the second kind of hydrothermal vents, namely alkalic vents or white smokers (Lane 2015; Baggott 2015, 215-219; Smith and Morowitz 2016, 121, 368-369). They are (presently) much rarer than black smokers and are formed at some distance from the spreading zone (several kilometers), they eject an alkalic solution of $pH > 7.0$ heated up to a temperature of some tens Celsius (up to 100°C), and are rich in molecular hydrogen (H_2) and calcium salts (hence their light color). They are created due to the deposition of calcium carbonate, exist for at least 100,000 years, and their height reaches 60 m.[21] H_2, rarely occurring on the earth, is used here as the reductive force (a source of electrons and hydrogen in general) for CO_2 assimilation and organic compounds production by prokaryotic (devoid of a nucleus) organisms: bacteria and archaebacteria (archaea) living in (the vicinity of) white smokers. The former are acetogens, which gain energy from the transformation of H_2 and CO_2 into acetate (acetic acid, CH_3COOH), while the latter belong to methanogens, in which the by-product of the processes leading to energy production from the reaction of H_2 with CO_2 is methane (CH_4). These organisms also constitute the basis of the food chain of local ecosystems associated with white smokers, although not so spectacular as the ecosystems in the neighborhood of black smokers. Nevertheless, the ecosystems driven by the chemical energy supplied by alkalic hydrothermal vents are totally autonomous in the thermodynamic and energetic sense, independent of solar light and other living organisms.

[21] An example is the so-called "Lost City" in the vicinity of the Mid-Atlantic Ridge.

Generally, presently most of the biosphere is energetically (thermodynamically) driven by the visible radiation coming from the sun and absorbed by plants (in a broad meaning, including cyanobacteria and eukaryotic algae) in the process of photosynthesis and transformed eventually (by plants, animals and other heterotrophs, including decomposers) into low-energy infrared radiation (far infrared radiation). On the other hand, a minute part of the biosphere relies upon chemical energy in the form of the "reductive force" (H_2 and H_2S) delivered from the interior (mantle and new crust formed from its material) of the earth by acidic and alkalic hydrothermal vents (black smokers and white smokers). It can be assessed that presently the energy supplied from the interior of the earth constitutes only one-ten thousandth (1/10,000, 0.01%) of the energy received from the sun in the form of visible radiation (Ozima et al. 2012) (a part of this radiation is reflected, and a part is absorbed). In the distant past, the proportions between the fractions of the biosphere driven by both energy/negentropy sources were probably different, as various tectonic processes were much more intense. Fig. 2-8 presents life as an assembly of dissipative structures sustained by either the dissipation of visible radiation into infrared radiation (Fig. 2-8 A) or by the interaction of the relatively reduced (containing more hydrogen) interior (mantle and oceanic crust) of the earth with the relatively oxidized (containing less hydrogen) surface (hydrosphere and atmosphere) of our planet (Fig. 2-8 B). The latter process requires closer discussion.

There is a redox (degree of reduction/oxidation) gradient between the earth's mantle and the oceans and atmosphere—the former is much more reduced, as it contains more hydrogens that easily donate electrons, while the latter are much more oxidized—they contain relatively fewer hydrogens, but much oxygen (in the form of molecular oxygen, O_2, and carbon dioxide, CO_2), which willingly accepts electrons (Smith and Morowitz 2016). Two processes sustain this gradient. First, by an escape of hydrogen (formed, e.g., due to photolysis—light-induced decomposition—of water) from the atmosphere and hydrosphere into cosmic space (hydrogen is the lightest gas, and therefore it is most difficult to keep it on or near the earth's surface by gravity), which makes them relatively oxidized. Second, the delivery of hydrogen (in the form of, e.g., H_2 or H_2S) from the earth's mantle to the oceans and atmosphere, replenishing its deficit caused by the escape into cosmic space. Due to this process, strongly reduced compounds (such as molecular hydrogen, H_2, hydrogen sulfide H_2S, or water H_2O itself) enter into contact with highly oxidized compounds (first of all, carbon dioxide, CO_2 and molecular oxygen, O_2). Formed in this way, the redox/chemical/thermodynamic gradients drive the process of chemosynthesis in prokaryotes living in the vicinity of acidic and alkalic hydrothermal vents (black and white smokers) and sustain the functioning of the entire ecosystems associated with them.

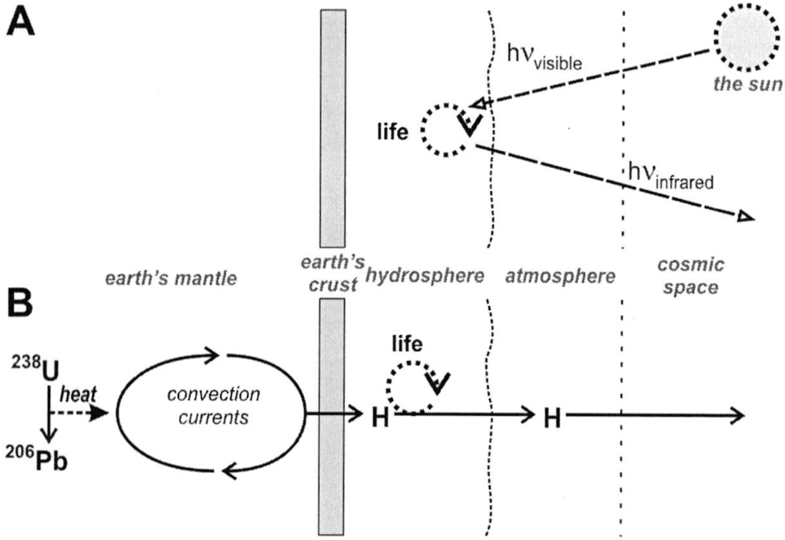

Fig. 2-8. Two processes presently driving life on the surface of the terrestrial globe. A. Transformation of visible radiation emitted by the Sun into infrared radiation sent away into cosmic space. B. Reaction of reduced (containing hydrogen, H, in various forms) content of the Earth's mantle with the, to a large extent, oxidized (containing less hydrogen) hydrosphere (oceans) and atmosphere, in particular with the entering into their composition carbon dioxide (CO_2). The mantle material issues outside through the crust and contacts hydro- and atmosphere as a result of the tectonics of the plates of the Earth's crust brought about by convective currents in the Earth's mantle, provoked in turn by heating up of the interior of our planet by the heat coming from the disintegration of radioactive isotopes, for instance of uranium ^{238}U into lead ^{206}Pb. Hydrogen (produced in the atmosphere, e.g., as a result of water photolysis) escapes into cosmic space, as the gravitation of the Earth is too weak to keep it on, and therefore its reservoir in the hydro- and atmosphere must be continuously replenished. As a result, a chemical/thermodynamic/redox gradient is formed between the relatively reduced mantle and relatively oxidized atmo-/hydrosphere, driving the ecosystems of living organisms existing in the vicinity of hydrothermal vents.

However, more important for the present life, including life directly driven by photosynthesis based on visible solar radiation near the surface of oceans and on land, is the constant supply of hydrogen, first of all in the form of H_2S and H_2, from the mantle to hydrosphere and atmosphere. Other phenomena associated with crustal plate tectonics play a much greater role than hydrothermal vents, first of all, the volcano eruptions in the subduction zone and lava effusions in the spreading zone. All these processes are grouped into one (the arrow from "convective currents" to "H" through the "earth's crust") in Fig. 2-8 B. If not for the convection of the mantle material and crustal plates' tectonics, the hydrogen lost due to hydrogen escaping into cosmic space would not be replenished, the earth's surface would become extremely oxidized, the hydrosphere (oceans composed of H_2O) would vanish because of the lack of hydrogen, most of the oxygen would bind with carbon and form carbon dioxide, and our planet would be enclosed by a dense atmosphere dominated by CO_2 which would exert enormous pressure, while the temperature would reach several hundred degrees centigrade due to a monstrous greenhouse effect. Life, of course, would not be possible, not only because of the prevailing high temperature but also because of the lack of water and reductive power in the form of hydrogen, which would allow for the synthesis of organic compounds. Such conditions are present now on the "twin" planet of the earth—Venus (where the dense atmosphere is mostly composed of CO_2, darkness prevails on the planet's surface, the pressure is almost 100-fold higher than on the earth, and the temperature equals about 740 K $\approx$ 470°C). This results not only from the relative proximity of the sun but, first of all, from the fact that for some reason, there is no crustal plate tectonics on Venus, and therefore most probably no convective currents in its interior. This is probably a consequence of the escape of hydrogen and lack of water that constitutes the "lubricant" for the convective currents in the earth's mantle and crustal plate tectonics. Therefore, the reason seems to mix with the result, and they probably mutually co-condition one another.

Consequently, we can decidedly state that we owe our existence to the crustal plate tectonics on the earth, although not for long (in the geological scale of time). Some calculations indicate that after slightly more than one billion years, the supply of hydrogen from the earth's interior will cease, the convective currents will wither away, the terrestrial atmosphere will become extremely oxidized because of the escape of hydrogen into cosmic space, oceans will vanish, life will die out, and the earth will begin to resemble the present Venus (Ozaki and Reinhard 2021). Because life has existed on our planet for about four billion years, this is quite a near perspective. In the face of this, we do not have to worry about the fact that in about five billion years, the sun will combust most of the hydrogen in its core, will distend itself to the size of a red giant (a type of star, in the stage of evolution of stars similar to the sun), and consume the earth, or at least burn its surface to naked rock. We will not have been around by then for a long time.

I previously mentioned not only "normal" (oxygen-related) respiration and "normal" (oxygen-related) photosynthesis but also sulfur-related respiration (desulfurization), sulfur-related photosynthesis and sulfur-related chemosynthesis (sulfurization). Perhaps, therefore, it is time to present a sketch of the processes leading to storing energy in a commonly-accessible form, e.g., ATP in different organisms on the earth. An overview of such processes is shown in Fig. 2-9.

As we see, we have here a set of processes in which electrons and protons (that is, overall, hydrogen atoms) are transferred from one chemical element to another, where the former is oxidized and the latter is reduced (for different aspects of respiration, photosynthesis and chemosynthesis see Voet et al. 1999; Nicholls and Ferguson 2002; Murray et al. 2012; Lane 2015; Smith and Morowitz 2016). Thus, in the process of *photosynthesis*, quants of visible radiation supply energy for the transfer of electrons/protons/hydrogens from the reduced form of oxygen (water, H_2O) or sulfur (hydrogen sulfide, H_2S) on the maximally oxidized form of carbon (carbon dioxide, CO_2), while a moderately reduced form of carbon (organic compounds) and an oxidized form of oxygen (that is molecular oxygen, O_2) or sulfur (elementary sulfur, S, or sulfates, SO_4^{2-}) is produced. *Respiration* is a process that is essentially the opposite of photosynthesis. In the presence of oxygen, electrons/protons/hydrogens are transferred from organic compounds, while CO_2 and H_2O are produced. In the absence of access to oxygen, electrons/protons/hydrogens are transferred from organic compounds to oxidized forms of sulfur (S or SO_4^{2-}) (sulfur respiration, desulfurization) or nitrogen (nitrates, NO^{3-}) (nitrogen respiration, denitrification), while reduced forms of these elements are produced: hydrogen sulfide (H_2S) or ammonia (NH_3). When the latter penetrate an environment rich in oxygen, they can be oxidized to S and SO_4^{2-} (sulfurization) or NO^{3-} (nitrification), respectively, while hydrogens are transferred to O_2, which undergoes reduction to H_2O. Electrons can also be transferred to O_2 from reduced iron Fe^{2+}, and as a result, oxidized iron Fe^{3+} and H_2O are formed (protons are taken up from the water phase). The above processes are classified as *chemosynthesis*. There is also the possible transfer of electrons/hydrogens from carbon to carbon—in which case, we have the *fermentation* process. Due to this process, moderately reduced glucose is decomposed to extremely oxidized carbon dioxide and strongly reduced ethyl alcohol (CH_3CH_2OH) (ethanol fermentation) or transformed into moderately reduced lactic acid ($CH_3CHOHCOOH$) (lactic acid fermentation). In the presence of oxygen, they can be oxidized by O_2 to CO_2 and H_2O. The discussed cyclic processes of transfer of electrons/ protons/hydrogens between chemical elements, associated with their alternating oxidation and reduction, represent, of course, a few of the huge number of hierarchically-organized dissipative structures in the terrestrial biosphere. Only iron participating in chemosynthesis is reduced due to geological and not biological processes (this also concerns the emission of H_2 and H_2S by hydrothermal vents discussed above).

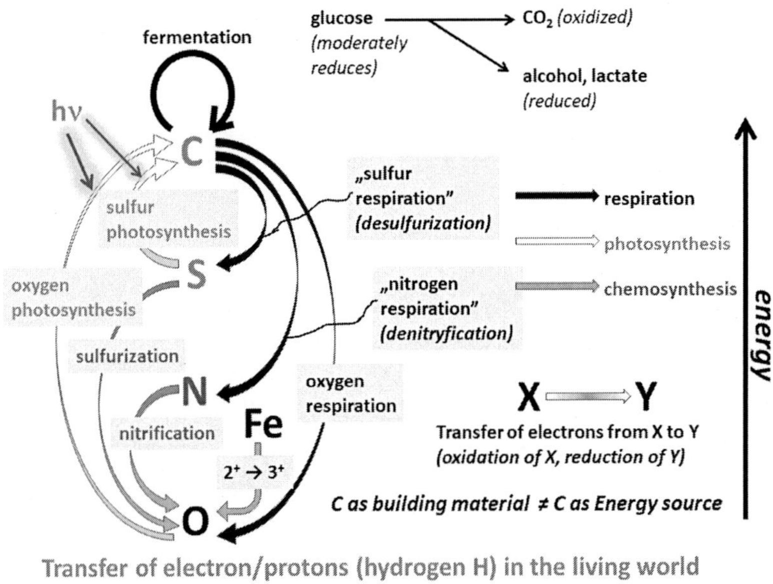

Fig. 2-9. General overview of reactions in the terrestrial biosphere (occurring in various living organisms) leading to gaining energy in an easily-accessible form, usually ATP. We distinguish here photosynthesis: oxygen-related photosynthesis (transfer of electrons/hydrogens from reduced oxygen, that is, water, H_2O, to oxidized carbon, CO_2, driven by the absorption of quanta of visible radiation, which leads to synthesis of moderately-reduced carbon compounds—organic compounds—and to molecular oxygen, O_2, release) and sulfur-related photosynthesis (transfer of electrons/hydrogens from reduced sulfur, that is, hydrogen sulfide, H_2S, to oxidized carbon, CO_2, driven by absorption of quanta of visible or infrared radiation, which leads to the synthesis of moderately-reduced carbon compounds—organic compounds—and to oxidation of sulfur to elementary sulfur, S, or sulfates, SO_4^{2-}). We also have chemosynthesis: sulfur chemosynthesis (sulfurization), nitrogen chemosynthesis (nitrification) and iron chemosynthesis, due to which electrons/hydrogens are transferred from the reduced form of sulfur (H_2S), nitrogen (ammonia, NH_3) or iron (Fe^{2+}) to molecular oxygen O_2, due to which oxidized forms of, respectively, sulfur (S or SO_4^{2-}), nitrogen (nitrates, NO_3^-) and iron (Fe^{3+}), as well the reduced form of oxygen (H_2O) are produced. In the process of

respiration, electrons/hydrogens are transferred from organic compounds (the maximally oxidized form of carbon, CO_2, is formed) to oxidized form of oxygen (that is simply O_2) (oxygen respiration), sulfur (S or SO_4^{2-}) (sulfur oxidation, desulfurization) or nitrogen (nitrates, NO_3^-) (nitrogen respiration, denitrification), leading to production of the reduced forms of these elements: H_2O, H_2S, and NH_3, respectively. Finally, in the process of fermentation, moderately reduced glucose ($C_6H_{12}O_6$) is decomposed to extremely oxidized CO_2 and highly reduced ethyl alcohol (CH_3CH_2OH) or transformed into moderately reduced lactic acid ($CH_3CHOHCOOH$)—electrons/hydrogens are transferred from some carbon atoms to other carbon atoms. All these processes with exception of oxygen-related photosynthesis, oxygen respiration and fermentation occur only in prokaryotes: bacteria and archaea (archaebacteria). It should be noticed that the transfer of electrons/hydrogens between elements in the biosphere, that is, alternate oxidation and reduction, has a cyclic character and represents an assembly of dissipative structures driven by energy dissipation/entropy production.

Many cyclic processes driven by the flow and dissipation of energy also occur on a smaller scale, e.g., in the cells of living individuals, their organs, or whole organisms. A good example is the circulation (cyclic course) of H^+ ions (protons) across cellular membranes associated with the transduction and storage of energy in the form of ATP, shown in Fig. 2-7 D. According to the famous (at least among biochemists and biophysicists) Mitchell's chemiosmotic theory (Nicholls and Ferguson 2002, 57-88), the intermediate in ATP synthesis in the process of oxidative phosphorylation (oxygen respiration) (but also sulfur and nitrogen respiration), photosynthesis, and chemosynthesis in animals, plants, and prokaryotes (bacteria and archaea) is the H^+ gradient (sometimes Na^+ gradient) across the inner mitochondrial membrane, thylakoid membrane in chloroplasts, or cellular membrane in prokaryotes. This gradient is composed of the osmotic (concentration) component associated with a greater concentration of protons on one side of the membrane (e.g., on the outer side of the inner mitochondrial membrane) than on the other side, and the electrical component, related to the gradient of electrical charge across the membrane: the charge is more positive on the one side (again, e.g., on the outer side of the inner mitochondrial membrane), and more negative on the other side. The discussed thermodynamic gradient is formed due to the transfer of protons and charges by *proton pumps*—large protein complexes containing numerous redox centers—from one side of the membrane to the other against the gradient of concentration and charge (e.g., from the inside to the outside of the inner mitochondrial membrane). There can be one to three proton pumps in an electron-transfer chain (for instance, three proton pumps in the respiratory chain in mitochondria or prokaryotes) (Nicholls and Ferguson 2002).

The pumping of protons is driven by (coupled with) a flow of electrons through a relevant electron transport chain (respiratory, photosynthetic, chemosynthetic), of which these pumps are elements. Electrons are transferred between redox centers, in which definite groups (frequently bound metal ions) can undergo alternating reduction (receive electrons) and oxidation (give away electrons). These can be cytochromes, containing the hem group binding an iron (Fe) ion, iron-sulfur centers (non-hem iron, Fe/S), or copper (Cu) ions, but also NAD, NADP, FAD, and UQ (ubiquinone) or PQ (plastoquinone): organic compounds not containing metal ions. The flow of electrons (eventually protons, and all together: hydrogen atoms) from respiratory substrates being organic compounds, for instance, sugars or fatty acids (they are oxidized to CO_2), on oxygen (H_2O is formed) in the case of oxidative phosphorylation, occurs spontaneously (is favorable energetically, has a negative ΔG value). On the other hand, an opposite process that occurs in photosynthesis, that is, knocking out of electrons (and protons) from water (O_2 is released) and their transfer on carbon dioxide (organic compounds are synthesized) cannot proceed spontaneously (is unfavorable energetically, has a positive ΔG value), and therefore must be driven by quanta of electromagnetic radiation in the visible range, being absorbed by chlorophyll (or bacteriochlorophyll), that deliver the needed energy and "force" the inverse (in relation to that in the respiratory chain) direction of the electron flow. As mentioned, in both cases, proton pumping against the gradient of concentration and charge takes place (from the compartment with a smaller concentration of protons and more negative charge to the compartment with a greater concentration and more positive charge). In one exceptional case (halophilic bacteria), only one type of proton pump (bacteriorhodopsin) is driven directly by quanta of visible radiation (Nicholls and Ferguson 2002, 189-193). For obvious reasons, the pumped-out protons tend (in agreement with the laws of thermodynamics) to return from the compartment where there are a lot of them, and which repulses them electrically (positive charge), to the compartment where their concentration is lower[22] and which exerts an electrical attraction on them. The lipid-protein membrane associated with energy transformation (and its storage in the form of ATP) is essentially

[22] Similarly, gas spontaneously flows from a container where a greater pressure prevails to the container, connected with it by a pipe, where the pressure is lower. On the way, the flowing gas can drive a turbine connected with a dynamo and generate electrical current. This is an analogy with ATP synthesis driven by the dissipation of the proton gradient.

impenetrable for protons.[23] Most protons return through an enzyme called ATP synthase, having a form of a "mushroom," whose "stipe" is anchored in a membrane associated with energy transduction (e.g., the inner mitochondrial membrane), and the "cup" protrudes into the compartment, where the proton concentration is smaller and the charge negative (e.g., mitochondrial matrix = interior). The spontaneous flow of protons through a channel in ATP synthase downstream the proton concentration and electrical potential gradient (the process with high negative ΔG) and is coupled with ATP synthesis from ADP and inorganic phosphate (P_i) (a process with positive ΔG). 2.7 H^+ ions (in higher organisms) are used for the synthesis of one ATP molecule inside the mitochondrion (three ATP molecules are formed for eight returning H^+ ions), and an additional one H^+ ion is used for the transport of ATP outside the mitochondrion, and the substrates for its synthesis: ADP and P_i inside the mitochondrion. In turn, the reverse reaction, hydrolysis of ATP to ADP and P_i, can be coupled with performing various kinds of work in the cell. Therefore, the energy transformation from one form into another (from the energy of respiratory substrates or visible radiation quanta—into the energy of the flow of electrons through redox centers from more "electronegative" to more "electropositive" centers—into the energy of the proton gradient—into the energy stored in ATP—into finally a certain kind of useful work), accompanied at each stage by energy dissipation (and entropy production) in the form of heat, is associated with a dynamic cyclic process (dissipative structure), that is, the circulation of protons across the lipid-protein membrane associated with energy transduction and ATP production. This is illustrated in Fig. 2-7 D. The horizontal bar represents the lipid-protein membrane associated with energy transformation, anchoring an exemplary proton pump and ATP synthase. In the upper part of the figure, there is the compartment with greater H^+ ions concentration and positive charge, while in the lower part—the compartment with lower H^+ ion concentration and more negative charge. The proton pump pumps protons upstream of the thermodynamic gradient of concentration and charge, while the ATP synthase couples the flow of protons downstream of the gradient with ATP synthesis. The transformation and dissipation of energy and production of entropy (energy loss occurs in each step: a part of the energy is released in the form of heat) drives the cyclic proton circulation. This process is very

[23] A relatively small proton leak through the inner mitochondrial membrane in our (and other animals') bodies is responsible for thermogenesis, that is, heat production, especially at rest (during physical exercise, the main source of heat are skeletal muscles at work).

common and constitutes the main source of ATP in almost all living organisms. Of course, the proton circulation is not so organized in space as convective currents, and it cannot be "seen" visually. However, from the formal point of view, these processes are fully equivalent (at least within the conceptual apparatus of thermodynamics).

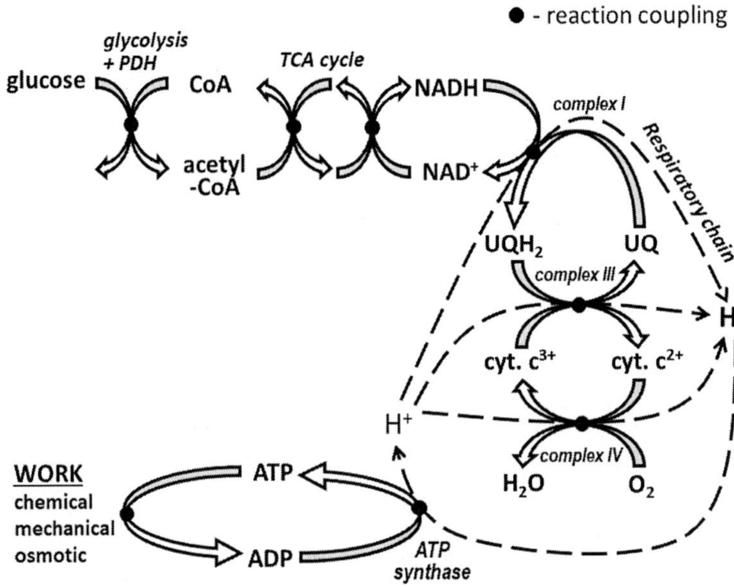

Fig. 2-10. A very simplified scheme of the bioenergetic system of the animal cell, composed of many alternating (cyclic) reactions of transformation of some substances into other substances, including alternate oxidation and reduction of redox centers. They represent a complex of dissipative structures transducing and dissipating energy (in the form of heat energy), and thus producing entropy. PDH, pyruvate dehydrogenase; CoA, coenzyme A; acetyl-CoA, "active acetate"; NAD^+/NADH (+ H^+), oxidized/reduced NAD (nicotinamide adenine dinucleotide); UQ/UQH_2, oxidized/reduced ubiquinone; cyt. c^{3+}/cyt. c^{2+}, oxidized/reduced cytochrome c; ATP/ADP, adenosine triphosphate/adenosine diphosphate.

Going further, the entire system transducing energy, for instance, in the animal cell, is largely cyclic, alternating transformations of some substances into other substances (including the alternating processes of reduction and oxidation of redox compounds discussed earlier). This is demonstrated in a very simplified way in Fig. 2-10 (see e.g., Marks et al. 1996; Voet et al. 1999; Murray et al. 2012 for details). Due to the decomposition of the 6-carbon glucose molecule, two 3-carbon pyruvate molecules are formed (through a chain of subsequent reactions). The enzyme pyruvate dehydrogenase (PDH) releases a 1-carbon molecule of CO_2 from a pyruvate molecule and transfers the remaining 2-carbon acetate group on coenzyme A (CoA), and thus an acetyl-CoA molecule ("active acetate") is formed. The acetate is transferred from acetyl-CoA to the Krebs cycle (tricarboxylic acid, TCA, cycle, citric acid cycle), free CoA is recreated to bind another acetate group, and the cycle closes. A similar process occurs during the transfer of the acetate group from the process of fatty acid β-oxidation (decomposition) to the Krebs (TCA) cycle (not shown). This cycle is composed of a sequence of reactions and intermediate metabolites. The net effect of its action is the complete oxidation of the acetate group to two CO_2 molecules, while the collected electrons (hydrogens) are used to reduce NAD^+ to $NADH + H^+$. The TCA cycle is a circular (closed) sequence of transformations (catalyzed by enzymes) of one into another of metabolites containing 6, 5, or 4 carbon molecules. In turn, complex I of the respiratory chain oxidizes NADH ($+ H^+$) to NAD^+, which is coupled with the reduction of UQ (ubiquinone, oxidized form) to UQH_2 (ubiquinol, reduced form) and pumping of four protons (per two flowing electrons) from the mitochondrial matrix outside mitochondria, as complex I is the first proton pump of the respiratory chain. Complex III, in turn, oxidizes UQH_2, transferring electrons (protons are released to the water phase) on the oxidized form of cytochrome c: cyt. c^{3+}, thus reducing it to cyt. c^{2+}. This process is again coupled with the pumping of four protons per two electrons (complex III is the second proton pump in the respiratory chain). Finally, complex IV (cytochrome oxidase) transfers electrons from cyt. c^{2+} on oxygen (O_2), which undergoes reduction into water (H_2O) (protons for formation of hydrogen atoms are taken up from the water phase). As the third proton pump of the respiratory chain, cytochrome oxidase pumps two protons per two flowing electrons. It is worth adding that the mentioned complexes contain several to several tens of redox centers (cytochromes, Fe/S centers, Cu ions, bound UQ or FMN): of course, they also undergo alternating reduction/oxidation. The pumped-out protons return, as mentioned above, to the mitochondrial matrix, which is coupled with ATP synthesis from ADP and P_i (omitted for simplicity in Fig. 2-10). In turn, hydrolysis of ATP to ADP (and P_i) becomes coupled with various processes in the cell which are unable to occur spontaneously: chemical work (the synthesis of different organic compounds, including proteins and nucleic acids), mechanical work (muscle contraction: relative movement in relation to each other of myosin and actin filaments, the transport of various cell elements along microtubules by kinesins and dyneins), osmotic work (the transport of various substances through cellular membranes against the concentration gradient), or electrical work (the transport of various ions, especially metal ions: Na^+, K^+, Ca^{2+}, through cellular membranes against the electric potential gradient). This entire system is presented, highly simplified, in Fig. 2-10.[24]

A subsequent example of the dissipative structures, in which a closed circulation of metabolites occurs, is the TCA cycle mentioned above; the Calvin cycle in the light-independent phase of photosynthesis that is responsible for CO_2 assimilation: acceptance of electrons from NADPH (an analog of NADP), the binding of CO_2 and synthesis of the primary product of photosynthesis: glyceraldehyde 3-phosphate; and the urea cycle, synthesizing urea from CO_2 and NH_3 (ammonia)—a by-product of, for example, amino acid degradation.

A cyclic character also has the circulation of the ions of sodium (Na^+), potassium (K^+), and calcium (Ca^{2+}) across lipid-protein cellular membranes. The Na^+/K^+ (sodium/potassium) pump (Marks et al. 1996, 141-142), driven by ATP hydrolysis, carries against the concentration gradient three Na^+ ions outside the cell for exchange with two K^+ ions transported to the cell interior. This process results in a gradient of ion concentrations (more Na^+ outside, more K^+ inside) and of electrical charge (there occurs a net transport of one positive charge outside the cell). Na^+ and K^+ ions can return spontaneously to the compartment, where their concentration is lower through special protein channels present in the cellular membrane that can be opened due to, for example, membrane depolarization (a change of its electrical potential from negative inside to positive inside), and next they can be pumped again outside/inside the cell by the Na^+/K^+ pump. The redistribution of the sodium/potassium ions across the cellular membrane of the axon (a process of the neuron that sends neural signals to other neurons) underlies the transduction of neural impulses—a sudden opening of sodium channels and, immediately afterward, of the potassium channels caused by membrane depolarization (a transient change of the electrical charge to more positive inside the axon) causes membrane depolarization in the next (neighboring) segment of the axon and consequently a translocation of the depolarization impulse along the axon (Alberts et al. 1998, 394-403). In turn, Ca^{2+} ions released from cisterns of sarcoplasmic reticulum (stores of

---

[24] The above example demonstrates why we eat and breathe. Namely, the transfer of electrons (hydrogens) from respiratory substrates (e.g., sugars, fatty acids, amino acids) on oxygen, that is, oxidation of the former and reduction of the latter, leads to building up of the proton gradient, and next to the synthesis of ATP, whose hydrolysis drives various processes in the cell keeping it alive and ensuring its proper functioning. Of course, the substances taken up with food also serve, after appropriate transformation (for which also ATP is needed), as the building material of our bodies.

calcium ions in the cell) to the cytosol of the muscle cell initiate its contraction (slipping in of myosin filaments between actin filaments). For muscle relaxation to occur, Ca^{2+} ions must be taken back up to the sarcoplasmic reticulum against the gradient of concentration and electrical potential by a special carrier (SERCA) anchored in the membrane of its cisterns, which is driven by ATP hydrolysis (Alberts et al. 1998, 536-542).

Within the cell is a ceaseless decomposition of damaged protein molecules and the synthesis of new molecules. This also concerns other cell components: RNA, lipids, carbohydrates, *etc.* (DNA is more durable, especially in non-dividing cells). Living organisms perpetually take up and excrete various substances, and their components are continuously exchanged. Over weeks, months and years, almost all atoms that enter our bodies' composition are replaced by new atoms (Toyama and Hetzer 2013) (this does not concern only tooth enamel and mineral components of bones, and DNA in non-dividing cells). Today, we are composed of almost completely different atoms than a year ago. What is (approximately) preserved is only the form (structure) of our bodies and their sub-systems and configurations, in which atoms, molecules, and higher-order elements are organized. This seems astonishing—the structure of (almost) all cells of our body remains (almost) unchanged, despite the exchange of all the atoms making them. This also concerns neurons, the connections between them (synapses), their excitation thresholds, (almost) the entire memory preserved in the functional pattern of these connections, and our personality, psyche, and self-consciousness that are epiphenomena of the functioning of our brain, that is, signal transmission within its neural network. This seems almost a miracle, but this is a truth resulting from the fact that, as dynamic structures feeding on the negentropy of the surroundings, we belong to the category of dissipative structures.

An example of a cyclic process representing a dissipative structure on the physiological level of the whole organism is the circulation of blood pumped by the heart, which delivers oxygen from the lungs and nutrients from the digestive system to each cell of the body, and receives carbon dioxide and dispensable by-products of the nitrogen metabolism that are transported to, respectively, the lungs and kidneys.

Other cyclic processes are the cell division cycle, the succession one after another of generations of living organisms, where in every generation, an adult individual develops from a zygote formed from the merging of an egg cell and spermatozoon, and afterward, it produces reproductive cells, or circulation of matter (carbon, nitrogen, phosphorus, but also electrons and entire hydrogen atoms, discussed above) in the biosphere. Life resembles, to a certain extent, a complex of hierarchically-organized whirls in a liquid whose flow has exceeded a certain critical speed (turbulent flow), or the discussed-above convective currents appearing at sufficiently large temperature gradients. Of course, the hierarchically-organized assembly of dissipative structures making up living organisms is much more complicated than mere convective currents.

Summing up, particular living organisms and the entire biosphere, whatever else they would be, are *also* a complex system of hierarchically-organized dynamic dissipative structures with locally low entropy, driven by sufficiently large thermodynamic gradients, whose "role" is the acceleration of the overall entropy increase. A continuous stream of matter, energy, and information flows through them, driving complex cyclic processes. Life in general and every concrete living individual must be a dissipative structure (or an assembly of such structures); this is one of the key elements of their essence. Using a little poetic metaphor, living organisms (or, more generally: living systems) are local islands of order lying in a big river flowing from order to disorder and, paradoxically, accelerating its flow.

The phenomenon of life can persist on the surface of the terrestrial globe as it dissipates energy and produces entropy faster than would be done by inanimate systems. In some important sense, the biosphere has its rationale because it is *darker* than the naked surface of our planet. Furthermore, the vegetation cover seen in satellite photographs seems almost black, significantly contrasting with the rather light rock surface or desert sand. How could such a simple and primitive property as the degree of darkness have any significant importance here? Well, the dark color means that a large part of visible radiation falling on an object having such color is absorbed and eventually transformed into infrared radiation, while a relatively small part is simply reflected, which is not associated with a change in the wavelength, and therefore with entropy increase. Of course, the inverse situation takes place in the case of light objects. As has already been discussed, the transformation of visible radiation into infrared radiation is associated with entropy production, and it becomes clear that the dark biosphere, for example, vegetation, dissipates the energy of solar light much more efficiently than the relatively light dead (inanimate) earth's surface. The possibly fast and efficient entropy production is just the "role" (from the point of view of thermodynamics) of the entire assembly of living organisms. Plants, in particular their photosynthetic apparatus and pigments, purposefully evolved to increase their efficiency in absorbing solar light in the possibly broad range of the wavelength of electromagnetic radiation (the color spectrum). This is the reason for their dark color in satellite photographs, and just because of this, they win the "competition" with the inanimate matter on the earth's surface for efficient energy dissipation. One should remember that black or almost black soil is composed mostly of dead organic substances that living organisms produce. Generally, an important role is played here by the relative fraction of the absorbed visible radiation, "processed" later into infrared radiation, and by the speed of this process. In addition, in this respect, living systems have an advantage over inanimate ones.

Theoretically, a stable biosphere could be composed of only autotrophic organisms (plants, algae, cyanobacteria) and decomposers—they would be sufficient to close the circulation of matter in nature and the transformation of visible radiation into heat radiation discussed above. Even decomposers are not necessary—the dead matter would finally undergo spontaneous decomposition by inanimate processes.[25] What, then, do herbivorous and carnivorous animals and parasites exist for? Of course, there is a biological-evolutionary answer to this question: living organisms exploit the available ecological niches. If there had become a possibility to exist at the cost of other organisms, why not profit from it? Many organisms chose (in the evolutionary sense) this way of living. And what about the thermodynamic answer? Well, the circulation of matter and production of entropy in the biosphere composed of only plants (in the ecological sense, including algae and cyanobacteria) and decomposers (or, all the more, of lone photosynthetic or chemosynthetic organisms) would occur relatively very slowly. The existence of a hierarchical food chain significantly accelerates this process—the eating of plants causes an intensified rebuilding and growth of the vegetation mass, increased production of carbon dioxide due to the combustion of plant organic matter, and, as a consequence, elevated assimilation of carbon dioxide and organic compounds produced by plants. Analogical reasoning can be carried out in relation to predatory animals and parasites. Generally, one could expect that large thermodynamic gradients allow forming diverse ecological niches, "supporting" ecosystems "containing" hierarchically-organized food chains and thus favoring a huge variety in the living world. In my opinion, there can exist in the universe (e.g., on relatively cold planets situated relatively far from their mother

[25] Similarly, archaea (archaebacteria) being methanogens that keep themselves alive based on chemosynthesis in alkaline hydrothermal vents (white smokers) could be self-sufficient and do without any other organisms. This also concerns the discussed above green sulfur bacteria that carry out photosynthesis at a depth of over 2 km, drawing energy from relatively high-energy near infrared radiation emitted by acidic hydrothermal vents (black smokers) and electrons/protons/hydrogens for CO_2 assimilation and organic compounds production—from H_2S, also supplied by these vents.

stars) environments where thermodynamic gradients are much smaller than on the earth, and the biosphere is composed of only photo- or chemoautotrophs and decomposers (or even without the latter) and is very poor.[26]

Nevertheless, it must be mentioned that in some biospheres, only half (one "arm") of the matter transformation is biological, while the other—purely physicochemical. This could be the case, for instance, in the time of the origination of life on earth, when organic compounds were synthesized abiogenetically and decomposed (consumed) by first (heterotrophic) living organisms. On the other hand, one could equally easily imagine an opposite situation—autotrophic (chemo- or photosynthetic) living organisms (for instance, archaea living in white smokers or green sulfur bacteria carrying out photosynthesis close to black smokers) building their biomass through the assimilation of carbon dioxide, and biomass would, in turn, be decomposed by "dead" physicochemical processes associated with CO_2 release. Nevertheless, in more advanced (thermodynamically and biologically-evolutionarily) biospheres, such a case probably could not occur.

As was mentioned above, living beings, as open systems, are not subject to the limitations of the second law of thermodynamics. Exactly the same amount of matter and energy leaves living organisms as was taken up by them (a small fraction of the matter can be used as the building material of a growing organism). However, because of metabolic transformations, both matter and energy leave a living system in a completely different form than it was taken up in. What an organism excretes into the environment possesses a much higher entropy than what was taken from it. Carbon atoms in the molecule of glucose serving as food are arranged, as was discussed,

---

[26] To continue this "biological fiction," one can imagine a biosphere composed of only one species, whose subsequent developmental stages would occupy various ecological niches. This species would live in the oceans. The youngest (e.g., unicellular) stage would carry out photosynthesis, and therefore fulfill the role of plants (unicellular algae). The next stage (first larva stage) would feed on the first stage, while the subsequent, still greater (larva) stages would feed on previous, smaller stages. The reproducing sexually adult stage would be situated at the top of the trophic pyramid and would devour the last larva stage. This cannibalistic life cycle would close in the moment of production by this stadium of numerous progeny representing the first, photosynthetic stadium. A similar, but composed of different species trophic pyramid exists in contemporary oceans (phytoplankton, zooplankton, still greater and greater fishes and so on). However, it is very probable that such a one-species ecosystem would be evolutionarily unstable – there would follow a specialization of particular stadia and their differentiation into distinct species.

in a much more ordered manner than carbon atoms in carbon dioxide produced due to its oxidation. The structure of glucose (information associated with it) that disappears during combustion of this sugar becomes, in a sense (quantitative, but not qualitative), transferred or transformed into the information associated with the hierarchical structure of the organism. Therefore, while the entropy of the organism actually decreases during its growth and development, the entropy of the surroundings increases even more. For this reason, the entropy of the whole system "organism + its surroundings" also increases over time.

Summing up, thermodynamics does not exclude the phenomenon of life. Just the opposite—the evolution of thermodynamic systems displaced from thermodynamic equilibrium enforces (in the appropriate, favorable conditions) the origination of dissipative structures, including living organisms, actively participating in efficient entropy production. The phenomenon of life constitutes a certain, important, from our point of view, although marginal in the overall scale, stage of the thermodynamic evolution of the universe. Whatever else life would be, it is also a dynamic structure shaped "for the purpose" of efficient energy dissipation. The whole biosphere and particular living individuals are, from the point of view of thermodynamics, nothing more than a complex of hierarchically-organized dissipative structures which participate in the dissipation of the energy of visible radiation into the energy of infrared (heat) radiation (alternatively, like the organisms living in the vicinity of hydrothermal vents, they "feed on" the redox gradient between the earth's mantle and surface). As mentioned, living organisms are a local "island of order" in the thermodynamic river flowing from order to disorder, paradoxically accelerating the overall disorder increase (Prigogine 1980; Prigogine and Stengers 1984; Schrödinger 1992).

Living systems are not systems of minimal entropy. They are characterized by entropy intermediate between minimal (the homogeneous distribution of matter in the gravitational field just after the Big Bang and inflation phase) and maximal (still larger and larger black holes) (Penrose 1990). From the point of view of thermodynamics, their most important property is the fact that they intensively dissipate energy and thus produce entropy, constituting tiny stop in the whole process of energy transformation in the universe.

The present chapter has demonstrated that the four physical forces, described in Chapter 1, not only are potentially able to participate in forming dynamic structures organized in space and time, but under some determined conditions, such structures originate in the physical world surrounding us. The difference between the potential ability for something and the actual realization of this possibility can be enormous. Let us recall once again the LEGO blocks and balls and sticks serving to build spatial models of chemical molecules. Something completely different is a chaotically scattered mixture of such blocks or balls/sticks, and something different again is a castle or DNA double helix model built of them. Of course, none of the discussed elements will assemble themselves into these complicated structures—an intervention of the human mind and human muscles is necessary here (the functioning of both the brain and muscles is driven by the transformation of matter, dissipation of energy, and production of entropy). The beauty of thermodynamics consists, among others, of the fact that it demonstrates the possibility of spontaneous self-organization of elements, which nature has at its disposal, into complex, frequently hierarchically-organized phenomena, such as convective currents, hydro- and atmospheric phenomena or living organisms, without any need of an intelligent action from outside. This forms a splendid bridge built between physics and biology.

Nevertheless, one should not forget about the other side of the coin. Living organisms differ fundamentally from inanimate physical systems, including non-biological dissipative structures like convective currents. Biological systems reach a degree of complexity of hierarchically-organized structure unavailable anywhere else in nature (I leave aside here some devices built by humans, being, after all, biological beings endowed with psyches and intelligence). The structure of living organisms differs from the structures of castles built from LEGO blocks or models of protein and DNA molecules composed of balls and sticks (and also from the structure of computers, robots and other devices constructed by humans) by several very fundamental properties (in fact, their nature is completely different) (Korzeniewski 2005, 2011). Namely, the structure of living individuals:

- is dynamic (happening in time);
- is based on a molecular level;
- can originate spontaneously in the process of self-organization;
- continuously exchanges its material, and only its form is (approximately) preserved;
- is sustained by the unceasing flow of matter, energy, and information;
- is capable of self-repair;
- can self-develop from some relatively simple “germ” (e.g., fertilized egg cell);

- realizes its superior purpose: sustaining its existence in time and its reproduction (more strictly: the possibly faithful multiplication of its identity in possibly a large number of copies);
- possesses numerous purposeful control/regulatory mechanisms responsible for this;
- is capable of self-modification and increasing the degree of its complexity (complexity of its identity) in biological evolution.

First of all, however, living individuals (and the biological, "valuable" information associated with them) are directed on a purposeful, intentional (in the operational sense) performing of a certain task: survival (at least to the reproductive period) and reproduction, and practically all (purposeful) details of their structure and function are subordinated to its realization. In order to understand the essence of life, it should be explained first how such a high level of complexity, functionality, and purposefulness can be spontaneously achieved, second, what differentiates the internally purposeful, associated with a certain new quality, endowed with a determined "value" information of living organisms from the purely thermodynamic, indifferent, having no "purpose" or "sense" statistical information of dissipative structures or, more generally, of physical systems significantly displaced from thermodynamic equilibrium. Third, finally, what is, from the formal (and therefore possible to be explicitly formulated within the terminology of strict sciences) point of view, the very essence of biological organisms? What can we learn about their animation? And how do we discriminate them from their physical surroundings? These questions can be answered by contemporary conceptions of the origin and evolution of life and the cybernetic definition of its decidedly most distinguished level—the living individual (Korzeniewski 2001, 2005, 2013a). These conceptions are presented consecutively in the following chapters.

CHAPTER 3

HOW COULD LIFE ORIGINATE SPONTANEOUSLY?

Attributes of life

Some configurations of atoms and chemical molecules on the surface of the terrestrial globe assembled themselves into objects that were not only congealed conglomerates of constituent substances, like rocks, or even dynamic dissipative structures, like convective currents. These configurations began to exhibit such properties, absent elsewhere, as self-copying, that is, the production of configurations similar to themselves (of similar identity), hierarchically-organized structure, and a set of functions forming a certain whole, directed purposefully for the preservation of themselves in a possibly good form to the reproductive period. In addition, the ability to take care of themselves purposefully, that is, the ability to neutralize (compensate) by special regulatory mechanisms various disturbances coming from the surroundings that could endanger their survival, the good condition and reproduction of these configurations by maintaining constant internal conditions. Finally, the possibility of evolution from generation to generation, that is, modification of their structure and function (their identity) over time, in particular, their improvement and increase in the degree of complexity. In other words, the origin of biological systems occurred, that is, life. The enumerated features of life chiefly concern its especially distinguished level of organization—the *living individual*. These are just living individuals that tend to self-copy, or strictly speaking—copy their identity in large amounts, possibly similar to the original. These individuals strive to endure in good condition, at least until the reproduction period, to which purpose their structure and set of functions, including purposeful regulatory mechanisms, are subordinated. These individuals (their identity, set of properties) undergo evolution from generation to generation. We will now see how purely chemical/ thermodynamic conglomerates of atoms/molecules could transform gradually, step by step, into living organisms.

The diversity of life forms existing presently (and in past geological periods) on the earth is stunning. All these forms originated in the process of biological evolution from simpler forms. Going back in time, we will eventually come to the moment when the first emerged somehow from inanimate matter. However, to investigate the process of the origination of life, it is necessary to specify first what life consists of and by which properties biological systems differ from physical and chemical systems.

Many characteristic properties can be attributed to presently known life forms. The set of these properties allows a rather univocal separation of this phenomenon from various inanimate phenomena. This results mainly from two reasons. First, despite the enormous diversity of forms, all living organisms on the earth possess astonishingly similar (in fact: identical) fundamental principles of structure and function (at least on the molecular and cellular level). Second, even the simplest contemporary (free-living) life forms exhibit a much higher degree of complexity than the first organisms that originated spontaneously from inanimate matter. Therefore, already the complexity itself seems a good determinant of life.

Nevertheless, a simple enumeration of properties attributed to known living organisms on the earth brings some dangers. On the one hand, we could potentially recognize as inseparable attributes of life its accidental features characteristic of present terrestrial life, but not that we like to regard as life in general. On the other hand, one should be cautious not to mistake properties characteristic for all living organisms with properties owned only by complex biological systems. In other words, we want our criteria of the division of phenomena (objects) on living and inanimate to be so deep and universal that they can be sensibly applied both to life forms existing on other planets in the universe, to the first living organisms on the earth, to possible cases of artificial or virtual life, and some presently known phenomena raising justified questions as to whether they should be classified as living systems or not. Speaking yet more simply, this is about reaching the very heart of the essence of life, its most immanent characteristics.

The living organisms on earth exhibit many properties that are characteristic of them. They are made of certain organic compounds, such as proteins, nucleic acids, sugars, lipids, *etc*. They possess a complex, hierarchically-organized structure; particularly, they are characterized by a cellular structure. They also exhibit a complex and hierarchical set of functions. They are equipped in metabolism (transformation of matter and energy), an assembly of biochemical reactions that convert certain substances into other substances, accelerated and directed by a set of enzymes (biocatalysts) of protein nature. They carry out redox reactions, that is, transfer electrons (and possibly also protons, that is, totally whole hydrogens) from certain compounds (elements) to other compounds (elements), oxidizing the former and reducing the latter. They store the record of their structural-functional identity in the form of the sequence of nucleotides in DNA. This record is being read, that is, translated into the sequence of amino acids in proteins (deciding about their structure and function) by the genetic apparatus composed of a few types of RNA,[1] and proteins, equipped with a "key" for reading the genetic record in the form of genetic code,[2] establishing the correspondence between the sequence of nucleotide threes (codons) in mRNA (complementary to nucleotide threes, anti-codons, in DNA and tRNA) and the sequence of amino acids in proteins.[3] They use a complex regulation system of gene expression (composed mainly of proteins and miRNA: micro-RNA, see below), which precisely decides where, when, and how much a given protein will be synthesized.

Living organisms possess many levels of organized structure and function. The level of organic macromolecules[4] (e.g., proteins, nucleic acids), level of complexes of macromolecules (ribosomes, chromosomes, cytoskeleton, protein-lipid membranes), level of cellular organelles (e.g., nucleus, mitochondria, chloroplasts, endoplasmic reticulum, Golgi apparatus, lysosomes), cellular level, level of tissues, level of organs, level of systems of organs, and finally—level of entire organisms (individuals)

---

[1] Messenger (matrix) RNA: mRNA, transport RNA: tRNA, and ribosomal RNA: rRNA

[2] The term "genetic code" is frequently confused with the term "genetic record." The latter refers to the sequence of (threes of) nucleotides in genes in DNA, determining the sequence of (threes of) nucleotides in RNA, and next the sequence of (single) amino acids in proteins synthesized based on mRNA. On the other hand, the former indicates the assignment of determined single amino acids in the protein chain to particular nucleotide threes (codons) in the mRNA strand, based on which this chain is synthesized, and which in turn are complementary to the nucleotide threes (anti-codons) in the gene in DNA (in one of the strands in the double helix) encoding this mRNA.

[3] For example, the three (codon) GGG (and three other threes with the same two first nucleotides: GGU, GGC, and GGA) in mRNA (and complementary to them threes, anti-codons, CCC, CCA, CCG, and CCT in a corresponding gene in DNA) correspond to amino acid glycine, the threes (codons) UAU and UAC in mRNA—to amino acid tyrosine, while the three UAC is a STOP mark: a signal of termination of the synthesis of a protein chain.

[4] And below it, of course, the level of simple organic molecules, atoms and elementary particles (discussed in Chapter 1).

can be distinguished.[5] Living individuals manifest functions specific to them, organized hierarchically, and therefore realized on different levels (e.g., biochemical, genetic, cellular, physiological). They exhibit a transformation of matter and energy not only on the biochemical level but also on the physiological level (of course, just those complex enough to possess this level). They are characterized by the ability for growth and autonomous, active movement, and the latter can mean a movement of the entire organism or its particular parts in relation to each other (e.g., movement of the cytosol within a cell, transport of various substances/objects along microtubules, including chromosomes during cellular nucleus division: karyokinesis, the flow of blood in animals driven by heart beating, the movement of cilia on the surface of epithelium cells). They exhibit the ability for self-repair: the neutralization of damage and fighting off infections. They can perceive stimuli from the environment and adequately react to them. Adult individuals spontaneously and independently build their structure and function from a certain simple germ, for instance, a zygote (fertilized egg cell). Living individuals reproduce, that is, produce individuals similar to them (of similar identity). This is closely associated with inheriting genetic information, in which the structure and function of the organism are recorded. They can evolve from generation to generation, that is, the gradual transformation of some forms into other forms can take place.

Of course, living individuals form higher-order systems: families, herds, populations, ecosystems, and finally—the entire terrestrial biosphere. However, this by no means changes the fact that, at least in my opinion, the living individual constitutes a decidedly distinguished level of organization in the whole hierarchy of living systems and the entire phenomenon of life (Korzeniewski 2001, 2005, 2013a).

Life on earth possesses many attributes, and in the face of this, one could question which of them is most important for explaining the phenomenon of life. It is well known that various organic compounds can originate spontaneously outside living organisms. On the other hand, there are no principal reasons that life on other planets could not be built with slightly different chemical compounds. Nevertheless, according to our knowledge, only organic compounds are sufficiently complex and diverse to constitute

[5] In my opinion, the concept of an organism is rather vague and indefinite; the organism can be identified with a spatially-separated, properly organized whole. On the other hand, the concept of an individual is univocal and precise, denotes a functionally-closed whole directed purposefully on its own survival and reproduction, possessing an identity that competes with other identities, constituting the unit of selection and subject of evolution—see more in Chapter 6.

the building material of spontaneously-originated life, although this must not necessarily concern artificially created life (Korzeniewski 2011; Supplement 2). Therefore, this is not a given concrete organic building material that decides the essence of life, although, generally speaking, spontaneously originated, without the participation of any intellect, living forms (individuals), whose functioning is based on the molecular level, seem doomed to be built of some organic compounds of carbon. It cannot also be excluded (and is quite probable) that physical and chemical conditions prevailing on planets or moons similar to the earth compel a concrete molecular basis of life, for instance, the frame (core) of metabolism, very close to this is known on our planet (Smith and Morowitz 2016).

It is true that, ultimately, both a living organism and a computer are composed of a dozen types of atoms. This would suggest that complicated systems (which undoubtedly a computer is) must not necessarily be built of molecules of organic compounds. However, in the computer, a device built with the hands of people, it is not single molecules that constitute the fundamental functional units but much greater objects, such as microprocessors or diodes, based on semiconductor crystals. Objects of this sort cannot originate spontaneously, like organic molecules, and their shape and function do not result at all univocally from the type of building material. A copper wire connecting two distant sub-systems represents one of an infinite number of possibilities of spatial shaping (arrangement) of x atoms of copper it is composed of, and the chance of an accidental origination of such a shape (leaving aside the connections of the wire within the system) is practically zero.

Therefore, for life to originate spontaneously, it must be composed of molecules representing one of the necessary, as determined by the nature of the chemical bonds, configurations of atoms entering into their composition (this does not contradict the astronomical number of possible sequences of nucleotides in nucleic acids and amino acids in proteins, as the structure of nucleotides and amino acids, and the general form of DNA, RNA, and proteins, are determined by quantum mechanics). The structure and function of such molecules must result from physicochemical laws, e.g., spatial distribution of atoms, their electron shells, and charges. Moreover, atoms can only form sufficiently complex and diverse connections in organic compounds to become a basis of spontaneously-originated life. Therefore, it seems that "natural" life is doomed to organic compounds of carbon, although not necessarily the same we know as components of life on our planet.

The conceptions concerning, for example, life based on silicon are highly speculative. Complicated devices can be made of this element (and many others), but they seem to have little to do with life. To return once again to this example, the functioning of a computer is based on the (designed by humans) shape of its elements (liquid crystal screen, microprocessor, wires) and a system of mutual connections between them. The structure of particular elements and the scheme of their connections were intentionally "imposed" in the technological process of their production. Without the human intellect, a computer could not originate spontaneously, as nothing in nature can "design" its functional constituent parts—they are in no way necessary, as they do not constitute a simple consequence of physicochemical laws.

On the other hand, the structure and properties of various organic compounds are determined by physicochemical laws (e.g., quantum mechanics determining the shape of atomic and molecular orbitals), which condition the specific features of their molecules. These compounds can therefore originate (and they do originate) without the participation of reason. Living organisms, being complex agglomerates of organic compounds, in a sense, designed their structure and function "by themselves" in the process of biological evolution. This was possible, as biological evolution could "start" from relatively simple assemblies of these compounds produced spontaneously by physical and chemical forces. A "spontaneous evolution" of a computer or any other system based (among others) on silicon would have nothing to start from, as physical processes by themselves do not lead to the production of sufficiently complicated and diverse compounds/conglomerates of silicon atoms, which could fulfill the function of progenitors of diodes or microprocessors, and all the more to a "sensible" connection of these elements into a coherent whole. The same is true of all other physical systems known to us apart from organic compounds.

The second property inseparably associated with the phenomenon of life is a complicated structure, organized hierarchically, that is, composed of many levels, where systems from a lower level constitute compound elements of systems on a higher level. Therefore, what on a higher level is regarded as a simple building unit, an element of some superior system, on a lower level itself constitutes a complex system containing elements of lower order (particular elements endowed with some degree of autonomy). Cells, for instance, on the one hand, form tissues and organs of higher organisms, while on the other hand, are themselves composed of a nucleus, mitochondria, Golgi apparatus, and other organelles. The hierarchical structure of living organisms contains many levels. Among the lowest, the level of atoms of various elements, and the level of simple chemical compounds and groups and the level of organic macromolecules, such as proteins, nucleic acids, polysaccharides (compound sugars), and lipids (fats) can be distinguished. They, in turn, enter into the composition of complexes of macromolecules, for instance, lipid-protein membranes, cytoskeleton, ribosomes, or chromosomes, which can then be components of various organelles (mitochondria, chloroplasts, cellular nucleus, and so on). The next level is the level of entire cells, whose building elements constitute the above organelles and cytoplasm "filling" the cell. This is followed by the level of tissues, constituted by one or more types of cells (e.g., epithelial, muscle, neural cells), the level of organs composed of an assembly of properly spatially-organized tissues, the level of systems of organs (e.g., the digestive tract divided into particular segments, equipped with glands producing digestive enzymes, for example). Finally, we have the level of entire organisms (individuals). The most complex organisms, which includes vertebrates together with humans, contain all the enumerated levels. On the other hand, the simplest life forms, for example, bacteria, "halted" on the level of a single cell, additionally "losing" on the way the level of cellular organelles (the lack of such organelles constitutes one of the basic properties distinguishing non-nucleated organisms, *Prokaryota*, to which bacteria and archaea [archaebacteria] belong, from nucleated organisms, *Eukaryota*, comprising protists, algae,[6] fungi, plants, and animals).

The distinction of particular levels of structure undoubtedly constitutes a convenient way of describing the structure of living organisms. Nevertheless, it should be stressed that those levels were conventionally separated to a certain extent, and it is impossible to delineate a sharp border between them. An example is a plasmodium, a cell containing many nuclei. In the case of the algae *Caulerpa*, the entire "plant" of the length of several tens of centimeters, divided into various "organs," including a "trunk" and "leaves," is just one big plasmodium. It seems that here we encounter something that represents a level intermediate between the level of a cell,

[6] Algae and protists are artificial groups, containing very different, unrelated to each other organisms. Generally, they could be characterized as photosynthetic (algae) and non-photosynthetic (protists) unicellular and simple multicellular nucleated organisms forming a thallus (body not differentiated into tissues). It is convenient to put such organisms into one sack, but this in no way represents the evolutionary kinship. For instance, charophytes (*Charophyta*) (group of green algae, *Chlorophyta*) are closely related to higher (vascular) plants, while classified as "protists" unicellular collar flagellates (*Choanoflagellata*) constitute a sister group of animals.

on the one hand, and the level of tissue, or even an organ, on the other hand. Similar relativity of the division on levels can sometimes also be observed in relation to other levels.

Of course, possessing structure in all certainty does not constitute a property characterizing only living organisms. A structure (understood as a certain differentiation of composition and the presence of the various constituent parts) is possessed by a huge majority of physical systems, both static (e.g., crystals, rocks) and dynamic (e.g., convective currents). In addition, our planet, stars, and galaxies are characterized by concrete, well-defined structures. The earth is an excellent example of an inanimate object with a strongly marked hierarchy of structure. The entire terrestrial globe is composed of an atmosphere, a hydrosphere (oceans, seas, and fresh waters), a planetary crust (oceanic and continental), a solid (although plastic) mantle appearing below it, and a situated centrally metallic core (liquid outer core and solid inner core). Among these layers, for instance, the terrestrial crust is composed of many types of rocks dislocated in relation to each other in a frequently very complicated way (resulting from the rich dynamics of the crust of our planet, which causes, for example, rock folding, formation of sedimentary rocks, an outflow of magma leading to forming basalts and so on). Particular rocks can also possess complicated structures containing many minerals, appearing as smaller or larger crystals and lodes. Of course, ultimately, all of them are composed of chemical compounds and atoms. Therefore, many levels of hierarchy can be distinguished in our planet's structure, which is hierarchical, as in the case of living organisms. One could argue that the structure of (especially the most complicated) living organisms exhibits a greater degree of complexity and regularity of repeatable structures (e.g., cells) than the structure of physical objects. In the face of the lack of any objective measure of complication (Gell-Mann 1995; Adami 2002), allowing a planet and bacterium to be compared as different objects, the estimation of the degree of complexity can be, at best, very arbitrary—I suppose that many people would be inclined to grant a higher degree of complexity to a planet. Apart from this, the first living organisms were probably much simpler than contemporary ones. Finally, if compared to living beings, human products such as computers, robots, or televisions, whose classing to living beings would cause our decided objection, are also characterized by a complex and hierarchical structure. For this reason, certainly, the possession of a (hierarchical) structure by itself does not constitute a principal determinant of the phenomenon of life.

Another property of living organisms is that their "sub-systems" (e.g., cells, organs, systems of organs) fulfill various functions that are, similarly to elements of structure, organized hierarchically. Out of several levels that can be distinguished here, the biochemical, genetic, cellular, and physiological levels comprising the functioning of tissues, organs, and systems of organs can be quoted. The assemblies of functions on a lower level of the hierarchy form partly autonomous blocks, serving together and carrying out some functions on a higher level. For instance, the complicated assembly of biochemical reactions (and also processes of reading genetic information and mechanisms of its regulation) forming the process of protein synthesis belongs to one of many blocks enabling the functioning of cells, for which a whole set of proteins, among other enzymes, is necessary. Actin and myosin, which are responsible for the contraction of muscle fibers, belong to such proteins in the muscle cells. Another assembly of biochemical reactions necessary for muscle contraction is the bioenergetic block responsible for the production of energy in the form of ATP (various blocks are, of course, linked also "horizontally"—the bioenergetic block supplies ATP to the block of protein synthesis, while the block of protein synthesis produces enzymes entering into the composition of the energetic block). Muscle contraction can fulfill different tasks on the physiological level: maintain a constant blood flow in the circulatory system and thus provide all cells of the body with oxygen and nutrients, as it takes place in the case of the heart, serves locomotion, manual operations, and speech generation, "driven" by skeletal muscles, or bring forth peristaltic movements of the intestines, causing the gradual passing of nutrients, and undigested food leftovers through subsequent segments of the digestive tract, realized by smooth muscles. Similarly, an action on the biochemical level, for instance, the synthesis of a given hormone, and subsequently on the cellular level, for instance, the secretion of this hormone, serves to maintain homeostasis on the physiological level, for instance, the control of the level of glucose in the blood by insulin and glucagon (Cryer 1991; Reece et al. 2011, 296-303). All these actions have their superior purpose: survival of the whole organism and production of progeny (reproduction).

An important feature characterizing the functioning of living organisms (individuals) seems to be a certain type of purposefulness. This purposefulness is expressed among others by a tendency toward preserving the organism as an intact whole, regardless of the conditions of the surroundings. The tendencies of this type, and chiefly their effect, are referred to as *homeostasis*. An example can be the gaining and assimilation of food to ensure a source of energy and building substances or the mentioned above, maintaining the constant glucose level in the blood. Another expression of the action of living creatures, a manifestation of a pursuit of a decidedly determined aim, constitutes the production of descendant systems possessing the properties (identity) of parental systems.

A whole system of inferior functions on all hierarchy levels—also characterized by well-expressed purposefulness—is subordinated to these superior functions: survival and reproduction (Korzeniewski 2001, 2005, 2013a). A huge majority of processes occurring in living organisms "serve" something (from the point of view of the superior aim—maintaining homeostasis, survival and reproduction) and do not simply take place. Most of Chapter 6 is devoted to this problem.

And what about the "functions" of non-biological systems? From the point of view of physics and chemistry, it is difficult to find any principal difference between the functioning of living organisms and processes occurring in inanimate nature. Physicochemical phenomena underlying the functioning of living systems are principally the same (categorically) processes that appear in inanimate systems. In both cases, the behavior of atoms and their assemblies is governed by the same laws—no qualitatively new factor, such as the vital force, decides the essence of life. In inanimate systems, especially in devices created by humans, one can discern a hierarchical ordering of various "functions" (processes) (which in turn appeared only in a rudimentary form in first living organisms). The only property specifically distinguishing the assembly of functions associated with the phenomenon of life could be the purposefulness discussed previously. However, a purposefulness very close to that present in living organisms can be met in the mentioned products of humans, such as a robot or computers intended for fulfilling strictly determined tasks.

Additionally, the functioning of these devices has a hierarchical structure, which further increases their similarity to living systems. The action (frequently complicated) of particular sub-systems, such as microprocessors or diodes, contributes to the whole's work, directed at realizing the aim imposed by the constructors of the device. Summing up, this is not in function as such, or at least not only where the essence of the phenomenon of life should be looked for.

The functioning of a living system on the biochemical level manifests itself in the form of metabolism, that is, transformations of matter and energy flowing through this system accelerated and directed by a set of enzymes. Organisms take up various substances from their surroundings and build their bodies. They also draw energy from various sources and subsequently transform it into some useful form (e.g., the energy of chemical bonds in ATP), in which it can be used to execute different types of work, for instance, chemical (synthesis of different chemical compounds), mechanical (muscle contraction, flagellum movement), electrical (transduction of neural impulses), osmotic (transport of substances across cellular membranes against concentration gradients), and so on. Part of this energy is always dissipated in the form of heat. Autotrophic organisms, mainly plants, algae, and some prokaryotes, synthesize the components of their bodies from carbon dioxide, water, and other inorganic compounds, while as the source of energy, they use the radiation of the sun or various inorganic compounds, or, strictly speaking, redox reactions, that is, the transfer of electrons from some such compounds to other, which was discussed in the previous chapter (see especially Fig. 2-9 and related discussion).

On the other hand, heterotrophic organisms use organic compounds produced by other organisms as both building material and as a source of energy. Overall, the transformation of matter and energy consists of leaving a living organism in a completely different form than they were taken up as. A large part of the accessible energy is dissipated by heat. Concerning matter, heterotrophs respiring with the participation of oxygen generally convert organic substances (e.g., sugars, fats, proteins) into carbon dioxide and water (a part of food serves after transformation as a building material, while undigested leftovers are expelled). In autotrophs conducting oxygen-based photosynthesis, the opposite transformation occurs—water and carbon dioxide serve to synthesize sugars and organic compounds, and oxygen is released. Sulfur photosynthesis, sulfur respiration, and nitrogen respiration were discussed in the previous chapter.

Transformation of matter and energy is also present in physical systems. An example is the crystallization of rocks under the influence of high pressures and temperatures or the release of various gases into the atmosphere during volcanic eruptions. The latter process constitutes at the same time an intermediate stage on the way, in which there occurs a transformation of energy released in the process of disintegration of radioactive isotopes into heat energy and the transport of the latter to the surface of our planet from where ultimately the heat energy is radiated into the cosmos in the form of infrared waves. A similar fate concerns visible radiation emitted by the sun and falling on the earth's surface, which transforms infrared radiation, as discussed in the previous chapter. Incidentally, perhaps the most dramatic example of the "transformation of matter and energy" in physical systems is thermonuclear synthesis in the interior of stars and the decay of radioactive isotopes. It is not only that lighter elements serve here as the substrate for synthesizing heavier elements or that heavier elements decompose into lighter elements, but also that a tiny part of matter (mass) is transformed into a huge amount of energy.

Another form of transformation of matter on the surface and in the crust and mantle of the earth is the so-called carbonate-silicate cycle (Berner et al. 1983). On the surface of our planet, carbon dioxide reacts with calcium or magnesium silicate, yielding silica and calcium or magnesium carbonates:

$$MeSiO_3 + CO_2 \rightarrow MeCO_3 + SiO_2,$$

where Me stands for Ca or Mg. The carbonates formed in this way enter into the composition of the earth's crust. On the other hand, in the subduction zone, where the rocks of the crust are melted, the opposite reaction occurs:

$$MeCO_3 + SiO_2 \rightarrow MeSiO_3 + CO_2.$$

Carbon dioxide is released into the atmosphere during volcanic eruptions, while silicates flow onto the crust's surface in lava.

Generally, life forms also exhibit the ability for growth and active movement. Bacteria cells grow before they undergo division into descendant cells, trees develop from seeds, and humans grow during their embryonic and individual development. As to movement, it can mean the dislocation of the whole organism (the running of an antelope, the creeping of an earthworm, the swimming of a euglena propelled by its flagellum) or the change of the mutual location of its parts in relation to each other (e.g., the heart beating, the movements of cytosol in the cell, the peristaltic movements of the gut, the transport of various objects in the cells by dyneins and kinesins along microtubules). A given system moves actively when the cause of the movement is located within the system and is not a result of a passive reaction to influences from the surroundings (for instance, the rolling of a jostled stone down a slope or the movement of a struck billiard ball).

The world of inanimate objects offers a broad gamut of growth and active (endogenous) movement. Salt crystals increase their size in a saturated solution of salt, stalactites and stalagmites grow, and the terrestrial globe was formed due to growth consisting of the joining of dust particles, pieces of rock, and planetesimals. Planets revolve around the sun, winds blow, and rivers flow to the sea. As a matter of fact, this is not an active movement, as it is caused by the external gravitational force in relation to the considered objects. However, the movement of fragments of the earth's crust, that is, the drift of tectonic plates, certainly should be acknowledged as active (if the considered system is the whole of our planet). Nevertheless, not to reach too far, an excellent example of objects moving actively are the various mechanisms produced by people: robots, cars, and planes.

The ability to receive stimuli from the surroundings and to adequately react to them is sometimes quoted as an attribute of life. The perception of a stimulus can adopt such different forms as the image of a mouse on the retina (and in the brain!) of a buzzard, on the one hand, and the activation of the production of enzymes decomposing lactose in bacteria by the presence of this sugar in the environment, on the other hand. An adequate reaction would be hunting down the mouse or the decomposition of lactose and using it as an energy source or building material. On the other hand, it is enough to consider a thermostat in a refrigerator detecting an increase in temperature, which prompts a response in the form of its fall, in order to reject the hypothesis concerning the exceptionality of a living system in terms of perception of stimuli from the surroundings and an adequate reaction to them. Another example is a computer "reacting" by a complicated system of operations to various combinations of signals introduced to it through the keyboard or an advanced robot carrying out the transformation of external signals on the input into appropriate actions on the output. If this analogy to perception seems too distant, it is enough to mention that there are artificial readers of writing already constructed.

Of course, all these properties and processes are much more complicated in contemporary life than inanimate systems (although not always: let us compare a bacterium and a supercomputer or the Internet), but they did not have to be such in the first primeval organisms. Therefore, the enumerated properties cannot decide the specificity of life, particularly concerning the essence and nature of living systems (individuals).

Life exhibits four properties, strongly mutually related to each other and constituting to a certain degree different faces of the same feature/phenomenon, which seem completely absent in physical and chemical systems. These are (1) the division of the entirety of the "living mass" into separate *individuals*, purposefully taking care of their own integrity and survival, (2) *reproduction*, that is, production by parental individuals of descendant individuals, associated with (3) *inheritance*, understood as the transfer of identity, that is, information about the structure and function of parental individuals, to descendant individuals, and (4) the ability of *evolution* of this individual identity from generation to generation (Korzeniewski 2001, 2005). Both reproduction, associated with the inheritance of traits, and evolution concern just *living individuals* and, strictly speaking—their identity. Therefore, we can accept that the transition from inanimate matter to living systems occurred when pre-organisms acquired the discussed properties and became the first living individuals. Because the reproduction of individuals, inheritance of their identity, and the ability to evolve from generation to generation are closely associated with the possession of genetic information (and the whole genetic apparatus responsible for its copying and reading) and separation, isolation from the environment, the origination of life (living individuals) consisted, in the first place, of the appearance of its genetic and structural (individual) aspects. The former is related, especially in more complex organisms, to the regulation of gene expression (the reading of genetic information). The real world also "imposes" other aspects of life (Korzeniewski 2005). Thus, we have the thermodynamic-energetic aspect, metabolic (including enzymatic) aspect, and redox aspect: transfer of electrons/hydrogens from some molecules/ions to other molecules/ions. The regulatory aspect is responsible for maintaining homeostasis and adequate reactions to disturbances coming from the environment (both external and internal). In higher organisms, composed of tissues, organs, and systems of organs, there is additionally the developmental aspect, that is, the self-creation of the complex and hierarchically-organized structure of an adult organism (that is, morphogenesis) occurring during embryogenesis (fetus development) and ontogenesis (individual development). Explaining the mechanisms of spontaneous and independent origin of all these aspects, which are mutually associated with each other and share many common elements, is equivalent to demonstrating how life emerged from inanimate matter. Nevertheless, I regard the genetic-structural aspect of life as primate and primeval, most related to its essence, and I will start my considerations just from this aspect. However, first, the questions should be posed: whether, and if so, in which way could life originate spontaneously from inanimate matter?

Could life originate spontaneously?

A huge majority of scientists presently accept the opinion that life originated spontaneously on the earth about 3.8 billion years ago (perhaps earlier). However, a fundamental question appears as to whether life at all could originate spontaneously due to a certain kind of self-organization of inanimate matter, independent arrangement of atoms and molecules into appropriate dynamic configurations. Even the simplest living prokaryotic free-living organisms capable of independent existence[7] are very complicated creations composed of thousands (of types) of proteins and other compounds, possessing genetic information in the form of the double (circular) DNA strand containing millions of pairs of nucleotide (nitric base) pairs, equipped with a complex and properly organized metabolism conducting transformation of matter and energy, and also copying and

[7] And therefore not being, e.g., parasites, unable to live independently.

reading of genetic information, finally, possessing a determined spatial structure (they are surrounded by the lipid-protein cellular membrane, equipped with various protein carriers and the system of energy transduction and ATP production, and also by a cellular wall outside the cellular membrane). The coming into being of such a highly-organized system spontaneously from inanimate matter seems completely improbable, more or less like the writing of Shakespeare's *Hamlet* by a chimpanzee striking at random the keys on the computer keyboard or the spontaneous assembly of a Boeing plane caused by the wind blowing through a junkyard. This argument is still quite frequently used against the possibility of the spontaneous origin of life.

However, the above argument should be regarded as completely ill-conceived. In order to explain this, I will use the example of the commonly known lottery. Winning the main prize consists of choosing the correct six numbers drawn by lot out of a possible 49 numbers (a square of seven by seven numbers). The probability of hitting six random numbers out of 49 in one shoot equals about 1:14,000,000 (one in 14 million). This is because it is necessary to choose any number from among the winning six out of 49, at the same time, one of the remaining five numbers out of 48, one of the remaining four out of 47, *etc*., so that all six numbers are hit in a single go. In order to have a 50% chance of winning the main prize,[8] we would have to "shoot" 0.5 * (49/6 * 48/5 * 47/4 * 46/3 * 45/2 * 44/1), that is, about seven million times. Such a minimal chance prevents many people from playing the lottery. At the same time, it ought to be added that the probability of the lucky hit in the lottery of life, that is, for example, a spontaneous coming into being in a moment of a bacterium cell from inanimate matter (a mixture of atoms or even simple organic compounds) is still far smaller, in fact, so astronomically small that it does not pay to take it into account at all. So, could life actually not spontaneously come into being?

The error in the above reasoning consists of the assumption that the process of the origination of life resembled the strategy of playing the lottery. Let us call it the "chimpanzee strategy." However, this was not so. Let us consider another strategy of this lottery game, which we will determine as a "stage strategy." Let us assume that if in a given game (still consisting of crossing of six numbers) we hit at least one number out of the six "winning" numbers correctly, then we are informed about this, and in

---

[8] I say here about 50% of chances, as no amount of shots gives certainty, that is, 100% of chances, for hitting the winning six. An exceptional jinx could shoot (mark sixes) practically *ad infinitum* and never win the main prize.

the next shot, we leave it ("put it aside"), knowing that it is good, and we try to hit the remaining numbers. After some time, we have already correctly hit two, three, or four numbers, and at the end, we come to all six numbers. How many shots would we need on average in this way of plying? In this case, in order to have a 50% chance for the main prize, we would have to make 0.5 * (49/6 + 48/5 + 47/4 + 46/3 + 45/2 + 44/1), that is, about 56 shots (in this case, we add, and not multiply, appropriate factors). This is a very small number compared to the seven million shots needed to hit all numbers at once in the previous "chimpanzee strategy." The second is, therefore, the much more efficient of the two described strategies. Furthermore, this is just the method by which life originated and evolved.

In other words, life could come into being as it originated *in stages*. It was not so that at once, in one step, even the simplest of the present free-living organisms came into being. Just the opposite, there had to be, and there were, many of such steps. In each stage, the originating life underwent various random changes — shots — and the external conditions (environment) decided which of these shots were correctly hit, that is, which accidental changes appeared beneficial for a given pre-organism, ensuring its survival and possibly the production of similar descendant systems. The cumulation of beneficial changes occurred step-by-step, remembering the numbers previously correctly hit. The structure and function of even the simplest cell did not have to come into being at once. Just the opposite — it was shaped during millions of years of evolution by accumulating many beneficial changes, each of which improved this pre-cell and sometimes increased its degree of complication. In other words, the configurations of atoms/molecules constituting the structure of living organisms (individuals), subject to normal physical and chemical laws, did not originate suddenly, in one step, but gradually, passing from relatively simple through still more and more complex configurations. Evoking the analogy with the LEGO playing blocks again would mean that the entire complicated castle would not come into being at once. Just the opposite, first, there are fragments of walls, battlements, and roofs, then whole walls, towers, and buildings, until the whole castle is formed only at the end. However, in contrast to the castle made of LEGO playing blocks and their complexities, the original and complicated structures of living organisms (individuals) at each stage were autonomous and functional and constituted certain closed whole units. This cannot be said about fragments of walls, battlements, and towers, which independently, in separation from other elements, do not fulfill any function, serve nothing, and have no sense.

To summarize almost everything said above, emerging life simply did not play the lottery. In other words, it started from a very simple version of the lottery, in which it was necessary to hit correctly, for example, two out of four numbers (a two × two square) instead of six out of 49 numbers. Afterward, the game became more complicated, and the square with numbers contained three times three (nine) numbers, four times four (16) numbers, and so forth. At the same time, the expansion of the square took place in such a way that the combination of the already correctly hit numbers within the initial (simpler) square remained to a large extent preserved, and the further random guessing occurred first of all within the added part of the enlarged square, while the old "core" remained relatively unchanged (this means the cumulation of changes). It appeared at the same time that different, frequently dissimilar to each other, combinations of numbers gave the main payout, that is, in other words, that life could successfully evolve from a certain primeval form in very divergent directions.

The origin and evolution of the earth and the earth's crust

It is commonly known that terrestrial life (and probably every spontaneously-originated life in the universe, which independently emerged from inanimate matter) is based on organic compounds (it has already been discussed why organic compounds are, as the only ones, predestinated to be the building material of spontaneously-originated life). Therefore, for life to originate spontaneously, there first had to occur abiogenic (not caused by living organisms) synthesis of the first organic compounds. As mentioned, life appeared on our planet at least 3.8 billion years ago, less than one billion years after its formation (Maynard Smith and Szathmáry 1995; Lane 2010, 2015; Baggott 2015; Deamer 2019, 2020; Knoll 2021). The earth was formed about 4.6 billion years ago from gas and dust thrown away into cosmic space during supernova star explosions (this is what our planet owes its composition to—all elements in the universe apart from hydrogen and helium originate in the interiors of stars or during supernova star explosions) (Macdoughall 1996; Ozima et al. 2012; Baggott 2015; Knoll 2021).

The entire solar system, that is, the sun, planets, moons, and smaller objects such as asteroids, meteorites, and comets, was formed from a gas-dust pre-solar nebula. It first underwent condensation from interstellar gas (a gas-dust cloud) into a gas-dust clump under the influence of a nearby supernova explosion, enriching the clump with heavier elements (Ozima et al. 2012; Baggott 2015; Sparrow 2016). This clump, also called the pre-solar nebula, condensed more and more around the common gravity center, and as it shrank, it began to rotate around the axis passing through this center and perpendicular to the plane of the flattened nebula. The sun was formed in its central region, which spun around its axis, containing almost 99.9%

of the system's mass. The remaining part of the nebula was further flattened (due to the rotation and centrifugal force related to it) into a rotating, proto-planetary disk. This is also known as the *accretion disk*, as conglomeration occurs in it, that is, the *accretion* of dust and ice particles into greater clumps. Within the *accretion disk* were separated compact rings, which gave birth to particular planets. Thanks to the initial disk rotation, all planets revolve around the sun in one plane and in the same direction. The rings in the proto-planetary disk rotating around the sun underwent condensation with time under the influence of the force of gravity.

First, innumerable small particles of fine dust (the size of a micrometer to a few millimeters) were formed. The rings in the proto-planetary disc containing these particles were gravitationally unstable, and they further condensed to form clusters approximately a few tens to a hundred kilometers in size called *planetesimals*. They fused into even larger planetesimals a few hundred kilometers in diameter. They grew by "consuming" smaller objects from the surrounding space. This resulted in forming moon-to-Mars-size planet embryos and, finally, full-sized planets. The entire planet formation process from the initial gaseous nebula took less than 100 million years. Meteorites are fragments of planetesimals or asteroids that collided and became pitched.

In the inner part of the proto-planetary disk, the temperature was high enough for water (and also hydrogen, methane, and ammonia) to occur in gas form, that is, water vapor. The water vapor and other gases were displaced by the pressure of radiation of the originating sun to the outer parts of the disc, where they could undergo condensation. The sphere, beyond which the temperature was low enough for this to happen is called the "snow line." Within this limit, planets with a rocky surface formed, which were practically devoid of gaseous hydrogen and helium (the so-called internal planets: Mercury, Venus, earth, Mars), while outside the snow line, gaseous and icy giants (external planets) composed mostly of hydrogen and helium (Jupiter and Saturn) or water, methane and ammonia ice (Uranus and Neptune) originated (Baggott 2015; Sparrow 2016).

The earth, like all other planets, originated from one of the rings in the proto-planetary disk. Gravitational disturbances caused the ring, the "ancestor" of our planet, to concentrate into a clump of matter, initially revolving around the sun on an orbit corresponding to the middle of the ring. The clump further clustered around the common center of gravity, as described above. As the spherical mass of dust, ice, and gas shrunk and concentrated, the pressure and temperature prevailing in it increased, leading to the material building melting. At this stage, the earth resembled a ball of melted magma. Heavier elements, first of all, iron, tended to accumulate under the influence of gravity in the center of the originating planet, which led to a metallic core forming. In other words, the earth's core underwent smelting from the primeval substance of the originating earth,[9] like iron undergoes smelting from ore in a smelting furnace and gathers at the bottom of the furnace (Ozima et al. 2012; Baggott 2015; Sparrow 2016).

On the other hand, lighter elements—for example, silicon, oxygen, carbon, magnesium, aluminum, and calcium—formed the outer layers of our planet: first, the mantle, and afterward, the crust. They are formed mostly from silica (SiO_2), metal oxides (especially oxides of magnesium, Mg, calcium, Ca, aluminum, Al, and iron, Fe), and metal silicates (the simplest of them have the general formula: $MeSiO_3$, where Me is a bivalent metal atom; frequently silicates are more complex and contain atoms of different metals). In the relatively heavier *mantle* of the earth, compounds (silicates) of magnesium and silica predominate; in the slightly lighter *oceanic crust* (of a mean thickness of about 7 km), the contribution of silica and compounds of aluminum and calcium increases, and the contribution of magnesium and iron compounds decreases (the rock called basalt predominates), while in the relatively lightest *continental crust* (of a thickness of 0-35 km), lying here and there on the oceanic crust, the contribution of silica is still greater, at the cost of metal (especially magnesium) compounds; the predominating metal is aluminum (the most common rock is granite). The process of delamination of the heavier and lighter elements under the influence of the force of gravity is called *gravitational differentiation*. It took place about 30 million years after the origin of our planet (Ozima et al. 2012).

Long afterward, the earth's gravitational attraction unceasingly pulled smaller pieces of rock from the surrounding space, which caused both an increment of its mass and a further rise of the temperature of its surface due to the continuous bombardment by these pieces. However, most of the nearby matter was eventually swept out by the force of gravity of our planet, because of which the frequency of collisions with smaller and greater crumbs of rock fell significantly. The terrestrial globe began to cool down, radiating heat energy into space. Its surface changed its physical state from liquid (or at least plastic) to solid, equivalent to forming the earth's crust.

As mentioned, while heavier magnesium rocks (containing silicon and oxygen) rich in magnesium (and to a smaller extent in iron) predominate in

---

[9] The composition of meteorites called carbonaceous chondrites very much resembles the composition of the homogeneous matter entering into the composition of the formed earth before its "gravitational layering" and smelting of the metallic core, as they are remnants of not-formed or crashed proto-planets (Ozima et al., 2012).

the mantle (essentially solid, although plastic in the prevailing high temperatures) lying under the crust (and much thicker than it), the oceanic crust ("lining" the bottom of oceans and lying under the continental crust) contains more much lighter silica and compounds (silicates and oxides) of magnesium, aluminum and calcium, while silica predominates in the continental crust, and the contribution of oxides and silicates of metals (first of all of the aluminum, and to a lesser extent of calcium, iron, magnesium, sodium, and potassium) is smaller. All the time, the melted surface of our planet, and after forming the earth's crust, numerous volcanoes released various gases, which formed the primeval atmosphere. Initially, nitrogen (N_2), carbon dioxide (CO_2), and water vapor (H_2O) most probably predominated in it.

Nevertheless, recent estimations suggest that degassing magma in the presence of iron (before most of it concentrated in the earth's core) also produced quite a lot of molecular hydrogen (H_2), methane (CH_4), carbon monoxide (CO), and therefore the primeval atmosphere was relatively reduced (Ozima et al. 2012). This was very important in the spontaneous abiogenic synthesis of organic compounds and, thus, the origin of life (see below). The earth's atmosphere was formed approximately at the same time as the gravitational differentiation of its interior took place—about 30 million years after its origin. One of the components of this atmosphere, water vapor, after a further cooling down of the earth's surface, began to undergo condensation, due to which the oceans originated[10]. At that moment, the basis for the birth of life was already prepared.

Incidentally, the origin of the earth's only satellite—the moon—is worth mentioning. Such a large moon is an exception among the rocky inner planets of the solar system (Mercury, Venus, earth, Mars), although moons of comparable size (or even greater) accompany the gigantic outer gaseous and icy planets (Jupiter, Saturn, Uranus, Neptune).[11] In the latter case, the satellites originated with their parent planets from the primeval proto-planetary (accretion) disc or were intercepted later by their strong gravitational field. On the other hand, the earth's moon was formed due to a collision of the young earth (some 4.5 billion years ago) with another proto-planet the size of Mars called Thea. This happened about 90 million years after the origin of the earth. The matter ejected into cosmic space due

---

[10] It is supposed that much of water on the earth comes from the Late Heavy Bombardment of the surface of the earth by icy comets and asteroids, carrying various organic compounds, that took place about 4.1-3.8 billion years ago (Ozima et al., 2012: Sparrow, 2016).

[11] Pluto, once regarded as the last planet of the solar system, is today classified as a so-called dwarf planet.

to this impact (70% coming from Thea) underwent aggregation under the influence of gravity around the common center of gravity and ultimately adopted a spherical shape, thereby forming our satellite. However, at the moment of the collision, both the earth and Thea had already, to a large extent, formed iron-nickel cores, whose material did not enter into the composition of the ejected matter. As a result, the Moon contains relatively little iron (it possesses only a very small iron core), and its specific weight is significantly smaller than earth. Immediately after its origin, the moon was located much closer to the earth and orbited it in a much shorter time than it does presently. It is also worth mentioning that initially, our planet spun around its axis much faster than today (a day lasted about 10 hours), and in the face of this, a year contained about 880 days, and the distance between the moon and the earth was much (about 6-fold) shorter (Ozima et al. 2012: Sparrow 2016).

Probably already from the very beginning, just after its formation, the earth's crust was divided into separate plates and exhibited (as mentioned before) plate tectonics; in particular, their mutual relative movement in relation to each other occurred. Until today, this process is the reason for most tectonic phenomena, first of all, volcanic eruptions, earthquakes, and forming of mountain ridges (orogenesis). It also causes the slow mutual movement of continents, "carried" by some of the earth's crust plates. Namely, these plates are made principally of a relatively heavy (basalt) and thin (7 km on average) oceanic crust, on which here and there lie a significantly thicker (up to 35 km) but lighter (granite) continental crust. Therefore, the relative movement of the plates of the earth's crust also results in the relative movement of the continents (Macdougall 1996; Ozima et al. 2012). 300-180 million years ago, all the continents were joined into one huge super-continent called Pangea. Afterward, its disintegration into recognizable continents took place, and the drift of the continents continues.

The drift of the plates of the earth's crust is driven by convective currents in the plastic (although principally solid) mantle of the earth, which is heated up by the heat energy coming from the disintegration of radioactive isotopes, especially uranium U^{238}. Where an ascending convective current reaches the surface of our planet, the crust is ripped, fragments on both sides of the crack move away from each other, between them a semi-liquid (low pressure prevails on the surface) matter of the mantle (lava) outflows, which then solidifies, and in this way, a new oceanic crust is formed. As mentioned above, this process is called spreading. In the place it occurs (the so-called spreading zone), a characteristic mid-oceanic ridge with a cleft in the middle, protruding about 1.5-2 km above the surrounding oceanic floor (lying at a depth of 3-4 km), is formed. This ridge exists because the rocks entering into its composition are extended due to the elevated temperature; the parts of this ridge that are moved away from the spreading zone as the oceanic floor expands because of new crust formation cool down and are lowered to the "normal" ocean floor level. The moving apart of the plates of the earth's crust due to spreading and forming new oceanic crust occurs in many places on the surface of our planet, for instance, in the middle of the Atlantic Ocean or around Antarctica. Europe is slowly receding from North America as the Atlantic Ocean becomes broader. The continental crust can also be ripped due to spreading, which occurs presently in the Great African Rift Valley. On the other hand, Baikal Lake in Siberia constitutes a leftover of an "unsuccessful," "abandoned" spreading zone.

Of course, the earth's surface area is constant, and therefore if a new crust is created in some regions, the old crust must be destroyed in other places. This happens in the so-called subduction zones (the whole process is called subduction), where plates of the earth's crust collide, and one plate descends under the other plate and submerges into the mantle. This occurs in places where a descending convective current subsides into the earth's mantle. The plate submerged in the substance of the mantle undergoes melting under the influence of high temperatures, and its material, lighter than the substance of the mantle, moves toward the surface of the mantle, punctures the oceanic crust, and outflows over its surface forming volcanoes and volcano islands, driving volcanic eruptions, and bringing forth earthquakes. For instance, Japan originated in this way and is still being formed. The pacific plate "submerges" beneath Japan, causing this tectonic phenomenon. Subduction zones, like spreading zones, appear in many places on our planet. The so-called "Pacific ring of fire," that is, a belt of intense earthquakes and volcano eruptions stretching along the western borders of both the Americas and the eastern Asia coasts (of which Japan is a part), is mostly associated with the phenomenon of subduction (and sometimes with transform faults, where tectonic plates slide past each other: see the famous San Andreas fault in California, USA) (Macgougall 1996; Ozima 2012: Smith and Morowitz 2016).

Additionally, it should be mentioned that the continental crust, lighter than the oceanic one, does not descend into the earth's mantle by subduction. When two continental plates collide, their material undergoes uplifting, forming mountain ridges (the process of orogenesis, that is, mountain formation), as took (and still takes) place in the Himalayas. The Indian subcontinent, an island for tens of millions of years, started to collide and still collides with Asia, exerting strong pressure on it and causing the Himalayas to continue to grow despite continuous weathering. Another example is the intrusion of the Apennine Peninsula colliding with Europe, leading to the uplifting of the Alps. The continuous destruction of the oceanic crust in subduction zones and its re-creation in spreading zones cause it to be generally relatively young: it is between zero (in spreading zones) to 180 (at some continent coasts) million years old. On the other hand, the continental crust does not undergo subduction; its new fragments are still being produced due to ongoing rudimentary gravitational differentiation (the smelting of lighter substances from the matter of the mantle), which can possibly be added to older fragments. Indeed, a few billion years ago, the area of the continental crust was much smaller than it is today, and the continents themselves were significantly smaller. However, the oldest fragments of the continental crust are about four billion years old.

Above, some attention was devoted to the plate tectonics of the earth's crust, not only because this phenomenon is interesting but also because it can be tightly associated with the genesis of life on the earth. Namely, as discussed below, an attractive candidate for an environment appropriate to the spontaneous origin of life is, as already mentioned, hydrothermal vents directly associated with mid-ocean ridges in the spreading zone or appearing in its neighborhood.

Abiogenic synthesis of organic compounds

At the moment of the origin of life, the conditions on the terrestrial globe were different from those prevailing presently. The salinity of oceans was much lower (its increase occurred due to the salts washing out gradually from the rocks of the earth's crust; Macdougall 1996). In the atmosphere's composition, which had a moderately reductive character, there was carbon dioxide, nitrogen, and water vapor, but also some methane, ammonia, and molecular hydrogen (Ozima et al. 2012, 89). On the other hand, it lacked oxygen and, therefore, the ozone layer. The lethal to life ultraviolet radiation, which is stopped by this layer today, could then reach the surface of our planet unobstructed. Volcanic and tectonic activity (e.g., earthquakes) was much more intensive, and there were probably more frequent and energetic atmospheric electric discharges (lightnings). Moreover, of course, the terrestrial biosphere did not exist yet.

It seems paradoxical that these conditions, decidedly hostile to contemporary life, enabled the coming into being of life. First, the gases entering the composition of the ancient atmosphere could constitute the substrate for the abiogenic synthesis of simple organic compounds. We should recall that organic compounds are, to a large degree, reduced (containing much hydrogen) compounds of carbon that can also contain oxygen, nitrogen, phosphorus, and sulfur, and, in smaller amounts, other elements, for instance, metal ions. In the present atmosphere, organic compounds could not originate, as it possesses a strongly oxidative character (much molecular oxygen and, simultaneously, a lack of hydrogen and its compounds if we ignore a certain amount of water vapor). On the other hand, the primeval atmosphere probably contained certain amounts of carbon, nitrogen, and sulfur in their reduced form (as methane, ammonia, and hydrogen sulfide), while oxygen was bound to carbon into carbon dioxide. Therefore, it "had at its disposal" all the most important elements necessary for forming organic compounds, constituting the chemical basis of life, and appropriate, moderately reductive properties.

In order to enter into chemical reactions leading to the production of simple organic compounds, the gases constituting the atmosphere at that time needed some source of energy, which could be transformed into the energy of appropriate chemical bonds. Such energy could be supplied by ultraviolet radiation, atmospheric electrical discharges, volcanic eruptions and the disintegration of radioactive isotopes. Therefore, a properly directed stream of energy could lead to the arranging of atoms of various elements, first of all, carbon, into configurations that afterward constitute the building bricks, the precursors of the structure of living organisms. Produced with the participation of these factors, organic compounds fell into the oceans and dissolved in water, forming the so-called "primeval broth" or "Urey's soup." Their concentration was rather small. However, presumably, they could frequently condense, for instance, through periodic evaporation of small water reservoirs or the adsorption on various minerals, for example, clays. Such organic compounds were relatively durable, as there were lacking the two main factors leading nowadays to their quick decomposition, namely oxygen and living organisms (Maynard Smith and Szathmáry 1995; Baggott 2015).

Summing up, what is today lethal for life, at the dawn of the history of our planet, those conditions enabled life to come into being. First, this is about the atmosphere containing moderate amounts of methane, ammonia, and hydrogen sulfide, a lack of oxygen and an ozone layer to protect from UV radiation, intense volcanic activity, and the absence of life itself. This explains why life does not originate spontaneously presently—for the same reason: the lack of appropriate conditions, for which we could not live on the earth over three billion years ago.

Numerous experiments concerning the synthesis of organic compounds in laboratories testify to the possibility of their spontaneous formation in the primeval atmosphere. When a mixture of gases with a composition to some extent resembling that atmosphere, although decidedly more reductive (containing first of all CH_4, NH_3, H_2, H_2O, sometimes also HCN, that is, hydrogen cyanide) was exposed to ultraviolet radiation or electrical discharges (first experiments of this kind were conducted by Miller, being then a Ph.D. student of the Nobel prize winner Urey), many simple organic compounds were obtained, including numerous compounds entering into the composition of living organisms, namely: amino acids (glycine, alanine, aspartic acid and others), nucleotides (ATP, ADP, AMP), nitrogen bases (adenine, guanine, uracil), simple sugars (including glucose, ribose, and deoxyribose), fatty acids, succinic acid, and many others. Among them are monomers of all the most important groups of complex organic compounds present in living organisms: proteins, nucleic acids, lipids, and complex sugars, and also the universal carrier of energy—ATP (Maynard Smith and Szathmáry 1995; Baggott 2015). Simple and even quite complex organic compounds were found in meteorites, especially in the so-called carbonaceous chondrites (Ozima et al. 2012, Baggott 2015: Deamer 2019). It was postulated that a substantial amount of various organic compounds was delivered to the earth's surface during the Late Great Bombardment (Ozima et al. 2012; Sparrow 2016). They were also detected (e.g., the glycine as mentioned earlier—the simplest amino acid) based on an analysis of the emission spectrum in comets and clouds of interstellar gas (Baggott 2015, 131-1320), which can testify to the common presence of organic compounds in the universe. The possibility of the spontaneous (abiogenic) coming into being of organic monomers seems not to give rise to doubt—therefore, the making of the first step toward the structure of living organisms, namely the independent formation of the configurations of atoms corresponding to simple organic compounds, does not meet any serious difficulties.

A spontaneous joining of monomers into polymers, in so far as appropriate conditions appear, was also encountered experimentally—the proteinoids discussed below can serve as an example. In Chapter 1, it was argued that linear polymers composed of various monomers are crucial for the complexity of living individuals. When conducive circumstances appear, the level of both simple and more complex organic compounds seems relatively easy to reach in spontaneous synthesis. These first stages leading to the origin of life (the synthesis of organic monomers and their polymerization) constitute a derivative of simple physicochemical processes that occurred in the conditions prevailing on the earth almost four billion years ago and that did not stand out as anything special from among other phenomena of this type. Nevertheless, it should be emphasized that it is not certain that life originated on the surface of the earth, in the upper layers of water reservoirs—an alternative conception is focused on hydrothermal vents in the oceans' depths, which is discussed below.

Where and how did life originate?

The first (historic) scenario of the origin of life on our planet is the self-organization of simple organic compounds (e.g., monomers) into more complex compounds (e.g., short polymers), and further into yet more complicated complexes on shallow sea shelves or in smaller water reservoirs, for instance, lagoons or ponds that underwent periodic evaporation, which led to a thickening of the organic compounds diluted in them (Oparin 1938; Haldane 1929; Maynard Smith and Szathmáry 1995). Organic compounds originated either in the atmosphere or in these water reservoirs under the influence of UV radiation, atmospheric electrical discharges, decay of radioactive isotopes, or volcanic activity. In this scenario, the first pre-organisms (likely not yet "alive"—resembling objects similar to coacervates or microspheres—see below) were heterotrophic and "fed on" the organic matter dissolved in water produced due to abiogenic synthesis. Initially, it was probably a simple absorption of chemical compounds, of which their "bodies" were built, for instance, amino acids, nucleotides, and simple sugars. Later, sulfur respiration (desulfurization) could emerge, as is discussed in Chapter 2 (in the face of the absence of oxygen in the atmosphere, oxygen-based respiration was, of cause, impossible). It cannot also be excluded that fermentation played a significant role at the dawn of life on earth. Finally, sulfur photosynthesis probably appeared, thus closing the cycle of alternate reduction and oxidation (transferring of electrons/hydrogens from one chemical element to another, see Fig. 2-9), constituting the main process energetically/ thermodynamically "driving" most of the terrestrial biosphere during about a half of the period of the existence of life on our planet. Oxygen-based photosynthesis, and consequently oxygen respiration, appeared with a more significant intensity relatively very late, about 2-1.5 billion years ago. The so-called Great Oxidative Event, that is an appearance of detectable of O_2 in the atmosphere, took place over 2 billion years ago (Baggott 2015), but only less than 1 billion years ago the content of O_2 exceeded 3% (today's value: 21%, maximum in Carbon: 35%) (Holland 2006). Therefore, life on earth was based mostly on sulfur photosynthesis and respiration for a long time.

The discussed scenario of the origin of life on the earth in a general energetic-thermodynamic aspect is presented in Fig. 3-1. On the left side, the well-known, present state is shown, in which the main processes of electron/hydrogen transfer are: (1) oxygen-based photosynthesis driven by quanta of electromagnetic radiation in the visible range emitted by the sun, which allows hydrogen (H) to be knocked out from H_2O (O_2 is formed) and transferred onto carbon (C) (its oxidized form CO_2), due to which organic compounds (that are indicated here as C-H) are formed, and (2) oxygen-based respiration consisting in the oxidation of organic compounds (C-H) with the participation of O_2 (CO_2 and H_2O are produced).

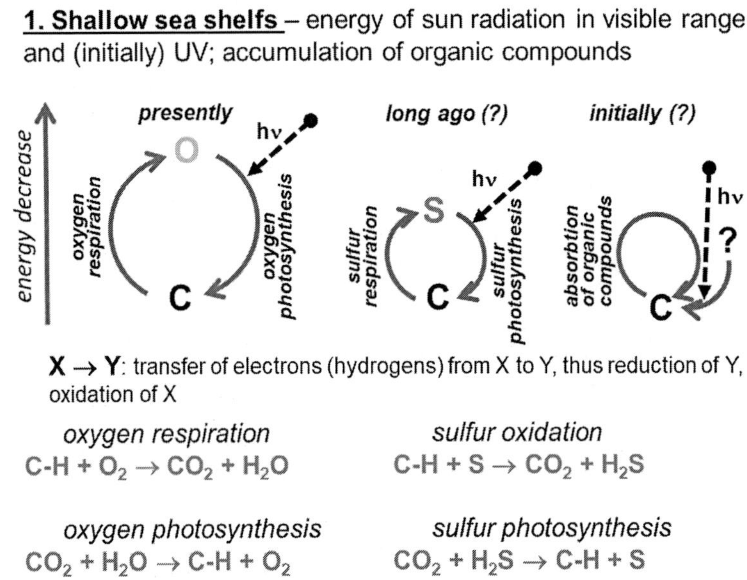

Fig. 3-1. The general energetic-thermodynamic aspect of life and a possible scenario of its origin in shallow water reservoirs.

In the middle of Fig. 3-1, the circulation of electrons/hydrogen between carbon and sulfur is shown, that is, the assembly of sulfur-based photosynthesis and sulfur-based respiration. This cycle probably drove most of life on earth for at least half its history. In the process of sulfur photosynthesis, quanta of electromagnetic radiation supplied energy for the transfer of electrons/hydrogen from the reduced form of sulfur (H_2S) to the oxidized form of carbon (CO_2), which resulted in the synthesis of moderately reduced organic compounds of carbon (C-H) and forming oxidized forms of sulfur (S, SO_4^{2-}). In the opposite process of sulfur respiration (desulfurization), electrons/hydrogen were transferred from organic compounds (C-H) to oxidized forms of sulfur (S, SO_4^{2-}), while the oxidized form of carbon (CO_2) and reduced form of sulfur (H_2S) was formed.

At the very beginning, presented on the right side in Fig. 3-1., organic compounds originated abiogenetically under the influence of, for example, UV radiation or lightning due to electrons/hydrogen transfer from some donor (marked as "?"), for instance H_2, NH_3, H_2S, Fe^{2+} or just CH_4 to carbon dioxide (CO_2). They were later absorbed by pre-organisms built of exactly those compounds, for instance, objects resembling coacervates (see below), illustrated by an arrow leading from C to C.

However, during the last few decades, alternative scenarios of the origin of life associated with the hydrothermal vents discussed in Chapter 2 have gained popularity (Lane 2010, 2015; Baggott 2016; Smith and Morovitz 2016; Deamer 2019). Acid "black smokers," which were discovered earlier and are much more frequent (at least presently), have a temperature reaching, and sometimes exceeding 400°C, and emit, first of all, hydrogen sulfide and metal sulfides (dissolved in water). As mentioned, chemosynthetic bacteria live in/near them, which reduce CO_2 using electrons/hydrogen taken from H_2S and form organic compounds (C-H). As food or symbionts, these bacteria constitute the first link of the food chain of the entire, surprisingly complex local ecosystems associated with "black smokers." However, for thermodynamic reasons (a positive change of the Gibbs free energy for real reagent concentrations), the reaction of CO_2 assimilation with the participation of H_2S cannot proceed spontaneously—it requires a supply of energy. This energy is gained in the process of H_2S oxidation using oxygen (oxidized forms of sulfur are produced: native sulfur or sulfates and H_2O). However, the only source of oxygen in the atmosphere and oceans is photosynthesis carried out by bacteria, algae, and green plants on the surface of water reservoirs and land. For this reason, the ecosystems associated with acidic hydrothermal vents are not self-sustainable or autonomous—they need other forms of life, producing O_2, for their existence. On the other hand, at the dawn of life and for a long time afterward, the terrestrial atmosphere was practically devoid of oxygen. For this reason, the scenario of the origin of life in the neighborhood of "black smokers" seems unrealistic (some trials to overcome this problem, for instance, Wächterhäuser 2007, were rather unconvincing).

A general energetic-thermodynamic aspect of acidic hydrothermal vents is presented in Fig. 3-2. In particular, it shows that chemosynthetic bacteria, which assimilate CO_2 using the "reductive force" in the form of H_2S, presently need the transfer of electrons/hydrogen from H_2S to O_2 in order to gain energy from the former process: both processes must be coupled with each other at the level of the bacterium cell. However, they simply could not function in anaerobic conditions in the distant past.

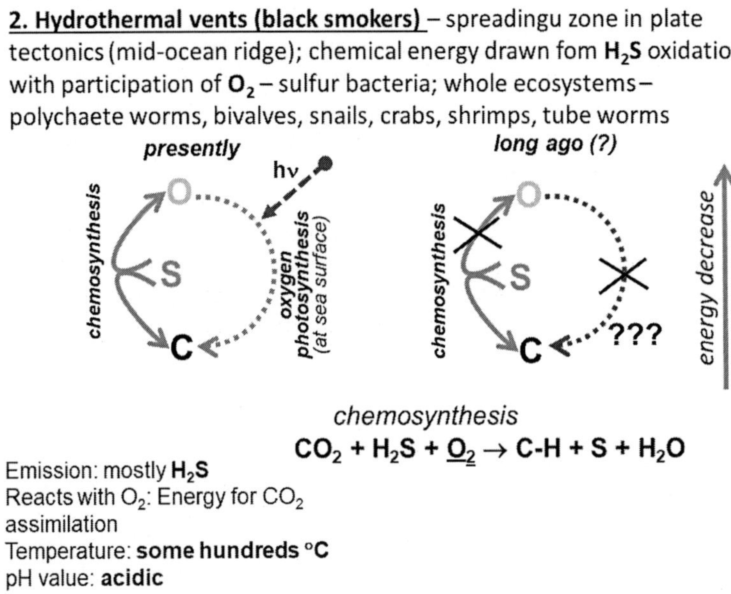

Fig. 3-2. The general energetic-thermodynamic aspect of life nearby acidic hydrothermal vents ("black smokers") and the inability of its existence (and therefore coming into being) in the absence of oxygen.

However, as mentioned in Chapter 2, "black smokers" offer the possibility of completely autonomous organisms, in the energetic-thermodynamic and "building substance" aspect, namely certain green sulfur bacteria (of the *Chlorobiaceae* family). They carry out photosynthesis in complete darkness (without electromagnetic radiation in the visible range) at a depth of about 2500 m. Their photosynthetic pigment, bacteriochlorophyll (a "relative" of chlorophyll), has one maximum absorption of electromagnetic radiation quants in the near-infrared range (that is, with a wavelength only slightly larger than visible radiation). Such radiation is emitted by the solution escaping from acidic hydrothermal vents of a temperature in the order of 400°C. This heat radiation (its quanta) has too little energy to knock out electrons/hydrogen from water and transfer them to CO_2 (and in this way produce organic compounds), that is, carry out oxygen photosynthesis. However, as a substitute for water, H_2S serves this purpose, which is also supplied by "black smokers," so here we deal with "primeval" sulfur photosynthesis. Today, CO_2 occurs commonly in oceanic waters (in the form of carbonates) and was even much more abundant on the early earth. Therefore, in contrast to the bacteria oxidizing H_2S by chemosynthesis in the neighborhood of "black smokers", they are self-sustaining energetically and regarding building material. Most probably, hydrothermal vents were previously much more common than they are presently and could "feed" a much greater number of organisms carrying out sulfur photosynthesis. However, photosynthesis is quite a complicated phenomenon, and it seems unlikely that it was the first process supporting life energetically/thermodynamically.

On the other hand, chemosynthetic bacteria and archaea exist that are fully autonomous and self-sustaining in the energetic/thermodynamic and building material aspect and the local ecosystems based on them. They carry out a relatively simple chemosynthetic process: the assimilation of CO_2 using molecular hydrogen (H_2) escaping from alkaline vents called "white smokers." The reduction of CO_2 by H_2 leading to the production of organic compounds occurs spontaneously and is favorable energetically. The terminal by-product here is methane in archaea (hence the name: methanogens) or acetic acid (or acetates) in bacteria (hence the name: acetogens).

The walls of the alkaline hydrothermal vents are porous like a sponge, and they contain a lot of roundish compartments the size of a bacterium or greater (Lane 2010, 2015). Additionally, between the alkaline solution escaping from "white smokers," filling out these pores, and slightly acidic or neutral oceanic water, a gradient of concentration of H^+ ions (protons) and electric charge (as it is commonly known, a solution of a high, alkaline $pH > 7.0$ contains relatively few positively charged protons, while a solution of a low, acidic $pH < 7.0$—relatively many) occurs (Lane 2010, 2015). An analogous gradient appears across the inner mitochondrial membrane and other lipid-protein membranes associated with energy transduction and constitutes the intermediate ATP synthesis according to the chemiosmotic theory described in Chapter 2 (Nicholls and Ferguson 2002). Once in the mentioned pores, the spontaneous, abiogenic synthesis of organic compounds could occur, and the gradient of protons across their walls could lead to the synthesis of some simple "ancestor" of ATP, for instance, pyrophosphate P-P (two connected inorganic phosphate groups) or acetyl-phosphate (acetyl-P). In such a case, such pores would be the "departure points" for the materializing and evolution of future cells (Lane 2010). Most certainly, "white smokers" or similar formations occurred much more frequently in the early stages of the evolution of our planet than they do presently. Therefore, they constitute a very attractive candidate for the cradle of life.

A general energetic-thermodynamic aspect of alkaline hydrothermal vents is presented in Fig. 3-3. Archaea and bacteria carry out a process of chemosynthesis, due to which they assimilate CO_2 using the "reductive force" in the form of H_2, thus producing organic compounds, the process of which is favorable energetically. In the face of this, the mentioned prokaryotes and ecosystems based on them are self-sufficient thermodynamically. At the same time, the two processes ought to be strictly differentiated. The first process, the reaction of CO_2 with H_2 leading to the production of CH_4 (methanogens—archaea) or acetate (acetogens—bacteria), serves the production of energy, which drives the second process, the synthesis, also from CO_2 and H_2, of moderately reduced organic compounds constituting the building material of a living cell.

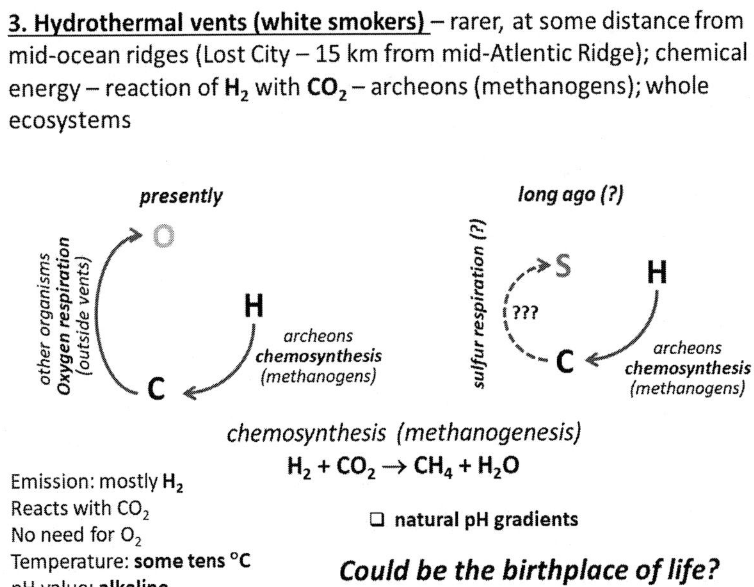

Fig. 3-3. The general energetic-thermodynamic aspect of life within and nearby alkaline hydrothermal vents ("white smokers") as a possible cradle of life.

If the first living organisms were actually born in and or nearby alkaline hydrothermal vents, they first based their energetics on chemosynthesis: they reduced CO_2 using H_2, and as a result, organic compounds were formed. Oxidized forms of sulfur: native sulfur (S) and sulfates (SO_4^{2-}) could serve other organisms as an acceptor of electrons/hydrogen received from organic compounds in sulfur respiration (desulfurization). In turn, yet other organisms could develop sulfur photosynthesis driven by infrared radiation and oxidizing hydrogen sulfide (H_2S), produced by sulfur respiration or delivered by "black smokers," to native sulfur and sulfates, and reducing CO_2 to organic compounds, thus supplying substrates for sulfur respiration. The next step would be transferring life to near the surface of the oceans and exploiting light in the visible range of the spectrum for photosynthesis. Finally, sulfur photosynthesis would be mostly replaced by oxygen photosynthesis, and after the accumulation of a certain amount of oxygen in the atmosphere, also dissolving itself in the oceans, oxygen respiration would appear.

Nevertheless, in my opinion, the scenarios of the origin of life of the type "shallow water reservoirs with accessible, visible radiation" type (Fig. 3-1) and the "alkaline hydrothermal vents with reductive power in the form of H_2" type (Fig. 3-3) do not exclude each other. In the early stages of the evolution of the terrestrial crust, formations resembling "white smokers" and "black smokers" could exist near the surface of shallow water basins. Both the visible radiation, needed for efficient photosynthesis, and UV radiation, leading to the abiogenic synthesis of organic compounds, could reach them there. In such conditions, the heterogenic mode of life (the absorption of organic compounds), chemosynthesis based on H_2, sulfur photosynthesis, and sulfur respiration could mutually support each other, creating a complex self-driving system composed of various pre-cells, from which there developed more complex and differentiated organisms in the process of evolution.

Indeed, it was recently proposed that life came into being in shallow water reservoirs (pools) on the surface of the land formed periodically around hot volcanic springs and geysers, similar to those appearing today in Yellowstone National Park in the USA and Iceland (van Kranandonk et al. 2017; Deamer 2029, 2020). A high temperature prevailed in them, and they underwent cycles of alternately drying out and filling with water. The volcanic springs, heated by "plumes of heat" (sub-crustal reservoirs of magma) or by lava pouring out in the spreading zones, emitted various substances of reductive characters, such as molecular hydrogen H_2 or hydrogen sulfide H_2S. In their neighborhood, porous rocks and clays appeared that possessed catalytic properties and could adsorb various compounds on their surface, favoring their high local concentration. Such rocks also contained much boron, facilitating the synthesis of ribose and phosphate, entering into the composition of nucleotides, RNA, and ATP. In hot, small volcanic lakes, formations also appeared, similar to hydrothermal vents. All this favored forming simple organic monomers (nucleotides, amino acids, primeval lipids), their polymerization into RNA and polypeptide (protein) chains, and the creation of bubbles surrounded by pre-lipids, which contained water with dissolved/suspended monomers and polymers of organic compounds. They constituted the yet inanimate "ancestors" of living cells.

The genetic aspect of the origin of life

Certainly, the most important aspect of both existing life and the process of its origination on the young earth is/was its *genetic aspect*. The genetic apparatus transcodes one kind of information into another. The chain of the transfer of genetic information in contemporary organisms from DNA (genes) to proteins is shown schematically in Fig. 3-4. As we know, in all living organisms, the set of instructions for forming their structure and function is contained in the genetic record, which has the form of a sequence of nucleotides in a nucleic acid (presently: in DNA or, in some viruses and viroids, in RNA). At the same time, it should be clearly stressed once again that this record has the form of a set of instructions and not a blueprint of the final product—the living individual that is to be constructed. It does not resemble, for instance, the plan of a computer or robot but rather a cookbook, that is, a description of the substances (components) that should be used and a set of operations that should be executed in order to obtain the "final product," that is, the living individual. This is, therefore, a dynamic, operational description not resembling the object that is to be produced by its structure, not a static, schematic blueprint similar by its structure (appearance) to the "ready product."

The genetic information can be duplicated by separating two DNA strands entering into the composition of the DNA double helix based on the complementarity of nucleotides (nitrogen bases entering into their composition) in pairs: A-T and G-C. This information can also be rewritten to an mRNA strand (transcription) in order to serve next as the matrix for protein synthesis in ribosomes (translation). The sequence of amino acids in the protein chain is determined by the sequence of nucleotide threes (triplets) (codons) in an mRNA strand ("messenger" RNA having the form of a linear chain of the length comparable to the length of the gene after removal of introns—non-coding gene fragments). Assigning various amino acids to particular nucleotide threes (triplets) is called the *genetic code*. The role of the agent (adaptor) in this assignment is played by the "key" to the genetic code, that is, tRNA (transfer RNA having the form of a chain organized into a structure resembling a clover leaf with segments of double-strand in the leaf "tail" and three leaflet "arms," and single strands in the three leaflets themselves [loops] and at the end of the leaf "tail"; nevertheless, this "clover leaf" in real conditions is folded into a three-dimensional structure). The "end 1" of tRNA (the end of the "tail" of the "clover leaf") (Fig. 3-4) directly recognizes the "end 2" (a niche binding tRNA) of the second adaptor, which is key to the genetic code, namely the enzyme aminoacyl-tRNA synthase, which catalyzes forming the tRNA-amino acid complex (there is more or less as many kinds of this enzyme, as there are different tRNA kinds). In turn, the "end 1" of this enzyme (a niche binding an amino acid) binds a proper amino acid. Due to these processes, the specific for a given kind of tRNA amino acid becomes ultimately assigned (bound) to the "end 1" of this tRNA. On the other hand, the "end 2" of tRNA, containing an anti-codon (nucleotide three on the top of the middle "leaflet" of the "clover leaf," Fig. 3-4),[12] recognizes a complementary codon on the mRNA strand. Within ribosomes,[13] amino acids bound to the "end 1" of tRNA molecules with appropriate anti-codons are connected into a protein chain in the sequence corresponding to the sequence of adequate codons in the mRNA strand. tRNA molecules "containing" complementary anti-codons and binding appropriate amino acids are bound (hydrogen bonds) for a short while to these codons. The synthesized protein chain is transferred from the previous tRNA molecule, to which it was bound, to a new tRNA molecule, with an amino acid specific for it, whereby the existing chain becomes elongated by just this amino acid. The "old" tRNA molecule is no longer bound to an amino acid or polypeptide chain and becomes released from the ribosome to cytosol, where it can bind a subsequent appropriate amino acid. This cyclic process is repeated to the moment of meeting (by the ribosome) of one of the three nonsensical (stop) codons in mRNA, indicating the termination of the synthesis of a given protein, that is, of *translation*. The proteins produced in the translation process, including enzymatic, structural, and other proteins, play a principal role in forming the structure and function of the living cell and the whole organism. The above-simplified chain of the transfer of genetic information from DNA to proteins is presented schematically in Fig. 3-4.

---

[12] For simplicity, only the middle "leaflet" of the "clover leaf" is show in Fig. 3-4, while the remaining two are ignored.

[13] Ribosomes are large complexes of macromolecules composed of two subunits, into the composition of which enter proteins and three (prokaryotes) or four (eukaryotes) kinds (strands) of rRNA—ribosomal RNA.

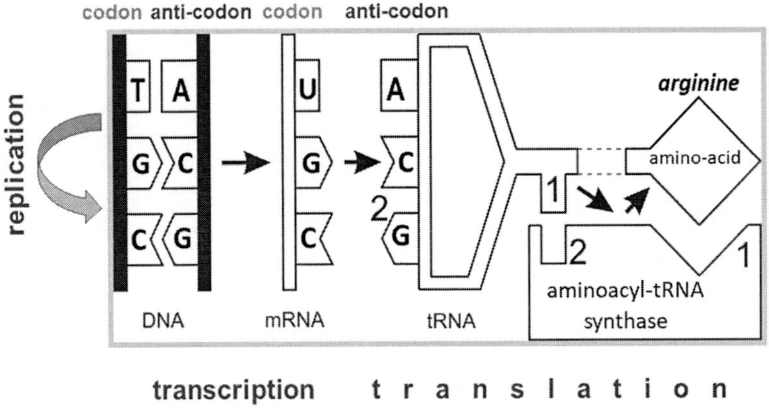

Fig. 3-4. The "chain" of the transfer of genetic information from the sequence of nucleotide threes (triplets) in DNA to the sequence of amino acids in proteins. The sequence of nucleotide threes (triplets) becomes translated, through codon threes (triplets) in mRNA, anti-codon threes (triplets) in tRNA, and appropriate aminoacyl-tRNA synthases into the sequence of amino acids in proteins. 1, 2, functional "ends" of tRNAs and aminoacyl-tRNA synthases: two "keys" to the genetic code. In this particular case, codon and anti-codon threes (triplets) for amino acid arginine were quoted.

However, proteins are not only the product of the very shortly-described process of the readout of genetic information. They fulfill, as enzymes, a very important function in catalysis and regulate particular steps of this process. For instance, DNA polymerase is responsible for DNA replication, RNA polymerase—for transcription, and aminoacyl-tRNA synthase and proteins that enter into the composition of ribosomes—for translation. Numerous proteins also fulfill supporting and supplementing roles in these processes. Therefore, nucleic acids and proteins record, copy and read genetic information. Regulatory proteins (together with miRNA—micro-RNA) constitute a key element of the whole system of the control of gene expression, that is, deciding which and how many proteins will be produced in a given place and time (regulatory proteins control rather the process of transcription, while miRNA—of translation, inactivating selectively some, partly complementary to them mRNA strands). The process of copying and reading genetic information is presented schematically in Fig. 3-5. It can be seen that these proteins are also indispensable for the reproduction and readout of genetic information, which ultimately serves the production of proteins. In other words, this is not nucleic acids (DNA and different types of RNA) alone, but the whole assembly: nucleic acids plus proteins that form the genetic apparatus of living organisms. Nucleic acids serve the coding of proteins, while proteins are responsible for synthesizing both nucleic acids and the proteins themselves (leaving aside the fact that they fulfill a huge number of other functions in the cell, for example, as enzymes or regulators of gene expression). Nevertheless, rRNA plays an important role in the catalysis of translation in ribosomes. Generally speaking, this is not only nucleic acids (in particular DNA) that undergo copying and transfer to descendant cells and individuals, but the whole self-reproducing system composed of DNA, different types of RNA, and proteins.

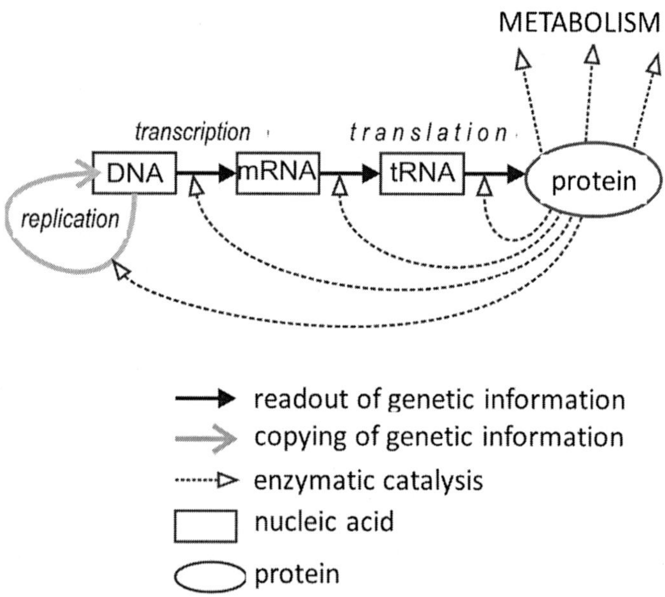

Fig. 3-5. The general scheme of the functioning of the genetic apparatus in present living organisms. Genetic information is both copied (the process of replication) as well as read in the multi-step process of protein synthesis (transcription, translation), in which mRNA and tRNA mediate. All these processes are catalyzed by proteins produced by these processes (rRNA also participates in the catalysis of translation). Proteins also regulate gene expression, that is, the production of appropriate proteins in an appropriate place, time, and amount (miRNA also participates in the regulation of gene expression). Additionally, proteins carry out, as enzymes, the whole metabolism and fulfill a broad range of other functions. Primarily everything that happens in the cell, the whole set of processes occurring in it, "begins" from proteins, and is conducted and regulated by them.

This system exhibits one principal property—the ability to propagate and reproduce itself (its own identity), producing similar systems. As we will see below, it also possesses the ability to evolve. Therefore, it can be said that the essence of the genetic apparatus of living organisms corresponds mainly to the essence of life in general. This leads to the conclusion that life originated at the moment of coming into existence of the first genetic apparatus capable of self-copying and reproduction (and an elementary structural-individual aspect, see below).

Even in simple live organisms, the genetic apparatus is very complicated—it possesses a complex structure and function, is operated by many enzymes, and undergoes complex, hierarchically-organized regulation. Because, as stated above, such a complicated system could not come into being at once in one step, it seems certain that initially, it looked much simpler. However, at least in my opinion, it had to be composed of nucleic acids and proteins from (almost) the very beginning.

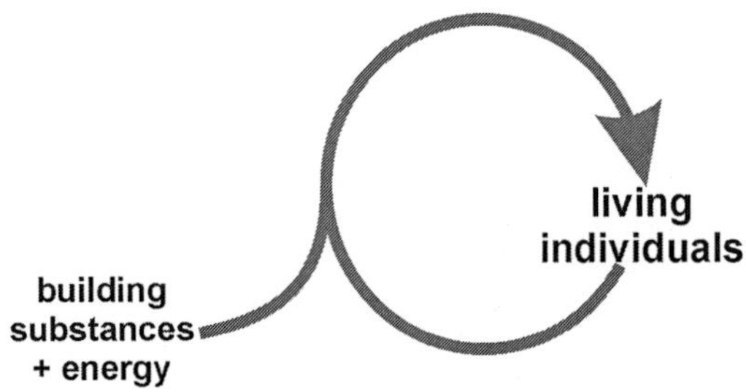

Fig. 3-6. The "auto-catalytic cycle" of reproduction of living individuals. Already existing individuals produce descendant individuals from building substances and energy drawn from the environment.

An inseparable property of living systems is self-copying (copying of own identity), that is, the production of descendant systems (reproduction). This drive toward quantitative increment is expressed in producing new systems similar to the parental system and not in a simple expansion of an indivisible whole, organic mass. Therefore, life is composed of particular, usually well-separated individuals. In reproduction, they produce individuals resembling them, deriving from the surrounding substances (and energy) needed for this. The "reaction" of the production of new individuals with the participation of already existing individuals can be written down in the following way:

***X* parental individuals + building substances →**
***Y* descendant individuals**

while the following condition is (at least potentially) fulfilled: $Y > X$. As over sufficiently long periods, more descendant individuals are always produced than exit parental individuals, the quantity of living individuals exhibits a natural tendency toward growth, limited only by the resources (capacity) of the environment. Self-copying life can undoubtedly be presented as an auto-catalytic system. Namely, living individuals produce descendants using matter, energy, and negentropy taken from the surroundings. The "auto-catalytic cycle" of the production of descendant individuals by parental individuals is presented in Fig. 3-6.

Nevertheless, simple chemical compounds are known, whose molecules catalyze forming additional such molecules from molecules of other compounds (Maynard Smith and Szathmáry 1995, 30-32). This phenomenon is called autocatalysis. It is presented schematically in Fig. 3.7.

Here, a molecule of compound X catalyzes the transformation of a molecule of compound A into the next molecule of compound X. Therefore, the simplest autocatalytic reaction can be presented as follows:

$$\mathbf{A + X \rightarrow 2X.}$$

From the formal point of view, such a reaction is identical to the reaction of "autocatalysis" for the production of living individuals presented in Fig. 3-6. Such purely chemical autocatalytic reactions apparently exhibit one of the main properties of living organisms (individuals) and their genetic apparatus—the ability to self-copy. However, as X is a simple chemical compound with a strictly determined composition and arrangement of atoms in its molecules, all these are identical. Therefore, no identity transfer to descendant molecules can occur; that is, no inheritance. Their structure cannot change and evolve. It is a univocal result of the elementary laws of quantum mechanics and chemistry. Therefore, it does not possess an identity or contain any information. Because of the uniformity of all such molecules, none are better or worse than other molecules, either more or less efficient in self-multiplication and the ability to survive. Therefore, any competition between them does not occur. As we will see in Chapter 4, the competition between systems to a different degree adapted to the survival and production of progeny constitutes the essence of biological evolution. However, no competition is possible between identical simple self-copying chemical molecules devoid of (strongly expressed) identity and not bringing with them any new information.

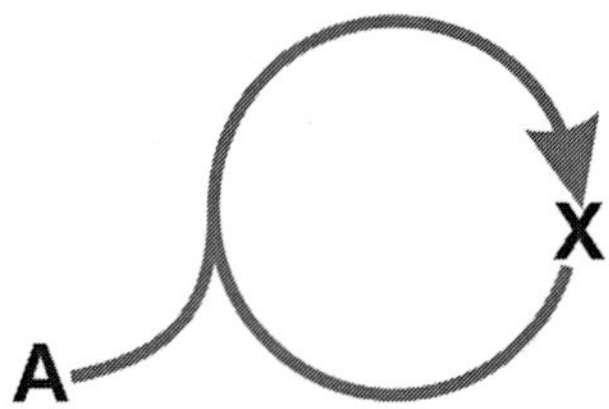

Fig. 3-7. The scheme of simple chemical autocatalytic reaction. Molecule X catalyzes the production of the next molecules X from molecules A.

On the other hand, living individuals have a very strongly expressed identity (set of properties), different from the identity of other individuals of the same, and especially—of other species, frequently as different as a bacterium, an oak, or a bat. An enormous, even unimaginable number of different identities of living organisms can potentially exist. Just this identity is inherited, transferred from generation to generation, and can undergo gradual changes in biological evolution. Ultimately, on the molecular level, it is based on linear heteropolymers: proteins and nucleic acids.

Proteins and nucleic acids, whose identity underlies the identity of living individuals, are large molecules that can differ between themselves in the sequence of, respectively, amino acids and nucleotides. This sequence deciding the specificity and identity of particular molecules of the discussed polymers can undergo tiny changes consisting of, for example, an exchange of one amino acid or nucleotide for another. For this reason, in principle, the structure of both proteins and nucleic acids can evolve. Systems composed of proteins or nucleic acids, or both these compounds, have, therefore, predispositions to fulfill the second main criterion of life—the ability to evolve (nevertheless, an indispensable condition of evolution also is the first criterion—the transfer of information about one's own identity to descendant systems).

The number of possible combinations of the sequence of amino acids in a relatively very short protein chain containing 100 amino acids is really astronomical (20^{100}, as there are 20 kinds of amino acids entering into the composition of proteins). This is still incomparably fewer (again—by a really astronomical factor) than the number of various living organisms that could potentially function on the earth (living organisms contain tens of thousands of types of proteins and millions to billions of pairs of nitrogen bases in DNA). Incidentally, all inheritable properties of living organisms are a derivative of the monomer sequence in the macromolecules. Particular living organisms possess a very strongly expressed identity—therefore, there is much to inherit. The structure of a horse does not result in any necessary way from the physical or chemical laws applied to the atoms entering into its composition. In order to describe it completely, it is necessary to deliver an enormous amount of information, among others, about the genetic record of the horse manifesting itself through the sequence of nucleotides in DNA. However, this will not suffice. The isolated DNA of a horse alone will never develop into a horse. Without the whole genetic apparatus, the "naked" sequence of nucleotides does not mean anything, has no sense, and constitutes only a passive record, which cannot do anything by itself.[14] In short, isolated DNA means nothing and can do nothing. Proteins are also needed, "taking care" of its protection and proliferation and "remembering" the genetic code (different types of aminoacyl-tRNA synthase), the whole fertilized egg cell (zygote), ensuring the appropriate conditions for the action of proteins, finally a mare carrying the fetus to the moment of birth, followed by feeding and protecting the colt. Therefore, the biological, inherited and evolving identity of a given individual of a horse is constituted by a whole complicated, purposefully directed on survival, and self-copying system, only one element of which is the genetic information recorded in DNA. Two horses' evolutionary identities (the set of inherited individual properties) are always slightly different. The identities of a horse and a rabbit differ in a much more principal way. Various identities can be, in given environmental conditions, to a different extent, efficient in ensuring survival and leaving progeny by individuals as their "carriers," and therefore, the transfer and possible propagation of themselves. Of course, in sexually reproducing organisms, only half of the identity of each of the parents is transferred to a given offspring. Nevertheless, there is no reason that an identity, understood as an assembly of biological properties, should be indivisible.

Summing up, the individual is not only a system producing systems similar to itself but also a system possessing a very strongly expressed identity (inheritable individual properties) and transferring this identity to its progeny. Identity is the information about the not necessary and, therefore, conventional (or arbitrary) structure and function of a given system, which nevertheless must fulfill the requirements associated with

---

[14] Similarly, a cookbook by itself will never bake a cake. A culinary recipe in English means nothing for somebody that does not know this language (and some fundaments of cooking). In order to bake a cake, a "cooking apparatus" in the form of a human is needed, who can cook and masters the language, in which the book is written. Analogously, in order to read the information recorded in DNA, the genetic apparatus "knowing" the genetic code is needed, and the whole set of mechanisms responsible for the regulation of gene expression.

functionality and fitness for the environmental conditions (expressing itself by the ability to survive and reproduce). However, many various solutions exist concerning the structure and function of an individual to ensure high fitness. The more possibilities there are to choose from, the more information is associated with the choice of one of them. A simple chemical molecule that catalyzes the production of further identical molecules has "to choose from" only one possibility concerning its own structure and therefore possesses zero information. Such a molecule cannot possess individuality, that is, identity. However, organic macromolecules, particularly linear heteropolymers, such as nucleic acids and proteins, that most probably gave rise to life and presently are the basic building material of this life possess a decidedly expressed identity. Living organisms, as complicated systems containing thousands of genes in DNA and protein types in the cell, are "equipped with" yet much more strongly expressed identity.

However, coming back to the topic of the first self-copying proto-individuals, systems containing only proteins or only nucleic acids cannot fulfill the four enumerated criteria characterizing living formations. They cannot (1) form individuals capable of (2) reproduction and (3) inheriting their identity (and therefore efficient production of their copies), which (4) can evolve in time, from generation to generation. This ability can (at least in my opinion) exhibit only systems composed of proteins and nucleic acids; therefore, the genetic apparatus of known living organisms is built of these two components. Let us see now why neither proteins nor nucleic acids can independently and efficiently proliferate and inherit their identity.

Protein (more precisely: polypeptide)[15] chains with a random sequence of amino acids exhibit a relatively broad range of non-specific (and not too strong) enzymatic activity; that is, they can catalyze many different reactions, although much less efficiently than "professional" enzymes appearing in living organisms, which is a result of millions of years of optimization in the process of biological evolution. One of the types of this activity is the catalysis of the synthesis of other protein chains with a random amino acid sequence through the polymerization of amino acids present in the solution. Therefore, the chains of primitive "proteins" of this kind cause the production of other chains of similar "proteins." Therefore, in this sense, they exhibit an autocatalytic activity (Kauffman 1986). This is presented schematically in Fig. 3-8, which shows the autocatalytic cycle of proteins.

---

[15] In the orthodox meaning, the term "proteins" is reserved for polypeptides (linear amino acid chains connected by the peptide bond -CO-NH-) produced by living organisms.

The example of the protenoids synthesized in laboratories (see below) (that are branched, non-linear polymers of amino acids) testifies to the fact that pseudo-proteins with a random amino acid composition and a broad spectrum of weak catalytic activity can originate spontaneously in appropriate conditions from a mixture of amino acids. Therefore, there are no fundamental barriers to the spontaneous formation of such configurations of atoms that can, specifically, preferentially accelerate various processes (chemical reactions, transport through membranes).

However, because of incorporating amino acids into the newly formed chain in an accidental sequence, the maternal chain cannot transfer the information about its structure and, therefore, its identity to descendant chains. There is no causal relation between the sequence of amino acids in subsequent generations in "parental" and "child" poly-amino acid chains. In this case, not a given concrete protein is proliferated, but only proteins in general, understood as some category of chemical compounds. If the information about the sequence of amino acids is not transferred to the next generation, one cannot speak here about inheritance, life, and evolution. Therefore, whereas primeval protein (polypeptide) chains were capable of catalytic activity and proliferation (as a type of organic compounds), they could not further transfer their identity.

Fig. 3-8. Autocatalytic cycle of proteins. Protein molecules with a random sequence of amino acids catalyze the production of further molecules of this type through polymerization of amino acids available in the solution. The system possesses some catalytic (including autocatalytic) activity, and therefore a potential possibility of reproduction; it cannot, however, transfer its identity to descendant systems, which in the face of this cannot inherit it.

The opposite situation takes place in the case of strands of nucleic acids. If a new nucleic acid strand is synthesized on the matrix of a given old single strand, this descendant strand possesses the sequence of nucleotides complementary to the parental strand. If, in turn, the newly formed strand serves as the matrix for the synthesis of the next strand, the latter will be identical to the initial strand—the negative of a negative gives a positive. In nucleic acid strands, the daughter is the mother's negative, while the grandmother is identical to the granddaughter (as long as the copying is faithful). Therefore, nucleic acids can potentially transfer their identity to "progeny." This takes place because, as already discussed, the spatial and electrical match of nitrogen bases in nucleotides forming pairs: cytosine (C) always appears opposite to guanine (G), and adenine (A) always appears opposite to thymine (T) (in DNA) or uracil (U) (in RNA). The autocatalytic cycle of nucleic acids is presented in Fig. 3-9.

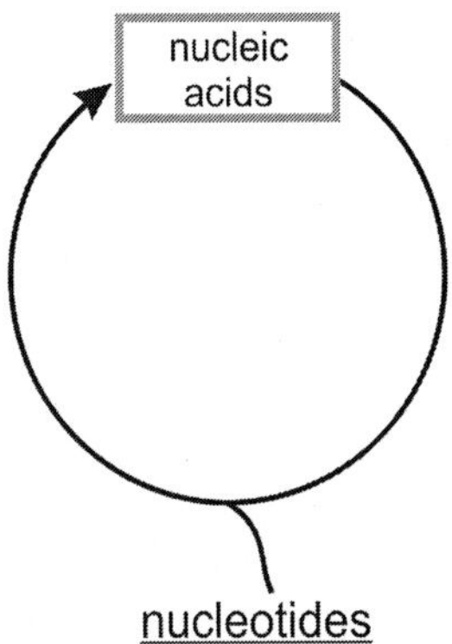

Fig. 3-9. Autocatalytic cycle of nucleic acids. On the matrix of existing nucleic acid strands are synthesized new strands complementary to them through joining of four types of nucleotides in an appropriate sequence. The system possesses a potential ability to inheritance of identity, but is devoid of efficient (auto)catalytic activity, and therefore the possibility of reproduction.

The case of nucleic acids demonstrates that relatively simple configurations of atoms are capable of, at least in principle, after creating appropriate conditions for them, the proliferation of their identity based on the sequence of monomers in a linear polymer.

However, nucleic acids either do not exhibit catalytic (enzymatic) activity, or this activity is rather weak and little differentiated (catalytic activity in short RNA strands, called *ribozymes*, was discovered; see below). Therefore, the production of the descendant strand on the matrix of the ancestral strand is a very inefficient process. The child strand can dissociate from the parental strand before completing the copying process. The replication of nucleic acids without the participation of enzymes is also very inaccurate—with every added nucleotide, the chance that an incorrect nucleotide will be incorporated increases, and therefore both strands will differ. For these two reasons, only extremely short nucleic acid strands (composed of few nucleotides) can potentially efficiently self-copy without the participation of proteins.

Additionally, if the rate of disintegration (due to different disturbances from the surroundings) of already exiting strands exceeds the formation rate of new strands, the entire population of strands will quickly become extinct. Until today, no RNA strand capable of self-replication in the absence of protein enzymes, which means the synthesis of a strand complementary to it (and next a strand complementary to the descendant strand) from nucleotides (A, C, G, U) present in the solution, have ever been discovered in nature or, despite numerous heavy trials, constructed in the laboratory.[16] Therefore, while strands of nucleic acids are potentially able, due to the physicochemical match of nitrogen bases in nucleotides, to transfer their identity to descendant strands, in the absence of appropriate catalytic activity of proteins, they cannot realize this potential.

This, which cannot be done by separate autocatalytic cycles of either proteins or nucleic acids, can be realized by *hypercycles*[17] (Maynard Smith and Szathmáry 1995, 51-58) of proteins and nucleic acids. In such a hypercycle, nucleic acids encode the sequence of nucleotides in the descendant strands and the sequence of amino acids in proteins, while proteins catalyze the copying of both nucleic acids and themselves. This is illustrated in Fig. 3-10. Specific "co-operation" takes place (Maynard Smith and Szathmáry 1995, 32-37): proteins offer their catalytic abilities, thanks to which the replication of nucleic acids occurs accurately and efficiently, while nucleic acids—a mechanism allowing the proliferation and transferring to descendant hypercycles of identity, their own and of the proteins encoded by them. Thanks to this, such a system can efficiently produce descendant systems that are identical (or almost identical).

In my opinion, the genetic apparatus of the first proto-organisms resembled the described autocatalytic hypercycle of proteins and nucleic

---

[16] Admittedly, there was once a grandly announced construction of self-copying RNA strands. However, they could only "splice" two previously prepared shorter strands complementary to them (Paul and Joyce 2002; Lincoln and Joyce, 2009). Therefore, this has nothing to do with the synthesis of a new strand from free nucleotides, and naming it "self-copying" constitutes a significant abuse. Unfortunately, the tendency toward prominence at any price is very common in contemporary science.

[17] A term introduced by Eigen and Schuster (1979).

acids. It is commonly supposed, for many reasons, that initially RNA (the equivalent of today's mRNA), and not DNA, constituted the carrier of genetic information, while the primeval proteins were simpler than the present ones and contained fewer (types of) amino acids (Maynard Smith and Szathmáry 1995; Copley et al. 2007; Lane 2010; Smith and Morowitz 2016; Deamer 2019). Apart from the carrier of information (nucleic acids) and catalytic factor (proteins), a connector (adaptor) translating the sequence of nucleotides in nucleic acids to the sequence of amino acids in proteins was needed. This connector would create the first genetic code, constituting the "key" to this code, something analogous to the present tRNA, appearing in the number of types corresponding to the number of functioning codons. Most probably, this code was initially "dyadic" (based on twos or duplets of nucleotides) and not "triadic" (based on threes or triplets of nucleotides), which means that only two out of three nucleotides in a codon were meaningful, recognized by tRNA, and participated in the determination of the amino acid corresponding to a given codon (Smith and Morowitz 2016). These coding twos were separated by one non-coding nucleotide (otherwise, it would be difficult to imagine the later transformation, in the course of evolution, of the "dyadic" into the "triadic" code). This would mean the existence of only 16 (4^2) codons, as opposed to 64 codons (4^3) in the "triadic" code, and at most, several encoded amino acids, that is, fewer than the twenty amino acids encoded today. Initially, only simpler amino acids were probably encoded, while more complex amino acids were added later when the "dyadic" code evolved into the "triadic" code (Smith and Morowitz 2016).

Nevertheless, contrary to what is commonly said, the present code is, in fact, "dyadic-triadic" and not "triadic." We have 61 codon triplets corresponding to amino acids (plus three nonsensical codons—stop codons—indicating the termination of translation) and only 20 encoded amino acids. Many amino acids are encoded by two, four, or even six codons (this lack of an unambiguous one-to-one assignment of nucleotide threes to amino acids is called the *degeneration* of the genetic code). Nevertheless, these are codons whose two first nucleotides (nitrogen bases) are identical and differ only by the third nucleotide (nitrogen base). Therefore, the nucleotide in the third position has no significance! (For instance, four codons correspond to the amino acid threonine: **AC**U, **AC**C, **AC**A, **AC**G). In the case of these fours of codons, differing only by the last nucleotide, even today the genetic code is strictly "dyadic." Because eight existing fours of codons, encoding one amino acid each, give a total of 32 codons (out of all 64, including 61 codons encoding amino acids), this "dyadic" character concerns at least half of the code. Therefore, the question arises whether a hypothetical genetic code allowing for encoding a greater number of amino acids (more than 20) would be more beneficial to enable greater diversity and complexity of the structure/function of proteins and higher efficiency and level of organization of organisms containing them.

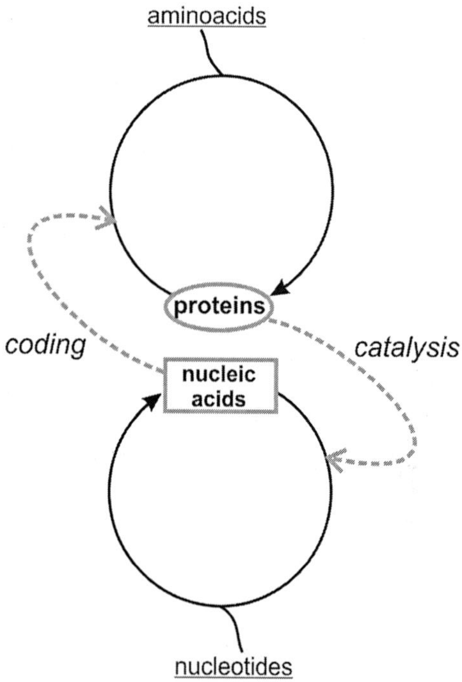

Fig. 3-10. A hypercycle of proteins and nucleic acids. Protein chains catalyze not only their own synthesis, but also copying of nucleic acid strands. Nucleic acids determine not only the sequence of nucleotides in descendant strands, but also encode the sequence of amino acids in proteins. The system possesses both (auto)catalytic properties, that is, the ability to reproduce, and the ability to transfer its identity (information about its structure) to descendant systems.

Furthermore, if so, why has such a code not evolved on earth: because of some physicochemical limitations, or maybe simply by chance? Perhaps we will never find this out. Nevertheless, we can imagine organisms on other planets based on RNA/DNA and proteins but containing more or fewer amino acids in their genetic code.

The genetic code is frequently regarded as totally arbitrary, with no causal relation between a given codon triplet and the amino acid assigned to it, or even just accidental. This is not completely true. First, as mentioned above, if four threes of nucleotides encode a given amino acid, the two first nucleotides in those threes are identical. In two cases (leucine and arginine), where six codons correspond to a given amino acid, at least the second nucleotide is common (of course, it cannot be so that six codons have the same first two nucleotides).

However, even more interesting is that the first position (nucleotide) in a codon corresponds strictly to the precursor (an organic compound being an element of metabolism), from which the encoded amino acid is synthesized. Thus, codons beginning with U correspond to amino acids, for which the precursor is pyruvate, A in the first place points to oxaloacetate as the precursor, while amino acids produced from α-ketoglutarate are assigned to codons with C in the first position (Smith and Morowitz 2016).

On the other hand, the second position (nucleotide) in a codon seems related to the properties, particularly polarity, of the encoded amino acids. Namely, A in the middle of the codon three characterizes the most hydrophilic (polar, "liking water") amino acids, while U—extremely hydrophobic (non-polar, "avoiding water") amino acids (Futuyma 2005; Smith and Morowitz 2016). Generally speaking, it seems likely that several properties of today's genetic code result from past evolutionary events and not pure chance. Therefore, this code is not fully optional and arbitrary.

One can (at least I do) feel the temptation to speculate that the most primitive genetic code was, in fact, "monadic," with single coding nucleotides separated by two non-coding nucleotides. Therefore, in such a case, only four amino acids would be coded, or, which seems more likely, groups of amino acids with a common precursor or similar properties, such as polarity. Perhaps the primeval tRNA did not distinguish particular amino acids but only groups of amino acids with similar physicochemical features, for example, hydrophobic, polar, acidic, and alkalic.

In my opinion, it is probable that the first genetic apparatus had the form of a hypercycle composed of RNA and primordial proteins. These proteins contained fewer types of amino acids than now and were much shorter (composed of fewer amino acids). The primeval "key" to the genetic code was some ancestor of today's tRNA (proto-tRNA), whose one (functional) "end" ("end 1"), containing an anti-codon, was bound to a codon in the ancestor of the modern mRNA (proto-mRNA), while the second "end" ("end 2," an analog of the present "tail" of the "clover leaf") recognized the proper amino acid (there was no aminoacyl-tRNA synthase yet, at least in the present form, recognizing by one "end," "end 2," the proper tRNA type, and by the other "end," "end 1"—the assigned to it amino acid).[18] Perhaps proto-tRNA molecules "carrying" amino acids assigned to them arranged themselves into a row along a proto-mRNA strand, binding their anti-codons to codons in the latter (as mentioned above, codons and anti-codons were initially "dyadic" or even "monadic"). Next, amino acids bound to proto-tRNA molecules neighboring each other in a row were also arranged into a line and connected into a polypeptide (protein) chain. This process was catalyzed by proto-rRNA with a small catalytic group, perhaps a short polypeptide joined to it (of course, ribosomes did not exist yet). Finally, proto-tRNA molecules. that were previously disconnected from the newly-synthesized protein, dissociated from the proto-mRNA strand. Such a system would be capable of self-proliferation (reproduction), inheritance of its identity (its transfer from ancestor systems to descendant systems), and evolution, that is, changes of this identity in time, from generation to generation. This hypothetical scenario was invented by me and is to some extent similar to that presented by Maynard Smith and Szathmáry (1995, 89-90, Fig. 6.4 therein). It seems that the described system could already come into being spontaneously from a mixture of organic compounds concentrated in objects resembling, to some extent, coacervates or microspheres (or bubbles surrounded by a proto-lipid membrane), discussed below in the context of the origin of the structural (individual) aspect of life, due to which it became the most fundamental unit of life—an individual (initially, it was, of course, a single-cell individual). Such a scenario is at least plausible. In agreement with the above stages of the origin of life, the genetic apparatus of contemporary organisms did not have to come into being at once, in one step in all its complexity. On the contrary, there are possible much simpler but still functional transitional stages capable of further development (evolution).

The postulated by me speculative early evolution and differentiation of RNA strands and primordial genetic apparatus are presented in Fig. 3-11. Its upper panel (A) shows that the proto-RNA (p-RNA) underwent differentiation into proto-mRNA (p-mRNA), proto-tRNA (p-tRNA), and proto-rRNA (p-rRNA). The lower panel (B) presents the possible functioning of the primitive genetic apparatus described above. There were no advanced proteinaceous enzymes.

---

[18] There could be simply one type of proto-aminoacyl-tRNA synthase, just binding appropriate amino acids to appropriate proto-tRNA types because of their physicochemical match ("affinity") that gradually disappeared in the course of evolution.

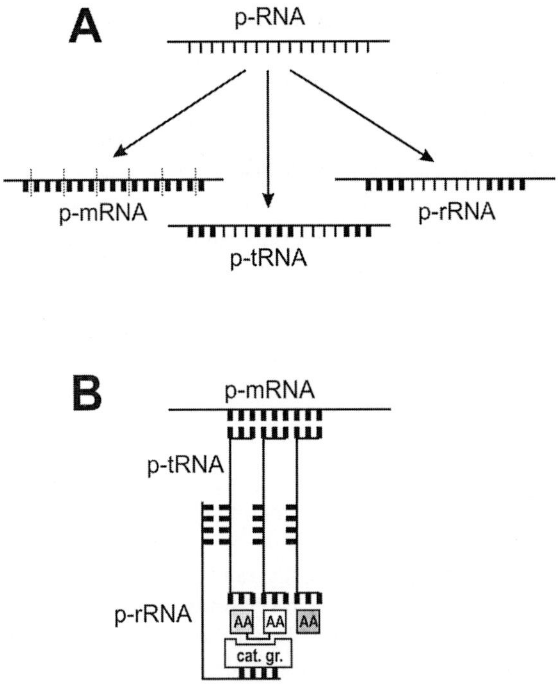

Fig. 3-11. A speculative scheme of the functioning of the primordial genetic apparatus based on RNA and proto-proteins (first of all, different RNA types are shown, assuming that particular steps of the process are catalyzed by primitive proteins encoded by p-mRNA). A, differentiation of proto-RNA (p-RNA) into proto-mRNA (p-mRNA), proto-tRNA (p-tRNA), and proto-rRNA (p-rRNA). B, mechanism of primitive translation (protein synthesis). p-tRNA molecules with appropriate amino acids (AA) bound to their "end 1" (see Fig. 3-4) (recognized by the specific spatial structure of this "end" without participation of amino-acyl synthase) arrange themselves into a row along a p-mRNA molecule: anti-codons (initially two-, or even one-nucleotide) bind transiently through hydrogen bonds to codons in p-mRNA in the sequence corresponding to the sequence of these codons. p-rRNA binds by its one "end" to p-tRNA, while a catalytic group (cat. gr.) (perhaps a short polypeptide) on its other "end" binds lying side-by-side amino acids into a polypeptide chain. After completing the whole process, p-tRNA molecules dissociate from p-mRNA and AA.

The RNA molecules of some catalytic abilities (*ribozymes*), which were discovered some time ago, led to the suggestion that autocatalytic RNA cycles preceded in time hypercycles of proteins and RNA in the history of life (Maynard Smith and Szathmáry 1995; Copley et al. 2007; Smith and

Morowitz 2016; Deamer 2019). This would be the so-called "RNA World." However, even if this actually happened (which, at least for me, is still highly problematic, although I do not deny that RNA played a much more important role in the early past than today), they were quickly and successfully replaced by the latter, as proteins are much more efficient and versatile catalysts than RNA. The known ribozymes catalyze, quite inefficiently, only a certain narrow set of processes, for instance, the self-cutting of introns[19] from pre-mRNA molecules or peptide formation during translation (rRNA in ribosomes; however, in this case, the support of ribosomal proteins is necessary). On the other hand, as mentioned above, scientists have not been successful in either the discovery or construction in the laboratory of an autonomously (in the absence of proteinaceous enzymes) self-replicating RNA strand (that would synthesize descendant strands on the matrix of parental strands from free nucleotides available in the solution).

Nevertheless, in my opinion, it seems likely that primitive protein chains from the very beginning already supported catalytically coexisting with them within coacervate-like bubbles (see below), proliferating RNA strands before they became encoded by these strands. Nucleotides and RNA could concentrate within coacervate-like formations for purely physical reasons because of their better solubility than in water, where an environment favoring the proliferation of RNA due to the non-specific catalytic activity of proteins additionally prevailed. In such a case, the autocatalytic cycles of proteins and nucleic acids co-existed for some time within coacervate-like bubbles (or lipid-bilayer vesicles, Maynard Smith and Szathmáry 1995, 97-106), whereas proteins "by chance" supported the multiplication of RNA catalytically, until encoding of proteins by RNA came about, that is, to the origin of hypercycles. The coacervate-like formations could not only constitute the ancestor of the primordial cell in the structural (individual) aspect but also lead to "striking up an acquaintance" between proteins and nucleic acids. This, in turn, directly enabled the origin of life, equivalent, in my opinion, to formation of the first primitive genetic code.

However, most probably RNA preceded DNA in the history of life (Maynard Smith and Szathmáry 1995; Copley et al. 2007; Lane 2010, 2015; Smith and Morowitz 2016; Deamer 2019). Therefore, at the dawn of life, this was RNA that underwent replication (copying): a descendant RNA

[19] Introns are non-coding fragments of genes lying between protein-encoding fragments—exons. They are cut out (by the so-called spliceosome containing proteins and snRNA, small nuclear RNA) or self-cut from the strand of pre-mRNA synthesized based on DNA, while exons are joined into a functional mRNA strand. Introns are common in eukaryotes and also appear in a small amount in archaea.

strand was synthesized on the matrix of a parental RNA strand. However, eventually, DNA became the carrier of genetic information, as it is much stiffer and more resistant to damage than RNA. Practically the only spatial structure that DNA can adopt is the stiff double helix, while elastic RNA strands can easily arrange themselves into different spatial conformations, both single- and double-stranded, variously folded in three dimensions, and therefore they can fulfill the role of tRNA, rRNA, mRNA, snRNA, miRNA, and ribozymes. However, initially, only RNA existed and fulfilled all the functions executed today by nucleic acids. Next, after getting on the stage of DNA, it was still RNA that was replicated, while DNA was synthesized on the matrix of RNA in the process of reverse transcription (presently, this process appears in retroviruses and retrotransposons)[20] and deposited as a genetic record.

On the other hand, RNA strands were, in turn, produced due to "normal" transcription, ubiquitous in present organisms, and through RNA replication (later RNA replication was replaced with DNA replication and reverse transcriptase became needless). It is supposed that this stage of evolution is represented by the last common ancestor of all the present organisms living on the earth, the so-called LUCA (Last Universal Common Ancestor) (Maynard Smith and Szathmáry 1995; Copley et al. 2007; Lane 2010, 2015; Smith and Morowitz 2016; Deamer 2019). Only later, presumably independently in the three main groups of living organisms (three "trunks of life"): bacteria (*Bacteria*, *Eubacteria*), archaea (*Archaea*, *Archebacteria*), and eukaryotes (*Eukaryota*), the origin of DNA replication and disappearance of, as already needless, RNA replication and reverse transcription (it endured in a rudimentary form in transposons) came about. Several cues testify to the supposition, among others, that while bacteria and archaea have principally very similar mechanisms of transcription and

---

[20] Retrotransposons are a type of transposons, that is, mobile elements (fragments) of DNA in the cell that can proliferate and incorporate their copies into other places in the genome (Schleif 1993, 531-563; Futuyma 2005). In their reproduction, the enzyme reverse transcriptase participates, encoded by the very retrotransposon and catalyzing the synthesis of a DNA strand on the matrix of the transposon RNA strand (first, "normal" RNA polymerase catalyzes the synthesis of an RNA strand on the matrix of DNA of the transposon). Most DNA in *Eukaryota* is the so-called "junk DNA," not encoding any proteins, not participating in the regulation of gene expression, and not fulfilling any other known function. It is supposed that a large part of it is active or already inactive due to accumulated mutations transposons. On the other hand, retroviruses are viruses that possess the genetic record in the form of RNA, also copied with the participation of reverse transcriptase: first, an appropriate DNA strand is synthesized in the host cell on the matrix of the virus RNA, and afterward, on its basis, RNA strands of new virus particles (virions) are produced.

translation, their DNA replication mechanisms are very dissimilar (Lane 2010). The main difference between the genetic apparatus of LUCA and contemporary organisms is presented in Fig. 3-12.

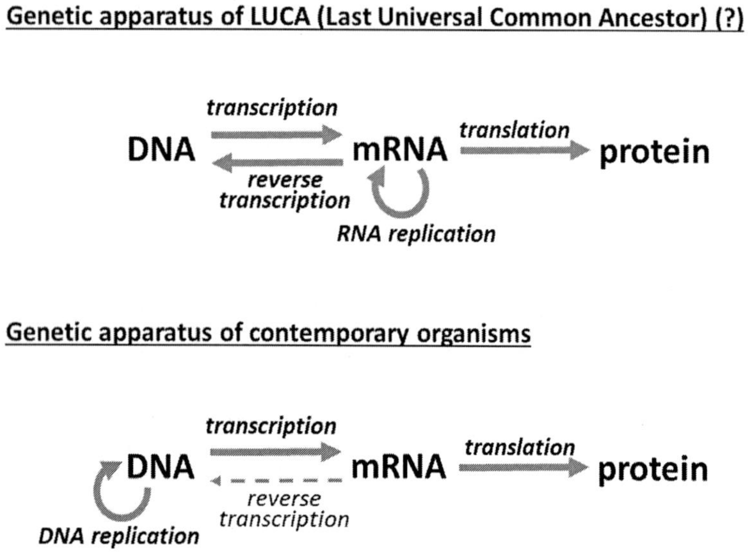

Fig. 3-12. Comparison of the general scheme of the genetic apparatus of the common ancestor of all contemporary organisms LUCA (Last Universal Common Ancestor) and genetic apparatus of the present organisms. In the former case, there underwent replication of RNA (first of all the ancestor of the present mRNA), and reverse transcription played an equally fundamental role, as "normal" transcription. In the latter case, DNA is replicated, while reverse transcription (and RNA replication) practically disappeared.

Nevertheless, it should be noticed that because of the universality of the genetic code (the same codon triplets are assigned, with very few tiny exceptions, to the same amino acids in all known living organisms), most probably LUCA already possessed a fully shaped genetic code. It cannot be excluded that the "complete" genetic code was formed before genetic information has become at all recorded in DNA (please notice that the genetic code is the assignment of amino acids to codons in mRNA).

The structural-individual aspect of the origin of life

As mentioned, one of the most important aspects of the originating life was the structural (individual) aspect (understood narrowly as a certain static spatial structure separated from the surroundings; in the broad sense, all aspects of living organisms represent an assembly of dissipative structures organized in space and time). Let us start with the fact that life does not form some continuous indivisible mass covering the surface of the earth (like the huge thinking ocean in the novel *Solaris* by Stanisław Lem 2002); but is divided into separate units, *individuals*. These are just individuals that reproduce and transfer their properties (their identity) to progeny, where metabolism, that is, the transformation of matter and energy occurs, whose structure and function are recorded in DNA, that are "equipped" with the genetic apparatus, that maintain internal homeostasis and "purposefully take care" of themselves, that, finally, (actually, their identity) evolve(s) from generation to generation (constitute the unit of natural selection and the subject of biological evolution). All self-dependent living entities, including a bacterium and each of us, are individuals. A fundamental property distinguishing an individual is its isolation from the environment, separation by some barrier from the surroundings (and other individuals). This barrier is constituted in the case of unicellular organisms by the cellular membrane (plus frequently the cellular wall), in plants—the epidermis or bark, while in the case of higher animals, including people—skin and mucous membranes. Most probably, at least in my opinion, life from the very beginning existed as discrete individuals. The coming into being of such individuals can be identified with the emergence of the *structural* aspect of life.

To digress, one could ask why life evolved from the "primeval broth" or pores in the walls of alkaline hydrothermal vents ("white smokers") as a set of individuals, and not some continuous mass, like a mush or jelly covering the whole surface of our planet. The answer seems simple. Generally speaking, biological evolution consists of, as we will see in Chapter 4, a generation of a diversity of features in competition with each other and the selection of the "best" of them (the criterion constitutes here the fitness in given environmental conditions). For the living mass to evolve and possibly increase its complexity, its different fragments would have to differ among themselves, compete for limited environmental resources, and propagate their identity (set of properties) in the possibly vast space of this mass. In turn, to compete with something, it is necessary to be strictly separated from this something and have clarity concerning the matter: "at this point, it is you; at this point, it is me." In the case of the hypercycles described above, it is necessary both to prevent them (proteins and RNA entering into their composition) from diluting and blurring in the surrounding solution and to avoid "mixing up" of two or more competing hypercycles with each other. Therefore, first, what would be done by the different areas in our hypothetical continuous living mass, would be creating a clear limit (border) around them, separating them from other areas and the surrounding environment. This would prevent the exchange or escape of matter, energy, and information (first of all genetic one), that would cause the dissipation of these areas' identity. For instance, the mycelia of individuals of some fungus species, in fact constituting continuous networks spread in soil, whose different fragments can fuse with each other, possess sharp defensive mechanisms protecting against merging with the mycelia of other, even similar individuals. In other words, the alive (and thinking) ocean from *Solaris* by Stanisław Lem (otherwise, an excellent novel) will forever remain in the domain of science fiction.

Therefore, life is doomed for individuals. For many reasons, a great role is played by a barrier separating the organisms of such individuals from the surroundings. The chemical composition of living beings differs fundamentally from the chemical composition of their environment. For instance, a bacterium living in water contains proteins, nucleic acids, lipids, sugars, many small-molecule compounds, and inorganic ions, which appear in relatively high concentrations and are essentially absent in the surrounding water or appear there in a much lower concentration. Therefore, it is necessary for something that would prevent the escape of all these compounds or simply "dissolving" of the bacterium in water. Every living organism also requires protection against the impact of various harmful environmental factors and disturbances of its metabolism by influences from the surroundings. The barrier, an envelope separating an individual from the surroundings, should also be capable of selective transport to the individual's interior of various substances serving as building material or energy sources and the removal of the unwanted by-products of metabolism. Such a sheath could also fulfill additional, secondary functions, for instance, participate in energy transduction in *Prokaryota*—respiring, photo- and chemo-synthesizing bacteria (and their descendants—mitochondria and chloroplasts), and in respiring and conducting chemosynthesis archaea (archaebacteria). As we saw above, this kind of ATP production depends on the proton gradient across the lipid-protein cellular membrane surrounding the prokaryote cells. This membrane, containing numerous specialized protein carriers, also plays an important role in transporting various substances inside and outside the cell.

The bi-layer lipid-protein cellular membrane, constituting the barrier separating from the surroundings in the simplest free-living organisms, just bacteria and archaea,[21] is presently a very complicated and specialized formation, composed of various lipids appearing in appropriate proportions and a great number of proteins fulfilling the function of enzymes, carriers, redox centers (elements of electron transport chains), receptors, *etc*. It constitutes a product of long-lasting evolution and certainly could not come into being spontaneously at once from inanimate matter. Therefore, the problem of the origin of the germs of the structural-individual aspect of life can be reduced to the question of how the first proto-organisms coped with the problem of creating a primitive barrier separating them from the environment and how they emerged as isolated units.

Some light on this problem can be shed by current model formations created in laboratory conditions, such as Oparin's *coacervates* (Smokers et al. 2022) and Fox's *microspheres* (Nakashima 1987), just like the experiments of Miller and his followers that facilitated us to understand the spontaneous synthesis of organic compounds on the earth's surface. Of course, this does not mean that the first organisms (individuals) were identical with coacervates or microspheres or that the first organic compounds were produced under the influence of electrical discharges or UV radiation in an atmosphere that had a chemical composition identical to that of the gas mixtures used in the discussed experiments. The point is only to show a certain general possibility and principle.

If we make a suspension in water of certain organic substances (in particular of gum Arabic and gelatin, that is, polysaccharides and proteins isolated from living organisms), the suspension particles will tend to merge into droplets. Oparin named them *coacervates*. These droplets will gradually absorb any remaining suspended molecules of gum Arabic and gelatin in the solution, increasing their volume. After exceeding a certain critical size, determined by a resultant of physical laws, the droplets become unstable and undergo division. The descendant droplets formed this way will grow, double their volume, and undergo subsequent divisions. This cycle of subsequent growth and division is presented in Fig. 3-13. At first glance, it resembles the growth and division cycle of living cells indistinguishably. Thus, we have here "feeding" (taking up particles from the suspension) that leads to the growth of the droplet and division into descendant droplets. Coacervates are also, like living cells, clearly isolated from their surroundings. In this case, a primitive barrier is simply constituted by the boundary of phases between water and the droplet's

[21] In fact, the cellular membrane of archaea is significantly different from that in bacteria (and eukaryotes)—its lipids contain isoprenes instead of fatty acids, and they are bound within lipid molecules to L-glycerol by ether linkages, and not to D-glycerol by ester linkages (Koga and Morii 2007).

interior. Strictly speaking, water molecules (which, as we remember, are dipoles, that is, electrically polar molecules) adhering to the surface of the coacervate undergo spatial orientation: their more electropositive side (that is, two hydrogen atoms) are directed toward the coacervate, while the more electronegative side (oxygen atom) are directed towards the water phase. This simplest-to-imagine barrier already forms a very primitive "membrane," selectively permeable for many substances and ions. It should be emphasized that both forming coacervates, their growth and division, and the emergence of the "water membrane" is a spontaneous physical process, an example of self-organization of inanimate matter, associated with entropy increase, and therefore with a decrease in thermodynamically-understood information and structure complexity. We can see again that the structure in the thermodynamic sense is not the same and frequently is completely opposite to the structure in the common meaning.

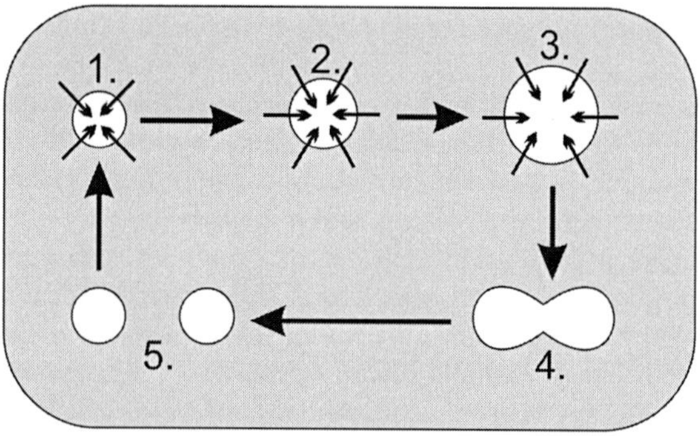

Fig. 3-13. "Life cycle" of coacervates. A coacervate droplet, absorbing remaining in the suspension particles of organic substances (gum Arabic and gelatin) gradually enlarges its size (1–3). After exceeding a certain critical magnitude, it undergoes narrowing (4), and eventually divides into two descendant droplets (5). They are capable of further growth (1…) and the cycle closes.

As it was already mentioned, while the structure and function of living organisms must be based on low entropy and a large amount of thermodynamic information and structure, the purposefulness, orientation on the realization of a concrete aim, or even "value" of information, structure, and function deciding about the essence of living individuals cannot be expressed in the language (and conceptual structure) of thermodynamics.

If an enzyme (phosphorylase) that converts (polymerizes) glucose (specifically—its high-energy form, glucose-1-phosphate) to starch, and another enzyme (amylase) decomposing starch to di-saccharide maltose, composed of two glucose molecules, is added to the suspension of coacervates, what follows will happen. These enzymes will concentrate within coacervate droplets because of better "solubility" in the substances forming coacervates than in water. When, in turn, glucose-1-phosphate is added to the suspension, this compound will penetrate the interior of the coacervate droplets and undergo conversion to starch, which is deposited in the coacervates, while inorganic phosphate separated from glucose will be released outside. Starch, in turn, will be decomposed to maltose, which will diffuse outside to the suspension's water phase (Smokers et al. 2022). Therefore, such modified coacervates exhibit something that can be regarded as a germ of metabolism. Formations resembling, to some extent, coacervates could, in the past, have been hosts of the first metabolic-genetic processes, such as hypercycles of proteins and nucleic acids. Coacervate-like objects not only isolate their content from the external environment and selectively capture some substances from the surroundings, excreting other substances outside, but also exhibit an ability for growth and division. Therefore, they seem ideal carriers of emerging life in many respects. As mentioned, many scientists consider the starting point of the origin of life to be the suspension in water of RNA molecules possessing some primitive catalytic properties (ribozymes) and capable of self-replication (the "RNA World"). In my opinion, however, this is an expression of the underestimation of the ability of inanimate matter to spontaneously self-organize into quite complex systems. In my opinion, life could start from much more complex formations that originated spontaneously, namely from autocatalytic cycles of proteins and nucleic acids (RNA) located in objects resembling coacervates (or microspheres, see below).

Nevertheless, these objects likely possessed rudiments of metabolism "driven" by non-specific catalytic abilities of proto-proteins of a random sequence of amino acids. For this reason, some germs of metabolism could appear before hypercycles came into existence ("metabolism first" v. "control first" concept, Smith and Morowitz 2016, 350, 428), and therefore before the system became "alive," and thus before the first living individuals emerged: hypercycles encapsulated in coacervate-like formations.[22] Instead of coacervates, they could be bubbles surrounded by a primordial membrane built of proto-lipids, that is, amphiphilic compounds (hydrophobic at one end, hydrophilic at the other) with a relatively short and little hydrophobic tail and therefore relatively permeable for various substances (see below)

[22] The "hard core" of metabolism concentrated around the tricarboxylic acid cycle is composed of only several tens of metabolites (Smith and Morowitz 2016).

(Cavalier-Smith 2001). Such a membrane would surround the water phase with autocatalytic cycles of RNA and proteins and the germs of metabolism contained in it. The ability of inanimate matter to self-organize is immeasurable.

The predispositions of coacervate-like formations for being the "ancestors" of the first living cells (individuals) are illustrated in Fig. 3-14. Such formations could concentrate and take up from the surroundings abiogenetically-produced ribonucleotides (nucleotides containing ribose, the building bricks of RNA), amino acids, and other organic compounds because of their better solubility in the interior of coacervate-like objects than in water. They could constitute a favorable catalytical environment for the first autocatalytic cycles of RNA and proteins and then hypercycles. Finally, they could constitute an excellent "starting point" for the evolution of more complex cells.

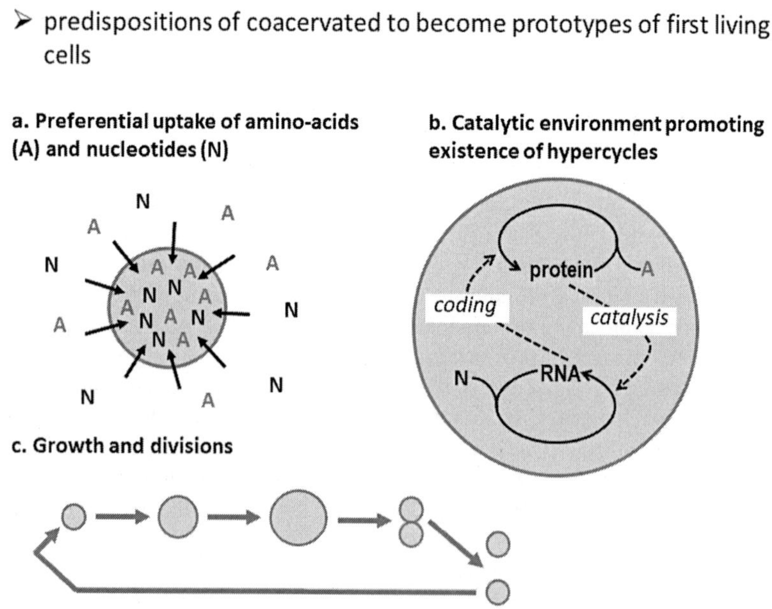

Fig. 3-14. Predispositions of coacervate-like objects to be ancestors of the first living cells.

Fox obtained objects similar to coacervates by heating up and drying a mixture of amino acids in the presence of phosphate as a catalyzer. The output of this procedure produced *protenoids*, complex polymers somewhat similar to proteins (chains of amino acids bound among others by a peptide bond), but branched (non-peptide bonds), which after suspending in water, formed bubbles called *microspheres* by Fox (Nakashima 1987). Microspheres, like coacervates, were capable of growth and division, that is, producing daughter microspheres. However, in contrast to coacervates containing proteins and polysaccharides produced by living organisms, they were composed of quasi-peptide chains synthesized exclusively due to physicochemical processes. More importantly, these chains with a random sequence of amino acids exhibited the ability for non-specific enzymatic activity, catalyzing various chemical reactions. This suggests that primeval protein chains arising spontaneously in primordial oceans could catalyze, for example, the production of other chains of this type or the copying of nucleic acids strands. This would have a fundamental significance for the origination of hypercycles of nucleic acids and proteins. Such proteins with a random sequence of amino acids could, when they had already become encoded on the matrix of nucleic acids, later constituted the starting point for the evolution of all presently known enzymes.

Additionally, microspheres are surrounded by a kind of membrane or wall, across which an electrical potential is present, and which exhibits osmotic properties. Therefore, microspheres seem closer to living cells in some ways than coacervates. However, the potential enthusiasm that we have found the protoplast of the living cells can be slightly damped because protenoids are branched polymers in which many non-peptide bonds appear. On the other hand, we can speculate that in the appropriate conditions prevailing on the young earth, there arose linear polypeptides that formed objects spontaneously resembling microspheres.

Coacervates and microspheres (or lipid-surrounded vesicles; Maynard Smith and Szathmáry 1995, 97-106) illustrate excellently the fact that, in principle, spontaneous, still purely physicochemical processes can lead to such configurations of atoms/molecules arising that exhibit an ability of a spontaneous taking up the building material from the surroundings, growth, division, separation from the surroundings by a clear barrier (and therefore constituting germs of individuals), and that can be the "residence" of metabolism. Self-organization can also lead to "nesting" in coacervate- or microsphere-like formations of autocatalytic cycles of nucleic acids and proteins. Moreover, from here, there is only one step toward spatially-separated hypercycles of RNA and proteins, equivalent, in my opinion, to *the first living individuals*.

It should be stressed that these creations did not need specialized proteins yet—they could appear later due to biological evolution and take over the functions realized initially by purely physicochemical processes.

This constitutes a subsequent example of the general rule "not everything at once," which was a predominating principle of life's origination (and evolution).

Therefore, already quite complicated structures could come into being not due to biological evolution in the way of natural selection but through self-organization of inanimate matter being a derivative of purely physicochemical processes. This self-organization was possible because of the potential ability of matter to create structure, discussed in Chapter 1, and realized because of fulfilling the appropriate thermodynamic conditions, as discussed in Chapter 2. Therefore, life did not have, at least in my opinion, to start from (almost) zero, from single (macro)molecules of organic compounds, for instance, self-replicating RNA molecules. On the contrary, its starting point could be quite complex structures, such as hypercycles "populating" coacervate- or microsphere-like objects. The crucial link that was lacking for life to originate and evolve, and more specifically—for the coupling of autocatalytic cycles of RNA and proteins into hypercycles—was the "keystone," an adapter establishing the correspondence between the sequence of nucleotides in nucleic acids and amino acids in proteins, conditioning the existence of the genetic code and genetic information, namely some ancestor of the present tRNA.

The formation of the cellular membrane surrounding the cell is strictly associated with the emergence of the structural-individual aspect of life. In contemporary organisms, it has a lipid-protein character. Its framework is constituted by the lipid bilayer, in which there are immersed various proteins: carriers (transporting different substances and ions), enzymes (e.g., producing the cellular wall), the elements of the electron transport chains, ATP synthase, or, in multicellular organisms, adhesive proteins connecting cells, and receptors (binding, e.g., hormones and turning on an appropriate response inside the cell).

Of course, the first organisms (individuals) did not yet possess specialized membrane proteins. The "pure" lipid bilayer alone is impermeable for most substances, particularly polar or electrically charged molecules (ions). The long hydrophobic tails of lipids constituting the "frame" (the "basis") of such a membrane are responsible for this (Cavalier-Smith 2001). However, a cell must intake various substances, serving as building materials and an energy source, and remove the harmful by-products of metabolism. For ATP synthesis according to the chemiosmotic mechanism to be possible, the circulation of H^+ ions (protons) (sometimes their role is played by sodium ions, Na^+) across a membrane associated with the production (strictly speaking—transformation) of energy is necessary: the cellular membrane of bacteria and archaea, the inner mitochondrial membrane of mitochondria, or thylakoid membrane in chloroplasts. Contemporary cells also maintain a gradient of sodium Na^+ and potassium K^+ ions across the cellular membrane through active transport: the concentration of K^+ ions is greater inside, while Na^+ ions—outside the cell (Alberts et al. 1998, 378-386). In present organisms, these tasks are executed by appropriate carriers that selectively transport various substances and ions, namely proton pumps in the respiratory, photo- or chemosynthetic chain and ATP synthase, and sodium-potassium pump driven by ATP hydrolysis. Of course, the first proto-cells lacked all these advanced specialized proteins. How, therefore, did they cope with the essential impermeability of the lipid bilayer?

Here, we must rely on speculation, at least at the present stage of knowledge. It can be supposed that the primordial cellular membrane was formed by amphiphilic[23] compounds that were less hydrophobic than lipids, with a much shorter hydrophobic "tail," and therefore, much more permeable for polar molecules and ions. A few candidates for such compounds (for instance, lipids with a much shorter "tail" of the fatty acids entering into their composition) have been proposed (Cavalier-Smith 2001). Additionally, some primitive proteins could inbuild themselves into proto-membranes and relatively non-specifically facilitate penetration of various substances. As the specialization and action efficiency of these proteins increased, the very "framework" of the cellular membrane could become more hydrophobic and thus impermeable for polar substances and ions, adopting the contemporary lipid character. This enabled the selective transport of only some substances, purposeful from the point of view of the cell functionality, by specialized protein carriers. Again, the "already ready", fully developed composition of specialized lipids and proteins of the cellular membrane did not have to originate at once, in one step, from inanimate matter. There are thinkable subsequent, fully functional, mediate stages in which an increase of complexity, optimization, and efficiency of the operation of the cellular membrane takes place. Also in this case, the emerging life did not play a lottery game.

The energetic-thermodynamic aspect of the origin of life

Of course, various other aspects of life had to develop more or less in parallel with the previously discussed genetic and structural-individual aspects. Their emergence will now be discussed, starting from the energetic-

[23] Amphiphilic compounds are compounds that are hydrophilic from one side, and hydrophobic from the other side.

thermodynamic aspect. Emerging life could not rely for a long time on organic compounds synthesized abiogenetically, as those underwent quick exhaustion because of their intake and usage by the first proto-organisms. Organisms needed a driving force that could allow them to function as dissipative structures. As we remember from Chapter 2, the necessary condition for the existence of dissipative structures is a sufficiently large thermodynamic gradient. The presence of such a gradient in a living organism (living cell) can be ensured by a chemical compound, whose decomposition can, under physiological conditions, supply a large amount of energy, which in turn is used to perform different kinds of work: chemical, mechanical, osmotic, electrical, and so on.

As it is commonly known, such a universal carrier of energy in the present cells is ATP. Earlier, this could be some simpler compound, for instance, pyrophosphate (P-P, two bound with other phosphate groups) or acetyl-P (acetate group bound with phosphate) (Lane 2010). The main mechanism responsible for ATP synthesis (which requires, of course, energetic investments, as the opposite reaction releases energy) in contemporary living organisms is the previously discussed membrane phosphorylation (including oxidative phosphorylation, photosynthetic phosphorylation, chemosynthetic phosphorylation, and ATP synthesis in halophilic bacteria) (Nicholls and Ferguson 2002). It is present in all prokaryotes (bacteria and archaea) and descendants of some respiring bacteria (α-proteobacteria) and photosynthetic bacteria (cyanobacteria), which entered into endosymbiosis with ancestors of eukaryotes and still today constitute, as mitochondria and chloroplasts, their power stations (Margulis 1981; Maynard Smith and Szathmáry 1995; Futuyma 2005; Lane 2010, 2015; Deamer 2015). As we remember, the mediator in ATP synthesis in membrane phosphorylation is the electroosmotic gradient (i.e., the gradient of both concentration and electric charge) of H^+ ions (protons) across an appropriate lipid-protein membrane, for instance, a bacterium cellular membrane. The proton pumps are responsible for building up the proton gradient, that is, for active transport of H^+ ions from the compartment (e.g., bacterial cytosol), where their concentration is lower, while electric charge negative, to the compartment (e.g., water phase outside bacteria), where proton concentration is higher, and electrical charge more positive. Therefore, this transport occurs against the gradient of proton concentration and charge. As such, it needs some thermodynamic driving force. Proton pumping can be driven by the flow of electrons received from respiratory substrates (e.g., glucose) through the respiratory chain (into its composition enter, e.g., proton pumps and cytochromes, compare Fig. 2-10) to oxygen, as it takes place in oxidative phosphorylation. Electrons can also be knocked out from the water by electromagnetic radiation in the visible range (chlorophyll mediates in this) and flow through the photosynthetic chain (also containing proton pumps) to ultimately reduce the carbon dioxide and synthesize organic compounds (e.g., glucose) (photosynthetic phosphorylation). In the case of chemosynthesis, electrons flow from reduced inorganic compounds (of sulfur, nitrogen, and iron ions) through the chemosynthetic chain to oxygen, which is invariably coupled with proton pumping. Finally, proton pumping can be driven directly by the absorption of quanta of visible radiation by the protein bacteriorhodopsin (halophilic bacteria). In all these cases, protons return to the compartment, where their concentration is lower and electric charge more negative (e.g., bacterial cytosol) through an appropriate channel in an enzyme called ATP synthase—compare Fig. 2-7 D. Just this return of protons is coupled with the synthesis of ATP from ADP and P_i, and thus drives the latter. ATP synthase seems a very ancient and primeval protein—it occurs and has a similar general form in all organisms conducting membrane phosphorylation. Therefore, it had to be already pretty complex in the common ancestor of these organisms, most probably LUCA (all known organisms devoid of membrane phosphorylation, exclusively eukaryotes, most probably secondarily lost it, as they get rid of mitochondria because of living in anaerobic conditions) (genetic analysis indicates that pre-ATP synthase was already present in LUCA; Weiss 2016).

It is supposed that such a system of energy production (transformation) developed relatively early in living organisms. The gradient of H^+ concentration and electric charge across a proto-cellular membrane likely appeared first, and only then was it adopted as the thermodynamic force driving ATP production (Lane 2010). The electrical concentration gradient could initially transport various substances/ions inside and outside the proto-cell. Its negatively-charged interior favored the concentration of molecules and metal ions endowed with a positive charge. The coupling of the transport of some substances through the proto-membrane inside the proto-cell with the backward flow of protons downstream of the thermodynamic gradient of proton concentration and charge could also lead to the condensation of these substances in the inner compartment. Negatively-charged molecules (ions) could be co-transported with protons,[24] which rendered the whole process electroneutral.

[24] This would not require any specialized proteinaceous carriers. A negatively-charged molecule could bind a proton and become electroneutral in the external environment with a lower pH (a higher H^+ ion concentration) and get into the proto-cell, wherein in a higher pH the proton would dissociate off from the molecule that

This constitutes a subsequent example of the rule: "not everything at the same time." The generation of the proton gradient across the (proto-)cellular membrane and the coupling of its dissipation with ATP synthesis did not have to emerge simultaneously. Just the opposite, first, the proton gradient could have started to be created, which "at once," instantaneously could be useful for transporting various substances. Only later, chemiosmotic ATP synthesis was added to this system—and a simple ancestor of the present ATP synthase appeared. Initially, it could use the Na^{2+} ion gradient, not the H^+ gradient, for ATP synthesis, as it still occurs in some prokaryotes (Nicholls and Ferguson 2002; Lane 2010).

Mutual co-conditioning of various aspects of the origination of life

It is worth mentioning that the origin of the energetic-thermodynamic aspect of life was most probably (at least in my opinion) strictly related to the origin of the structural-individual, genetic, metabolic, and redox aspects. An impermeable membrane for protons surrounding the cell is indispensable for membrane phosphorylation in prokaryotes. Therefore, for this energetic-thermodynamic process to occur, the strongly-expressed presence of the structural-individual aspect in the form of a lipid membrane "equipped" with proton pumps and ATP synthase (and protein carriers transporting various substances) is of key importance. The circulation of protons across the membrane, ATP synthesis, and hydrolysis, coupled with driving various processes in the cell, are examples of physical and chemical dissipative structures. Therefore, here, the structural-individual aspect meets the energetic-thermodynamic aspect. Proton pumps are driven by electron flow between redox centers in an electron transport chain—thus, in this case, the redox aspect crosses with the energetic-thermodynamic aspect. The proteins participating in the process of ATP synthesis were initially encoded in RNA and then in DNA and produced due to translation carried out by the primordial genetic apparatus. All these processes could not occur without metabolism, which in turn was driven by the hydrolysis of ATP to ADP and P_i, while the enzymes conducting it were encoded by RNA/DNA. Therefore, the genetic and energetic-thermodynamic aspects were also necessary for the coming into existence of the metabolic aspect, and they, in turn, needed this aspect for their functioning. The synthesis of (the components of) the lipid-protein membrane (crucial for the structural-

would again become negatively charged. In this form, it could not permeate back through the proto-cellular membrane and would be trapped within the proto-cell.

individual aspect) required the metabolic aspect (lipid synthesis), genetic aspect (proteinaceous carries synthesis), and energetic-thermodynamic aspect, driving the latter two, while all of them could not do without the structural-individual aspect, and so on.

The adenylate (ribo-)nucleotide, that is, the complex: adenine (purine, nitrogen base) + ribose (pentose, 5-carbon monosaccharide) + P (phosphate group), is common for many aspects of the origin of life. It enters into the composition of RNA—this is just the already familiar to us nucleotide A.[25] ATP (adenosine triphosphate) is nothing else than just the adenylate (ribo-)nucleotide with attached two additional phosphate groups (altogether, ATP contains three phosphate groups). The adenylate (ribo-)nucleotide also enters into the composition of NAD, a redox compound that transfers electrons and hydrogen between respiratory substrates and the respiratory chain (this was discussed above), of FAD (auxiliary in relation to NAD electron/hydrogen carrier), and a compound similar to NAD, NADP (NAD with attached P group), transferring electrons and hydrogen from the light (light-dependent) and dark (light-independent) phase of photosynthesis, that is, between the photosynthetic chain and CO_2 assimilation in the Calvin cycle, for which its reduced form ($NADPH + H^+$) constitutes the "reductive force." Finally, the adenylate (ribo-)nucleotide is a component of coenzyme A (CoA), which can transfer the acetate group as acetyl-CoA (compare Fig. 2-10) (CoA can also bind fatty acids and acyl-CoA is formed). This enables the "metabolic communication" between glycolysis (the decomposition of sugars) and the fatty acids β-oxidation (decomposition of fats), on the one hand, and the tricarboxylic acid (TCA) cycle, on the other hand. Therefore, the adenylate (ribo-)nucleotide constitutes the "common denominator" of the genetic aspect (as a component of RNA), energetic-thermodynamic aspect (as the "core" of the ATP molecule), redox aspect (as a component of NAD, NADP, and FAD), and the metabolic aspect (as a part of coenzyme A; CoA also participates in fatty acid synthesis) (Murray et al. 2012). It is also worth recalling that adenylate nucleotide and ATP were produced spontaneously in the sequence of experiments conducted by Miller and his followers. No wonder, therefore, that many aspects of emerging life exploited this nucleotide. Therefore, life is likely based not on "optimal" chemical compounds but on compounds that originated spontaneously on the young earth.

---

[25] The adenylate (deoxyribo-)nucleotides entering into the composition of DNA differ in this way from the (ribo-)nucleotides in that, instead of sugar ribose, they contain a very similar sugar deoxyribose.

The common element of the genetic aspect and the structural aspect of the origin of life was, as mentioned, the encapsulation (resulting in its separation from the surroundings and its harmful influences and other autocatalytic cycles) of autocatalytic cycles of proteins and nucleic acids in bubble-like objects resembling, to some extent, coacervates or microspheres, which favored the mutual acquaintance between these cycles, and the emergence of hypercycles resulting from this.

We can see, therefore, that various aspects of the process of the spontaneous origination of life mutually supported and interweaved with each other and most probably reciprocally co-conditioned one another (at least in my opinion). This is another manifestation of the causal closure property, so characteristic of living beings.

The enzymatic aspect of the origin of life

Another aspect of the origin of life that should not be omitted is the enzymatic aspect. Enzymes, or biocatalysts, are characterized by two fundamental properties (Voet et al. 1999; Murray et al. 2012). First, they hugely accelerate (catalyze) various (bio)chemical reactions. Second, they exhibit a high specificity, that is, conduct strictly determined, just such-and-not-other reactions. Together, these two features direct the whole process of matter and energy transformation by the complex of enzymes in a given cell (that is, metabolism). Of course, metabolism serves both the functioning of the living cell and the construction of its structure through synthesizing various, mainly complex, polymeric organic compounds. Essentially all enzymes appearing presently in living organisms are (simple or complex) proteins (although they may contain non-protein prosthetic groups), and their structure, being a derivative of the sequence of amino acids in a polypeptide chain, is encoded in the form of the sequence of nucleotides in DNA. The improvement and quantitative growth of genetic information were inseparably associated with an increase in both the number of enzymes and the degree of their specificity and efficiency, which in turn led to a broadening of the repertoire of the processes entering into the composition of the metabolism. Enzymes are crucial for all aspects of emerging life: the genetic aspect (e.g., RNA polymerase, and next DNA polymerase, reverse transcriptase, aminoacyl-tRNA synthase, proto-ribosomes), the energetic-thermodynamic aspect (e.g., ATP synthase, proton pumps), the metabolic aspect (catalyzing of all biochemical reactions), and the redox aspect (electron transferring complexes can be regarded as broadly-understood enzymes), *etc.* Of course, the genetic apparatus is responsible, among other things, for enzyme synthesis, while the regulation of gene expression–for their production in a proper place, time, and amount. This is a subsequent example of the mutual co-conditioning of various aspects of the origin of life discussed in the previous subchapter.

What does the principal property of enzymes, their ability to catalyze chemical reactions, consist of? If we come back to Chapter 1 and Fig. 1-3, we will recall that this, which limited the rate of chemical reactions and prevented the instantaneous joining of all reagents, was the "hump" in the dependence of the system energy on the distance between reacting atoms/molecules, corresponding to the activation energy (Marks et al. 1996, 104-108; Voet et al. 1999, 288-289; Murray et al. 2012, 57-69). As mentioned, the probability of a molecule jumping over this bump, that is, in the case of a huge number of reacting molecules, the reaction rate can be elevated through an increase of temperature, and thus confer to a significant fraction of atoms/molecules kinetic energy higher than the activation energy. (In the case of such complex molecules as organic compounds, not only the translocation energy of whole molecules related to the rate of their movement matters here but also the intensity of vibrations of the "springs" of chemical bonds—vibrational energy and the rate of rotation around the bond axis—rotational energy). However, this method would not work in the case of living organisms, as a high temperature would destroy their structures, and apart from this, the energy of chemical bonds would be dissipated in the form of heat and could not serve to perform any useful work.

Additionally, a high temperature would elevate non-specifically the rate of all, not only the selected reactions, due to which there would emerge not an ordered and strictly directed metabolism but a chaotic mixture of chemical processes. For this reason, enzymes had "chosen" (in the process of biological evolution) another strategy. Through their appropriate structure and spatial distribution of electrical charges in their active centers, they significantly lower the hump (reaction activation energy) shown in Fig. 1-3 (Marks et al. 1996, 104-108; Voet et al. 1999, 288-289; Murray et al. 2012, 57-69). Thanks to this, that hump can be jumped over by many molecules in the unit of time, even at room temperature, by molecules with relatively low kinetic energy. Therefore, ultimately, the structure of living organisms, constituting a derivative of the huge assembly of enzymatic reactions, owes its existence to the fact that the discussed hump is not too high, which allows the reactions to occur at a sufficiently high rate, and not too low, which enables selective acceleration by enzymes of selected reactions that are purposeful from the point of view of the whole metabolism, and exists at all, and thus all reagents do not react with each other instantaneously. Generally speaking, the last property enables the existence of such happening in time dynamic (chemical) structures as living organisms. Here we once again can conclude that the existence of the structure of living organisms is possible thanks to a really completely improbable coincidence of physical laws and appropriate parameter values resulting from them, for instance, those determining the magnitude of the discussed hump.

It is also worth mentioning that many enzymes, apart from a huge acceleration of strictly determined chemical reactions, couple spontaneously occurring reactions (processes) with reactions that alone are not favored energetically (thermodynamically). In this way, the absorption by chlorophyll of electromagnetic radiation quanta in the visible range becomes coupled (through the photosynthetic electron-transport chain and its proton pumps) with ATP and NADPH synthesis and, finally, with sugar production in plants. This way, the oxidation of sugars and fats is coupled (through the respiratory electron transport chain and its proton pumps) with ATP synthesis in mitochondria in oxidative phosphorylation. Finally, in this way, the hydrolysis of ATP drives various types of work in the living cell (chemical, mechanical, osmotic, electrical). Generally, this is the system of thousands of such couplings on the molecular level that causes the dissipation of visual radiation into infrared radiation and drives the hierarchically-organized assembly of dissipative structures constituting the terrestrial biosphere, described in Chapter 2.

Apart from enzymes proper, there also occur in cells proteinaceous carriers "anchored" in lipid-protein membranes transporting various molecules and ions selectively (Marks et al. 1996, 104-108; Voet et al. 1999, 288-289; Nicholls and Ferguson 2002; Murray et al. 2012, 57-69). Such transport can be passive transport, occurring down the concentration gradient (from the compartment with a greater concentration of the transported substance to the compartment with a lesser concentration) or charge gradient (e.g., the transport of positive ions from the more positively charged to a more negatively-charged compartment) or active transport, proceeding against the thermodynamic gradient, coupled, for example, with ATP hydrolysis or flow of H^+ ions from a compartment with higher H^+ ion concentration (lower pH) and more positive charge to a compartment with lower H^+ ions concentration (higher pH) and more negative charge. For instance, the transport against the concentration gradient of Na^+ ions outside the cell and K^+ ions inside the cell by a sodium-potassium pump is driven by ATP hydrolysis (Alberts et al. 1998, 378-386). In turn, the transport of inorganic phosphate (P_i) and some metabolites inside mitochondria is coupled with the spontaneous flow of H^+ ions in the same direction (the so-called co-transport or symport). The ATP/ADP exchange (anti-transport, antiport) through the inner mitochondrial membrane, conducted by an appropriate proteinaceous carrier, is driven by the electric gradient across this membrane (Nicholls and Ferguson 2002). Exploiting the H^+ concentration gradient and or electric gradient across a lipid-protein membrane, or the chemical gradient related to ATP hydrolysis, for driving the active transport of substances/ions is associated with the energetic-thermodynamic aspect, structural-individual aspect, and the metabolic aspect (an appropriate concentration of metabolites in various cellular compartments and the whole cell) of the origin of life. Additionally, the enzymatic (-carrier) aspect is tightly interweaved with the metabolic aspect, as enzymes and carriers constitute the functional building bricks of metabolism. Undoubtedly, all these aspects were inseparably related to each other during the spontaneous origin of life.

Transduction and processing of information in living organisms

In addition, the processes of transfer and processing of information in living organisms developed and underwent complications during the initial steps of the evolution of life. The most primeval process of this kind was the reading of genetic information with the help of the primordial (still dyadic, or maybe even monadic) genetic code, that is, translation of the sequence of nucleotides in RNA (proto-mRNA) to the sequence of amino acids in proto-proteins. Afterward, RNA was replaced by DNA in the function of the record of genetic information, and the information was transferred from RNA to DNA (reverse transcription) and back again (RNA polymerase) (LUCA stage). Of course, the genetic record had to be copied (first, RNA was replicated by RNA-dependent RNA polymerase, and next, DNA was proliferated by DNA-dependent DNA polymerase). Subsequently, there appeared and developed the regulation of the reading of genetic information consisting in the fact that various proteins and short regulatory RNA strands could, binding themselves to DNA nearby the regulated genes or to mRNA strands, stimulate or, just the opposite, inhibit gene expression, that is, synthesis of particular proteins (Brown 2012, 157-175). The intercellular signaling that controls the fluxes through various metabolic pathways (e.g., the activation of various processes, including ATP hydrolysis by actomyosin-ATPase and several steps of oxidative phosphorylation during muscle contraction (Korzeniewski 1998, 2017b) and of other metabolic processes by Ca^{2+} ions and or cAMP—cyclic AMP (cyclic adenosine monophosphate, that is, the complex: adenine + ribose + one phosphate group, closed into a ring) developed (Martin 1987; Alberts et al. 1998, 482-

513). In multicellular organisms, inter-cellular signals appear between neighboring cells (as part of a paracrine system), and inter-organ signals encompass the whole body in the form of hormones transported by the blood (the endocrine system) (Hill and Wyse 1989; Hopkins 1995). Finally, many vegetative functions, and first of all, the behavior of animals, began to be steered by neuronal information transfer (the appearance of the neural system, including the brain) (Hill and Wyse 1989, 385-418).

The common ancestry of known life

It is worth mentioning something else when discussing the process of the origination of life. Despite the enormous diversity of the forms of life on the macroscopic level (there is no need to prove that a bacterium, an oak, and a bat differ very significantly from one another), the most general principles of the functioning of terrestrial living organisms on the microscopic level seem identical. This identity concerns the carrier of genetic information (DNA, earlier RNA), fundamental elements of the cell machinery (enzymes of protein character), a universal carrier of energy (ATP), and a barrier surrounding the cell (lipid-protein membrane). This same is (in general terms) the highly complicated apparatus for reading the genetic information (protein synthesis), the complex mechanisms of membrane phosphorylation, and lastly, the genetic code (the property of the universality of the genetic code) (Maynard Smith and Szathmáry 1995; Copley et al. 2007; Lane 2010, 2015; Smith and Morowitz 2016; Deamer 2019). The indicated similarities (in reality, there are many more of them) lead directly to one extremely important conclusion: all living organisms known to us on the terrestrial globe come from *one common ancestor* (the *last* common ancestor was the one discussed above LUCA, but it was certainly not the *first*).

Additionally, this ancestor had to be already quite complicated; it possessed at least all the previously enumerated properties common to the known living world. It probably (at least in my opinion) already had the form of a primitive, simple cell surrounded by a lipid-protein membrane or by something that was its ancestor, containing a relatively advanced genetic apparatus, specialized enzymes, and a fully shaped genetic code. Therefore, it must have passed quite a long period of evolution from the stage of a simple hypercycle encapsulated in a coacervate-like bubble described above.

The fact that all presently living organisms come from a single ancestor does not imply that life originated only once in the initial stages of the earth's existence. However, even if separate life forms with slightly (or very) distinct fundamental principles of structure and function emerged from the "primordial broth," most became extinct, and only one gave rise to the entire living world. It is improbable that the same genetic code was shaped twice due to pure chance.

Reference to the next chapters

We saw in the present chapter that the capabilities of inanimate matter, resulting from physical and chemical laws, can, at least potentially, lead to the spontaneous emergence of surprisingly complex structures (the specific configurations of atoms and molecules) capable of growth and division, and recording and transferring to daughter configurations of their identity as a set of characteristic properties. In particular, purely physicochemical forces enabled the creation of proto-organisms (proto-individuals)[26] complex enough to constitute the "starting point" for the origin of life and its development in biological evolution.

Well then, inanimate self-organization processes could create some very primitive proto-cells but nothing above them. For the present living organisms, with their entire complexity, diversity, ingeniousness, and purposefulness of structure and function, to develop, there had to appear something more. This is the process of biological evolution in the way of natural selection. This is just natural selection, an extraordinarily simple and elegant mechanism that has led, in confrontation with the potential ability of the physical reality to form the structure discussed in previous chapters, to create the whole wealth of the living world with humans at the front. The fundamental mechanisms, principles, rules, and course of biological evolution will be discussed in the next two chapters.

Three "trunks of life"

As mentioned, three main "trunks" (domains) of life arose from LUCA: bacteria (*Bacteria*) = eubacteria (*Eubacteria*), archaea (*Archaea*) = archaebacteria (*Archaebacteria*), and eukaryotes (Eukaryota) (Maynard Smith and Szathmáry 1995; Copley et al. 2007; Lane 2010, 2015; Smith and Morowitz 2016; Deamer 2019). The two former groups were once classified as nucleus-less organisms—prokaryotes (*Prokaryota*). They are characterized by a lack of a nucleus, other organelles, and intracellular

---

[26] In my opinion, these were autocatalytic cycles of proteins and nucleic acids, followed by hypercycles, encapsulated in proto-cell bubbles resembling to some extent coacervates or microspheres.

membranes. They possess a small circular DNA molecule. Their cellular wall (primitive property) is associated with a lack of phagocytosis (food intake from its surroundings by a bubble due to the cellular membrane concaving inside the cell). They lack a cytoskeleton (microtubules, microfilaments, mediate filaments) that maintains the shape of cells devoid of the cellular wall and participates in transporting substances and structures within the cell. They have small ribosomes. Finally, their cells have tiny dimensions. However, despite the similarities regarding the size and structure between bacteria and archaea, the latter are considered more related to eukaryotes (some scientists even classify eukaryotes as a sub-group of archaea; Lane 2010; Hug et al. 2016). Archaea have a mechanism of DNA replication completely dissimilar to that in bacteria (Lane 2010), and therefore they most probably independently evolved from LUCA and "learned" DNA replication (compare Fig. 3-12). Archaea also differ from bacteria by other fundamental properties (Koga and Morii 2007). They have significantly different ribosomes, resembling more eukaryotic ribosomes; also, the sequence of nucleotides in their rRNA is more similar to the eukaryotic than the bacterial sequence. They possess characteristic tRNA and genes with specific nucleotide sequences. Some of their genes contain introns, which brings them nearer to eukaryotes. They have an apparatus of transcription similar to eukaryotic. Histones occur in some archaea, which is also a eukaryotic feature. Their cellular wall lacks peptidoglycans, including murein, common in bacteria. Finally, they have a different cellular membrane composition than bacteria and eukaryotes (just this property differs archaea from the latter). Phospholipids (the main lipids of the cellular membrane) in bacteria are composed of two molecules of fatty acids with a linear hydrophobic hydrocarbon "tail" bound with L-glycerol by the ester bound: -C-CO-O-C- ;[27] glycerol additionally binds a phosphate group (hence the name "phospholipids"). On the other hand, the phospholipids of archaea contain branched isoprenes, instead of linear fatty acids, that are bound through ether bonds: -C-O-C- to D-glycerol.[28] Additionally, in many

---

[27] This bond is formed due to the reaction of the carboxylic group -COOH of organic acid and hydroxyl group -OH of alcohol.

[28] L-glycerol and D-glycerol are *enantiomers* of glycerol, that is, its chiral forms constituting the mirror image of each other. Enantiomers appear in compounds, in which a least one carbon atom has four different radicals (bound chemical groups). "L" means "left-handed" and "D" means "right-handed," which refers to the direction of the optical rotation of polarized light (L—anti-clockwise, D—clockwise), although for many compounds these are arbitrary names—it is important that the forms L and D rotate polarized light in the opposite direction (Patrick 2000, 49-54; Baggott 2015, 207). No transformations (displacements, turns) in space can

archaea groups, the isoprene "tails" of the lipids lying opposite each other in the lipid bi-layer of the cellular membrane are bound by a covalent bond, which never occurs in the case of fatty acids in bacteria (Koga and Morii 2007). This dissimilarity of the cellular membrane structure probably results from the fact that LUCA had a simpler form of membrane, from which the membranes of bacteria and archaea evolved. Alternatively, the cellular membrane of the latter constitutes a derivative adaptation to the conditions in which many live because most archaea are organisms existing in extremely hostile (from our point of view) conditions: in high temperatures (thermophiles), environments of high salinity (halophiles), or environments which are very acidic (acidophiles).

Nevertheless, most closely located to the main "stem" of the family tree of all living organisms (that is LUCA), elaborated based on the nucleotide sequence in rRNA, are various thermophiles belonging to bacteria and archaea. This suggests that the common ancestor of all living organisms was also a thermophile (Weiss et al. 2016). This makes perfect sense, as the earth's surface was much hotter than it is now during the initial stages of life's evolution. In particular, the terrestrial crust plates' volcanic activity and tectonics were much more intensive. Therefore, it is possible that the cellular membrane of contemporary bacteria (and eukaryotes) is secondary in relation to the cellular membrane of archaea and constitutes an adaptation to "normal," less extreme conditions (e.g., temperatures).

Eukaryotes differ from prokaryotes, chiefly bacteria and archaea, by a series of properties. They possess a nucleus surrounded by a double lipid-protein membrane with characteristic pores. Their cells are (usually) much greater than prokaryotic cells and contain, apart from the nucleus, other organelles, such as mitochondria, chloroplasts (in plants), endoplasmic reticulum, Golgi apparatus, lysosomes, and so on. Their DNA is composed of linear segments and contains more nucleotides (and genes) than the circular DNA of prokaryotes. Eukaryotic genes contain numerous introns, while eukaryotic DNA is bound to histones and undergoes condensation (coiling on a few hierarchy levels) into chromosomes during cell division. Eukaryotes are originally devoid of a cellular wall (the cellular wall of, e.g.,

cause enantiomers L and D to overlap, such as it is not possible to overlap spatially left and right gloves. All amino acids entering into the composition of proteins are L-amino acids, while all monosaccharides entering into the composition of terrestrial living organisms (e.g., glucose) are D-saccharides. This results most probably from an accident event, from the fact that primordial enzymes started by chance to catalyse synthesis of just such-and-not-other amino acids and monosaccharides (of course, enzymes, as a whole, are also chiral). Nevertheless, most probably, life based on D-amino acids and L-monosaccharides is also possible.

plants or fungi is derivative) and capable of phagocytosis. They possess a cytoskeleton, for instance, microtubules, maintaining the shape of the cell, forming the mitotic or meiotic spindle, and the "skeleton" of eukaryotic flagella, serving as "rails" for transport proteins: kinesins and dyneins, or actin filaments, also giving shape to the cell and serving as "rails" for myosin heads, which is important for the transport of various substances/objects in the cytosol and for muscle contraction.

Mitochondria and chloroplasts originated due to endosymbiosis (intracellular symbiosis) of respiring bacteria (α-proteobacteria) and photosynthetic bacteria (cyanobacteria), respectively, with ancestors of eukaryotes (according to some authors—a group of archaea). Today, they are produced only by parent mitochondria and chloroplasts, contain a small circular DNA molecule of their own that encodes, for example, subunits of the respiratory chain complexes, and mitochondrial/chloroplast tRNA and rRNA, possess their own ribosomes and the whole genetic apparatus, and are all similar to those occurring in bacteria (Margulis 1981; Maynard Smith and Szathmáry 1995; Futuyma 2005; Lane 2010, 2015; Deamer 2015). It is a matter of debate whether the "host" of, for example, proto-mitochondria was a proto-eukaryotic cell, something similar to an amoeba, possessing all the properties of eukaryotes apart from being "equipped" with mitochondria, or simply an archaeon cell, which only after the "absorption" of mitochondria acquired all the features characterizing eukaryotes.

It should be stressed that in the initial stages of the evolution of life, very intensive (and to some extent also present today) so-called *horizontal gene transfer* took place, that is, a transfer of DNA segments between different organisms, frequently unrelated to each other (Futuyma 2005; Lane 2010, 2015). Therefore, the genetic information underwent a significant mixing between particular "branches" of life, and the family tree of terrestrial organisms can be equally well presented as vigorously bifurcating (branching) or as joining separate branches with horizontal "connectors" (and therefore, the "family tree" resembles a "family network"). For this reason, the family trees elaborated based on nucleotide sequences in genes encoding proteins, rRNA, and tRNA entering into the composition of the genetic apparatus can look completely dissimilar to family trees based on nucleotide sequences in genes encoding enzymes "conducting" cellular metabolism. Because of the very intensive horizontal transfer of "metabolic genes," the priority in forming the systematics of main groups of organisms is given to "genetic apparatus genes." Nevertheless, one should remember that living organisms constitute genetic hybrids ("chimeras"), as they contain genes of different origins. For instance, eukaryotes have genetic apparatus similar to that in archaea, while their metabolism resembles more bacterial metabolism (Lane 2010).

CHAPTER 4

HOW HAS THE COMPLEX STRUCTURE AND FUNCTION OF LIVING INDIVIDUALS EVOLVED?

Introduction

In my opinion, the first living individuals were bubbles that resembled, to some extent, coacervates or microspheres, containing hypercycles of RNA and proteins and equipped with the germs of metabolism. They were much simpler objects than the last common ancestor of all presently living organisms—LUCA. Nevertheless, they had already exhibited the ability to evolve, perfecting and optimizing their structure and function, and a possible increase of complexity (and, therefore, overall, a modification of their identity) from generation to generation.

Living organisms are, among others, dynamic configurations of atoms, which can produce (form) configurations of atoms very similar to them. In order to do this, they must possess some receipt that is transferred to daughter configurations; based on this, these configurations can re-create their structure and function. As already mentioned, this prescription does not resemble at all, for example, a static project of a building, a plan (blueprint) of a car or computer, but rather a recipe for baking a cake with an extremely important difference: this recipe can self-realize. The prescription for a complex living organism can be more adequately compared to a little robot (fertilized egg cell), which possesses a handyman "Do-it-yourself" set, that is, a basic full set of tools (e.g., enzymes), production projects (matrixes) (DNA segments—genes) of these and other tools, and some structural elements (that is, various proteins fulfilling structural functions, e.g., cytoskeleton, lipid-protein membranes containing carriers and receptors, and so on), finally—devices for reading and realization of these prescriptions/matrixes (that is, the whole genetic apparatus = machinery for protein synthesis, and the system of control of gene expression). The robot is also provided with many building materials and takes up the rest of the needed materials from its surroundings. Due to the realization (by itself) of the possessed record (set of instructions), the self-constructing little robot grows, differentiates and complicates its structure (during the embryonic and individual development) in order to eventually become a fully-shaped robot—a configuration of atoms representing an adult individual. This configuration can proliferate the possessed constructive prescription in many copies (each of them will pass to every descendant cell), some of which are transferred to the "precursors" of future "little robots"—reproductive (sexual) line cells (egg cells and spermatozoa). At the same time, the egg cell brings to the little robot-to-be a half (strictly speaking—one copy) of the constructive prescription and practically the whole handyman workshop (set of tools), while the spermatozoon contributes practically only the second half of the prescription (in principle: the second, slightly differing from the first, copy of this prescription). After fertilization, a new little robot capable of self-constructive development arises, and the cycle closes. Of course, the "robot" corresponding to the living individual, with an independent complex structure and function determined by the self-production recipe it is equipped with, has nothing to do with any robots built by people. One of the most important differences is that the functioning of living organisms occurs at the molecular level, particularly the biochemical and genetic levels. Biological systems are also dynamic structures—they exchange the components of their bodies incessantly, and a continuous stream of matter, energy, and information flows through them. It is worth emphasizing that for the production of daughter configurations corresponding to such "biological robots," the continuity of transfer of not only the construction prescription but also the whole set of tools for its realization, that is, the executive part of our little robot, is necessary. In the case of viviparity and progeny care (e.g., in mammals), the newly shaping little robot is strongly supported by the executive machinery of the robot-mother, first during pregnancy and next during feeding, protection, and teaching. One should ponder the fact of the continuity of the transfer of not only the production recipe but also of the complex set of tools indispensable for reading it before uncritically accepting the gene-centrism proposed by Dawkins, according to which bodies are only carriers or transient containers for "selfish genes" (Dawkins 1976).

Nevertheless, it is true that the structural-functional identity of a given individual, that is, a such-and-not-another dynamic configuration of atoms that ensures the similarity between parents and progeny, is to a greater extent a derivative of the copies of the production prescription (set of instructions for development procedures) transferred from the former to the latter. The form of this prescription can change from generation to generation. As a result, the dynamic configuration of atoms representing a living individual can undergo modifications, which can (although it does not have to) be associated with the increased complexity of the structure and function of the individual. The fossil record of life on the earth excellently confirms this. This kind of more or less continuous sequence of changes in the identity of living individuals in subsequent generations is called *evolution.*

Natural selection

The main mechanism of biological evolution is *natural selection* (Darwin 1859; Futuyma 2005). This process results from the reproduction of living organisms and inheritance, that is, identity transfer from parent individuals to progeny. In the case of the fulfillment of these two conditions, natural selection, and thus evolution, becomes practically inevitable when fundamental properties of the physical world are taken into account. As discussed above, this is just the reproduction of living individuals and the inheritance by them of the set of properties of their parents that best distinguishes life from among various phenomena of inanimate nature. These processes became possible at the moment of the appearance of the genetic and structural-individual aspects of life discussed in the previous chapter. Almost simultaneously, other aspects appeared that supported them and had numerous elements in common. Therefore life, once it has come into existence, must evolve. Natural selection constitutes a logical consequence of the very nature of living organisms—in fact, the latter can be defined as systems that can evolve. Now let us analyze where this inseparable association between life and evolution comes from.

Living organisms (strictly speaking—individuals) *reproduce*, that is, produce descendant organisms (individuals). In the simplest case, for instance, in bacteria, a bacterium cell undergoes division into two cells. These divide further, giving in subsequent generations 4, 8, 16, 32, 64... cells. In optimal conditions, and therefore being provided with a sufficient amount of food, place, lack of pollution of the environment by harmful by-products of metabolism, *etc.*, the number of bacteria would grow as a function of time in a geometric progression, tending to infinity. After a relatively short time, the progeny of a single bacterium would form an enormous ball expanding beyond the surface of our planet with the speed of light.[1] Similar reasoning can be applied to all other organisms, also those reproducing sexually and producing many more than two progeny.

---

[1] If somebody thinks I am exaggerating here, then they must realize that in the case of the exponential increase of some parameter value, both the parameter value and the rate of its increase tend to infinity.

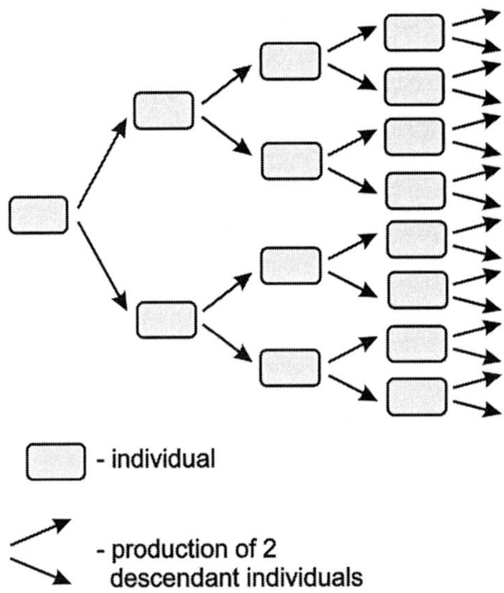

Fig. 4-1. Reproduction of organisms in the absence of any environmental restrictions. Every individual (represented by a rounded rectangle) produces two daughter individuals (in the case of bacteria; other organisms can produce much more progeny). The number of individuals increases in geometric progression, tending to infinity.

This situation in the case of an asexually-reproducing bacterium cell is illustrated in Fig. 4-1. Each bacterium individual divides into two descendant individuals that survive and divide further. In subsequent generations, we have 1, 2, 4, 8… individuals (cells). Their number grows unrestrictedly in an exponential way to infinity, and this growth continuously accelerates in time. In the case of sexually reproducing organisms, an analogous scheme would be slightly more complicated, although this has no significance for the course of our reasoning. Therefore, in subsequent schemes, we will also restrict ourselves to asexually reproducing organisms.

It seems clear, therefore, that in appropriate (ideal) conditions, the numerical force of any living organism's population would grow faster and faster (exponentially, in geometric progression) to infinity. However, in the real world, of course, such unhampered reproduction of organisms never happens—the reason is the *limited capacity of the environment.* Every organism occupies some place in space and takes up various substances from its surroundings that are indispensable for its life. Space and natural resources on the surface of the terrestrial globe are limited. Additionally, all organisms live in some concrete environment suitable for them, whose capacity is frequently still far smaller than the capacity of our whole planet. Many organisms also need hiding places, nesting places, *etc*. Every environment offers the possibility of only a limited number of living organisms of a certain kind. When this number is reached, and therefore the environmental capacity becomes exhausted, some of the produced progeny must die. Only such part of it survives and reproduces, which is necessary to replace parent individuals, for which the resources (capacity) of the environment allow.

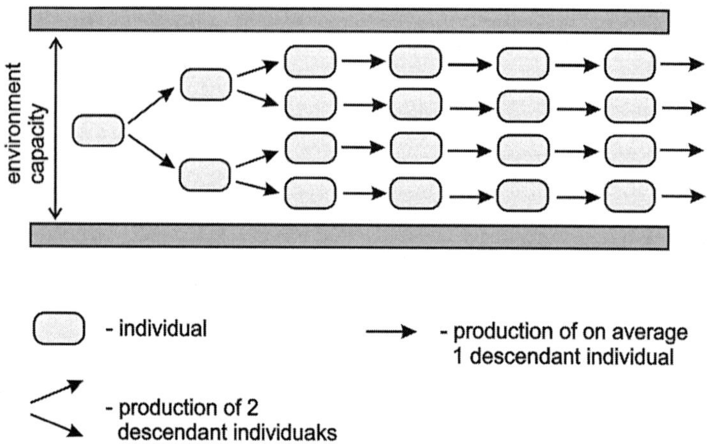

Fig. 4-2. Reproduction of organisms at limited capacity of the environment. After exhaustion of this capacity (artificially low in the scheme, amounting only four individuals), only such amount of progeny can survive that is necessary to replace the parents. In the case of bacterial cells dividing into two daughter cells, statistically only one of the latter will survive and undergo the next division. The other will die without reproducing itself. Of course, chance can cause two descendant cells of some bacterium will survive and divide, while both descendant cells of another bacterium will die. However, in principle, all bacteria cells have equal chances to leave on average one descendant cell.

Fig. 4-2 corresponds to just such a situation. The environment capacity limits here the numerical force of the population to four individuals (this is, of course, an artificially small number so that it is suitable for a clear presentation in the scheme—in a real population, it can amount to thousands, millions, and billions of individuals). If we start from one individual, then already in the third generation, the environmental capacity becomes saturated. From this moment, as the number of individuals cannot exceed four, each time (in every subsequent generation), only one of two descendant individuals can survive (on average), while the other must die without leaving progeny. Therefore, as a result, each individual produces now (statistically) one surviving and reproducing descendant individual, illustrated by single arrows in the scheme (in the case of sexual reproduction, two parents produce, on average, two children that survive and reproduce). Overall, half of the individuals produced in each generation must die without producing progeny. If, unlike in bacteria, some organisms would produce more than two descendants for one parent, then correspondingly, a larger part of this progeny would have to become extinct childless. A fundamental question for biological evolution is this, which individuals will survive and reproduce, and which will die without leaving progeny?

Undoubtedly, a role will be played here by chance. However, this will depend to a large extent on the organisms themselves. It was assumed in Fig. 4-2 that all individuals are identical, so each has an equal chance to leave the same number of descendants that will survive and reproduce, namely one. Nevertheless, particular individuals in a population of some species are never identical in a genetic respect—*genetic variability* prevails among them. Descendant individuals can differ from parent individuals, while individuals in the same generation—between them. Genetic variability is unavoidable and results from the physical nature of the carrier of genetic information, that is, DNA. For purely thermodynamic reasons, in particular, due to the second law of thermodynamics, it is not possible to faithfully store and copy information recorded in any carrier, regardless of whether it is papyrus, clay boards, medieval manuscripts, a CD, pen drive, or DNA.

As it is commonly known, copying genetic information consists of separating the double DNA helix into two single strands and synthesizing on each of these strands' matrix—a complementary strand. However, copying the genetic record cannot be accurate, at least because of the chaotic thermal movements of molecules, always appearing above the temperature of absolute zero. They disturb the process of DNA copying (replication) and can lead to the incorporation of a wrong nucleotide (not complementary to the lying opposite nucleotide in the matrix strand). The genetic record can also be damaged due to, for example, the action of ionizing radiation, various chemical compounds, or free radicals. In this way, the sequence of nucleotides in DNA, and therefore genetic information, will be changed. This change will, of course, be transmitted to all DNA strands synthesized on the matrix of this modified strand. The change of this kind is called mutation (the described type of mutation is one of many mutation kinds).[2] If a mutation occurs in the genome of a unicellular organism or the genome of the germ line cell of a multicellular organism, then it will be *inherited*, like the remaining unchanged part of genetic information. A mutation in a gene means an origin of a new *allele* (variant) of this gene. An individual endowed with a mutation will differ genetically from unmutated individuals. Many mutations, consisting of a modification of the genotype, also exert an important effect on the individual's *phenotype*. Such mutations will cause it to differ from other individuals also in regard to macroscopic appearance, structure, and function. It should be emphasized that mutations arise randomly and do not tend to change genetic information in any particular direction.

The frequency of mutations in the DNA of various species can be different (Futuyma 2013, 199-200, 270-273). It depends on many factors, to a large extent, on the efficiency of the DNA replication apparatus, whose fundamental element is DNA polymerase, and also on the system of repair of damages and correction of errors in the genetic record. The optimal frequency of mutations depends significantly on the stability of the environmental conditions. This results from the fact that the frequency of the appearance of mutations determines, at least potentially, the rate of evolutionary changes. In very stable environments, unchangeable in time and homogeneous in space, such as oceanic depths, organisms do not need to fit themselves to frequent changes and variability of external conditions; therefore, a low pace of evolution will suffice for them. Because, due to the non-directionality of mutations, a large part of them is harmful (more on this shortly), frequent mutations would cause unnecessary impairment of many descendant individuals. On the other hand, in environments quickly

---

[2] This is the so-called point mutation, that is, a mutation concerning a single nucleotide and consisting in *substitution* of one nucleotide by another, *insertion* of a nucleotide between two already present nucleotides, or *deletion* of one nucleotide from the exiting sequence of nucleotides. There also exist *chromosomal mutations*: *deletions* (removal a fragment of a chromosome), *duplications* (doubling of a chromosome fragment), *inversions* (reversal of a chromosome fragment), and *translocations* (transfer of a chromosome fragment between non-homologous chromosomes). They can lead to a change in the number of chromosomes: merging of two chromosomes and cutting out of one centromere decreases their number by one, while a duplication of a centromere and a split of a chromosome between two arisen centromeres—increases their number (also by one). Finally, *genomic mutations* are associated with multiplication of the set of chromosomes in the cell: diploids (organisms with a double set of chromosomes, to which we also belong) can mutate into tetraploids (a quadruple set of chromosomes) or even hexaploidies (a sextuple set of chromosomes).

changing in time and diversified in space (periodically drying out water reservoirs, mountain areas), the high frequency of mutations becomes a necessity so that the rate of evolution can keep pace with those changes and genetic diversity could allow for adaptation to various conditions. Nevertheless, the frequency of mutations is different not only in different species but also for different genes of the same species.

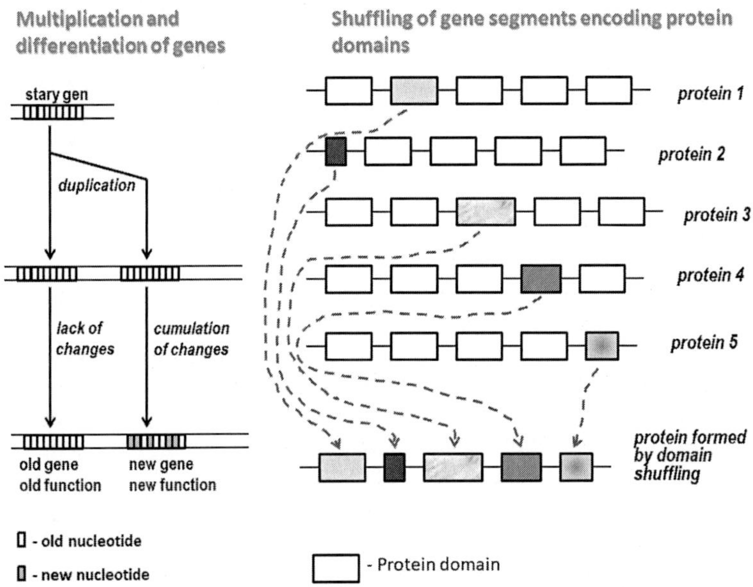

Fig. 4-3. Origination of new genes. Left: creation of new genes due to copying and differentiation of old genes. Right: creation of new genes through "shuffling" (merging into various combinations) of gene fragments encoding various protein domains (structurally and functionally distinguished fragments of proteins) coming from already existing genes/proteins.

The described mutations lead to the origin of alleles of a given gene located in the same place (*locus*, pl. *loci*) in a chromosome as the original gene. However, the forming of new genes is also possible. Duplication (doubling) of the initial gene can occur—due to this mutation, two identical copies of the gene arise. One of these copies encodes a certain protein fulfilling a useful function in the organism. On the other hand, the other copy may undergo numerous mutations and thus accumulate genetic changes within it. This is not threatened with a loss of activity by the produced protein, as it is still synthesized based on the first, unchanged copy of the gene. On the other hand, it can appear that the protein produced by the second modified copy will serve a different function than the original protein. This way, the second copy would become a new gene encoding a new protein. Due to the described process of multiple duplication and differentiation, so-called *gene families* originate, that is, their groups coming from one ancestor gene. Such a family is constituted, for example, by the genes encoding α subunit and β subunit of hemoglobin (hemoglobin, carrying oxygen in the blood, is a tetramer, that is, a protein composed of four subunits: two α and two β), and myoglobin (myoglobin, transporting oxygen within muscle cells, is a monomer, that is, contains only one subunit). Another example of a gene family is regulatory genes with the homeobox sequence (see below) belonging to the HOM/Hox family, discussed in Chapter 5. The origination of new genes through the multiplication of the existing genes in two or more copies, and their subsequent differentiation, is presented on the left side of Fig. 4-3.

New genes can also originate through "shuffling" (an arrangement in different combinations) of fragments of other genes, in particular their coding segments, that is, exons (that sometimes, though not always, encode structurally and functionally separated parts of proteins, that is, *protein domains*). For instance, the family of serine proteases (Krem and Di Cera 2001)—which includes prothrombin, tissue plasminogen activator, urokinase, factor IX, factor XII, fibronectin, and protein S—originated through an arrangement in different combinations of gene fragments encoding various protein domains: the catalytic domain of serine protease, the calcium-binding domain, the *kringle* domain, the different domains of the epidermal growth factor, and different domains of fibronectin, for example. Forming new genes by shuffling the gene fragments of the encoding protein domains is presented on the right side in Fig. 4-3. Of course, the emergence of new genes frequently leads to an increase in the complexity of an organism.

A secondary, in relation to mutations, source of genetic variability of individuals in a population, is a *recombination* of genetic information. In bacteria, the aforementioned horizontal gene transfer frequently occurs, that is, a relay of DNA fragments between cells, for instance, in the process of conjugation (conditioned by plasmid F[3]), a substitute for sexual reproduction. In the case of sexual reproduction in *Eukaryota*, the recombination manifests itself by the production of individuals being

---

[3] Plasmids are small circular DNA molecules, whose copies can be transferred between bacteria cells, responsible for instance for resistance to antibiotics.

heterozygotes in respect of certain genes (possessing slightly different copies of the same gene, that is, alleles, in the same place, *locus*, in homologous chromosomes), and by the creation of new combinations of alleles of various genes. A diploid child individual formed due to sexual reproduction receives half of its chromosomes (one haploid set of chromosomes) from the mother, and one from the father, which can meet in its genome mutations originated earlier (in the same or different genes) independently in different individuals (parents or their ancestors). This greatly increases the number of combinations in which alleles of various genes can appear.

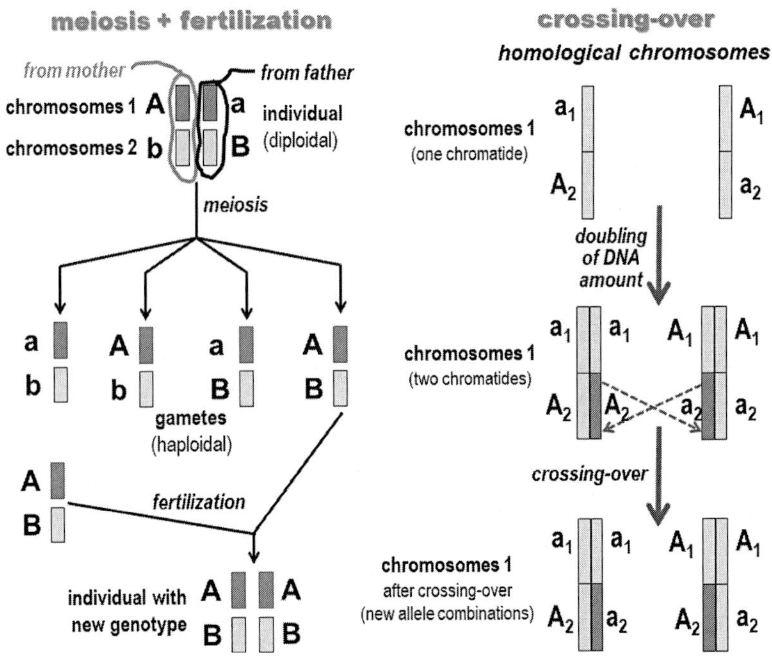

Fig. 4-4. Recombination of genetic information. Left: shuffling of entire non-homologous chromosomes due to random separation to gametes (egg cells and spermatozoa) during meiosis of chromosomes with different gene alleles coming from the mother and from the father. From one maternal cell there are formed either four spermatozoa or one egg cell, to which a random combination of chromosomes and alleles are distributed, and three microscopic so-called polar bodies that do not fulfill any further function in reproduction. While the maternal cells of gametes are diploid (possess two homologous sets of chromosomes), gametes are haploid (possess a single set of chromosomes). After fertilization, a diploid organism is re-

created, whose set of alleles can differ from those in its ancestors. Right: shuffling of homologous chromosome (in fact chromatid) fragments in the process of crossing-over (exchange of chromatid fragments between homologous chromosomes). Just after fertilization, chromosomes contain a single amount of DNA, single chromatids (during meiosis two homologous chromosomes containing two chromatids each divide ultimately into four single chromatids, which are transferred to four cells: either four spermatozoa or one egg cell and three polar bodies). The amount of DNA is doubled before the maternal cells of gametes are ready for meiosis, and therefore each chromosome contains two chromatids. During meiosis, the corresponding to each other fragments of chromatids can be exchanged between homologous chromosomes, which leads to the creation of new combinations of gene alleles (and of mutations within alleles) within a given chromosome. A, a and B, b—alleles of different genes in different chromosomes; A_1, a_1 and A_2, a_2, alleles of different genes in different places (loci) in the same chromosome.

On the other hand, in organisms with an asexual manner of reproduction, for two mutations to meet, they would have to arise within the same clone, which is much less probable. For this reason, sexually reproducing organisms are much more diversified genetically, can evolve much faster, and reach a much higher degree of complexity in their structure and abundance of forms than asexually reproducing ones. An additional source of genetic variability (apart from the "shuffling" of entire chromosomes) is an exchange of homologous fragments between corresponding (homologous) chromosomes (strictly speaking: between their chromatids) coming from the father and mother (i.e., *crossing-over*) during meiosis (the "reductive" division of the cell and nucleus during the production of gametes), leading to the shuffling of alleles of genes within chromosomes (and also of various mutations within a given gene) (Brown 2012; Futuyma 2013). Therefore, we deal with mixing genes (alleles) occurring in different, non-homologous chromosomes and within particular homologous chromosomes. Both the "shuffling" of the gene alleles (and of the mutations within them, that is, the origination of new alleles) due to crossing-over and forming new combinations of gene alleles due to random segregation of chromosomes and chromatids in meiosis. Bringing together the various combinations due to fertilization is shown in Fig. 4-4. Generally, sexual reproduction constitutes a very efficient source of genetic variability. However, as recombination only mixes and shuffles the genetic diversity created earlier by mutations, it is called a secondary generator of variability.

Therefore, the aggregate effect of mutations and recombinations causes genetic differentiation of individuals within the population. Let us come back to Fig. 4-2 and see what happens when a mutation occurs in the genome of one of the individuals. Three types of mutations are distinguished in regard to their effect on the fitness of the individual carrier of these mutations: neutral, harmful, and beneficial. *Neutral* mutations do not bring either harm or profit. This can happen when, because of the degradation of the genetic code, an exchange of a given nucleotide in a codon for another nucleotide does not change the coded amino acid (as we remember, some amino acids are coded by several codons). In other cases, even if, due to a mutation, a different amino acid is coded, this does not influence the structure and function of the entire protein, especially if this amino acid is located in the protein chain outside important regions of this protein, for example, outside the catalytic niche of an enzyme. Finally, a neutral mutation can cause some visible phenotypic effect, not impacting the functioning of the organisms—for instance, for the distribution of colorful spots on a snail's shell.

Generally speaking, neutral mutations do not affect the chances of survival of the individuals carrying them and the chances of leaving progeny by them. These individuals reproduce equally efficiently as individuals without neutral mutations, and therefore their progeny will exist beside the latter's progeny. This is illustrated in Fig. 4-5. It can be seen there that a neutral mutation changes nothing—a mutated individual produces descendant individuals equally efficiently as before the mutation and therefore has statistically the same chance for transferring the mutated gene to the next generations as other individuals—for the transfer of unmutated genes. Nevertheless, the accumulation of neutral mutations can change the population's genetic composition and increase its diversity. However, this kind of "evolution" is completely random and directionless, although frequently, the genetic material it prepares may lose its neutrality and appear advantageous in the new conditions.

Many mutations are harmful and negatively affect the chances of survival and production of progeny. In the extreme case of *lethal mutations*, the death of an organism carrying them inevitably follows before achieving the ability to reproduce (the reproductive age). As an example, a mutation in a gene encoding an enzyme indispensable for metabolism can occur, causing the loss of catalytic properties by this enzyme (for instance, a "knock-out" of cytochrome oxidase, an enzyme transferring the electrons coming from respiratory substrates to oxygen, causes automatic suffocation of the organism). Frequently, the effect of a harmful mutation is less drastic, only slightly or significantly lowering the efficiency of survival and progeny production. These are so-called *sub-lethal mutations*. In both cases, however, sooner or later (in the case of lethal mutations—instantaneously), the mutated individuals are "ousted" from the population by unmutated individuals, more efficient in survival and reproduction. This situation is presented in Fig. 4-6. One can see that in the place of a possible descendant of an individual with a harmful mutation, a descendant of another, unmutated individual "jumped in," filling the gap (a complete saturation of the environment capacity) which arose due to the death of an individual "endowed with" this mutation.

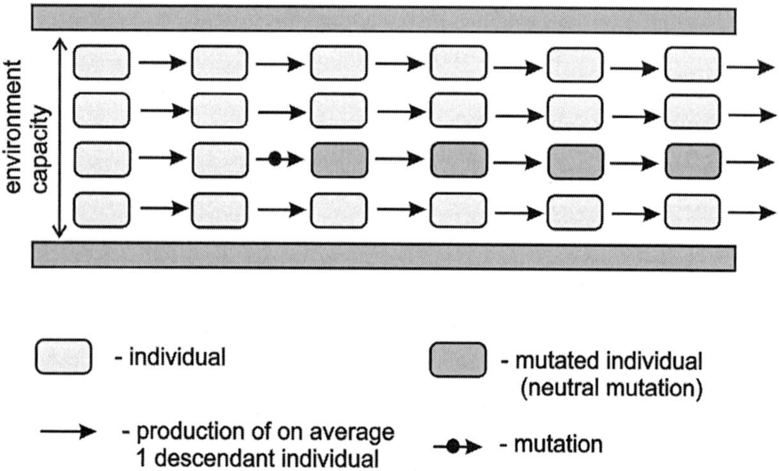

Fig. 4-5. Effect of neutral mutation. Neutral mutation (black dot) causes an appearance of a mutated individual (dark rounded rectangle), which has a statistically identical chance to survive and leave progeny as not mutated individuals (light rounded rectangles). Neutral mutation will be neither eliminated nor spread in the population.

A few mutations may appear *beneficial*; that is, they give the mutated individual an advantage over unmutated individuals regarding the efficiency of their survival and or progeny production. Suppose, due to a mutation, a new enzyme appears to decompose a hitherto non-assimilable nutritious substance present in the environment; the mutated individual will gain obvious supremacy over individuals not possessing this enzyme, especially under the conditions of a lack of other nutrients in the environment. It can more easily avoid starvation, grow faster, and produce more progeny. Its progeny will gradually oust the progeny of other individuals, and at the end, the entire population will be composed of individuals who are carriers of the beneficial mutation and possess the useful enzyme. This case is illustrated in Fig. 4-7. Here, after two generations from the moment of the appearance of an individual endowed with this mutation, the whole environment will be filled with individuals being its carriers (a beneficial mutation can additionally enlarge the environment capacity—more individuals will be able to feed themselves). A beneficial mutation causes an organism to become better adapted to its environment (the ecological niche; nevertheless, such a mutation also slightly modifies this niche). The described situation can be repeated, and a subsequent beneficial mutation will become consolidated into the population. Therefore, the genetic changes of individuals in subsequent generations will have a cumulative character. After a sufficiently long time, descendant individuals will differ significantly from their ancestors. Eventually, these differences will be so large that a new species will originate. This process is called *evolution*.

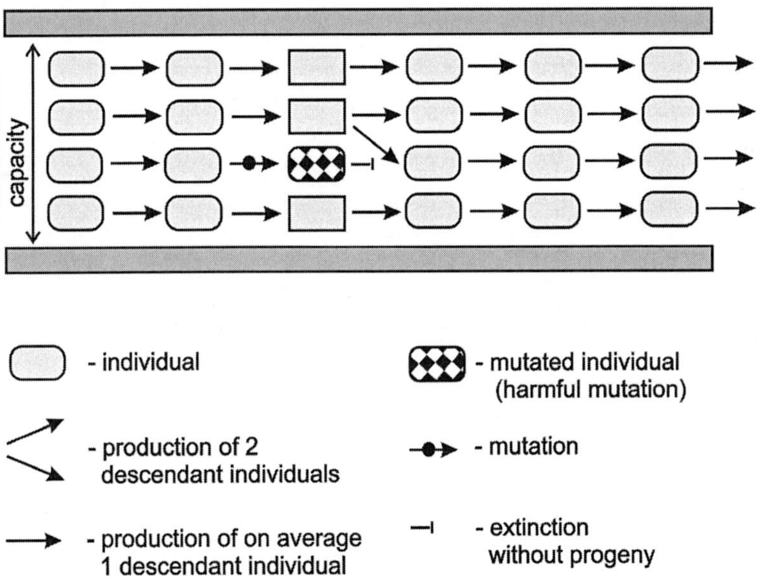

Fig. 4-6. The effect of harmful mutation. A harmful mutation (dot) diminishes the chances of survival and leaving progeny, and therefore the (potential) progeny of a mutated individual (patterned rounded rectangle), also possessing this mutation, will be earlier or later ousted by the progeny of not mutated individuals (light rounded rectangles). The harmful mutation will be eliminated from the population.

The described mechanism of evolution is *natural selection*. It expresses itself in the survival and reproduction of individuals best fitted to the environmental conditions (ecological niche) due to the accumulation of beneficial mutations. The competition for environmental resources can take place between individuals of the same species or of different, frequently very little related to each other species (e.g., the displacement of herbivorous Australian marsupials by rabbits and sheep brought from Europe) (it can be said that different identities compete with each other based on different sequences of nucleotides in DNA and amino acids in proteins, or, reaching even deeper, various, immensely complex dynamic configurations of atoms corresponding to living individuals). Mutations that are beneficial in one environment can be harmful in another environment—therefore, this environment constitutes the reference frame for the fitness of organisms. Non-avian dinosaurs were perfectly adapted to live in the tropics (and not only) of the Mesozoic era, but they could not survive, contrary to primitive mammals, the asteroid impact (and related to it, the climate cooling and huge dust clouds cutting out the sunlight) at the turn of the Cretaceous and Tertiary eras.

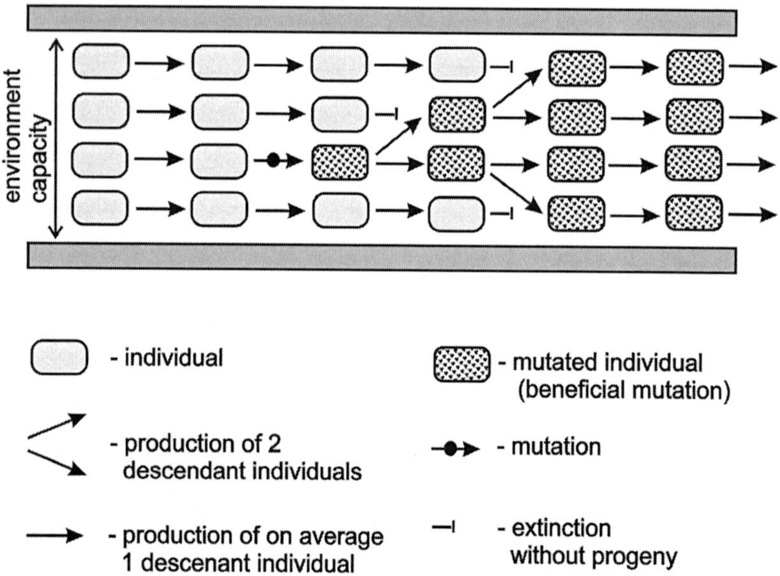

Fig. 4-7. Effect of beneficial mutation. A beneficial mutation (dot) will increase the chances of survival and leaving progeny by individuals endowed with this mutation (patterned rounded rectangles) in relation to individuals without the mutation (light rounded squares). The progeny of mutated individuals will gradually replace not mutated individuals, and eventually all individuals in the population will be carriers of the beneficial mutation.

On the other hand, some mutations, elevating the overall efficiency of an organism, can increase its chances of survival in many different environments. For example, mammals, equipped with a relatively large brain, a four-chamber heart, viviparity, and homeothermy, conquered many environments on the earth (land, sea, air, the tropics, polar zones, deserts). It should be stressed that evolutionary success can be ensured not only by an appearance of a beneficial mutation but also by meeting in one individual due to recombination of two (or more) different mutations, which alone are not particularly beneficial, but whose cooperation brings about synergistic positive effect for their carriers.

To sum up, the general scheme of natural selection is based on four principal premises:

- in the process of *reproduction*, more descendant individuals are produced than can survive in the *limited capacity of the environment*;
- descendant individuals inherit the properties (identity) of parent individuals;
- among descendant individuals, accidental variability prevails, caused by directionless mutations (inevitable for thermodynamic reasons) and recombinations;
- only individuals *best fitted* to the present environmental conditions survive and reproduce.

Natural selection constitutes an inevitable result of the existence of living organisms (reproduction and inheritance) and the realities of the physical world in which they live (the limited capacity of the environment, inevitability of errors in copying any information, properties of the environment, to which living organisms are to adapt). It is even possible to formulate, half-humorously, a "logical proof for evolution," analogically to mathematical proofs, where the final theorems are derived in the logical, necessary way from the assumed axioms. Such a "proof" is presented in Fig. 4-8. However, otherwise than in logic and mathematics, the "axioms" represented here are not arbitrary assumptions but most elementary and unshakable facts. The physical reality imposes the limited capacity of the environment (which is trivially true) and the physical nature of the carrier of the genetic information, DNA—everything in the universe is (also) physical.

On the other hand, at the core of living beings (individuals) is reproduction (due to which, at least potentially, more progeny are produced than there are parent individuals), associated with the inheritance of identity (a set of properties based on the genetic record). It results from the limited environmental capacity and reproduction that some progeny must die out before reproduction. The (also) physical nature of genetic information and the fact of its inheriting imply an inevitable necessity of the appearance of damage to the genetic record and errors in its copying (the second law of thermodynamics), which leads to mutations. In association with their recombinations, mutations result in the differentiation of (the identity of) individuals in populations. This, in turn, in association with the necessity of the extinction of some progeny, leads inevitably to the conclusion that only the "most efficient" individuals survive that are best fitted to current environmental conditions and reproduce. Beneficial mutations, increasing this fitness, will be consolidated in a population, and the identity (the genetic record and the structure and functions resulting from it) of individuals being their "carriers" will be changed. Therefore, their *evolution* will take place *quod erat demonstrandum*. There is no sense in deliberating whether natural selection and evolution proceed. They *must* proceed. They are an elementary and necessary consequence of the existence of living organisms (individuals) and fundamental properties of the physical environment in which these organisms live.

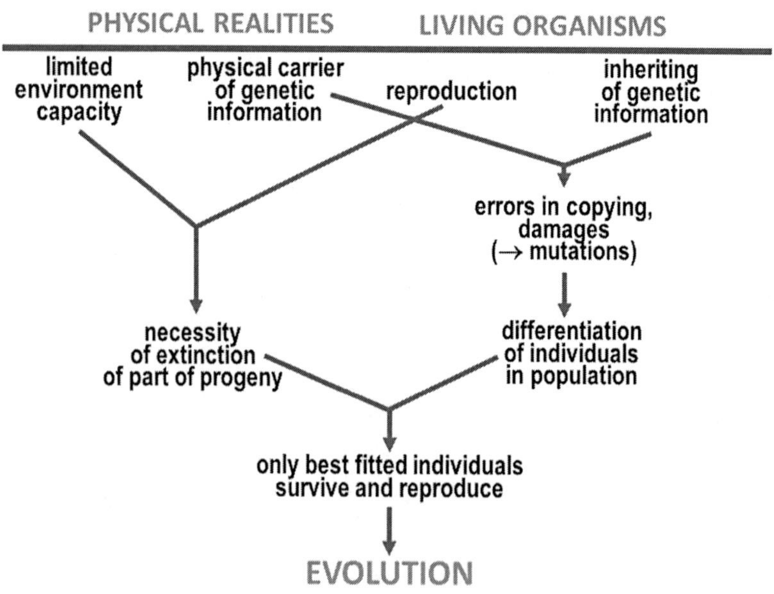

Fig. 4-8. Logical proof of evolution.

However, although natural selection constitutes the main mechanism of evolution, it can cooperate or even be replaced by supplemental mechanisms, such as sexual selection (a certain specific form of natural selection), artificial selection, or genetic drift, as discussed below. It is also worth emphasizing that natural selection only constitutes a selector or sieve for the variability offered, but not the generator of the variability itself. The latter, especially in the case of complex tissue organisms, such as plants and animals, is constituted of mutations in regulatory genes supervising the development of the structure (and function based on it) (morphogenesis) of an organism during fetal development (embryogenesis) and individual development (ontogenesis). They can lead to significant, macroscopic changes in the general plan of the structure and set of functions of organisms and also to increased complexity. They will be discussed more closely in Chapter 5.

It should be stressed that natural selection does not realize any advanced sketched plan. It appears in systems possessing certain properties and, from an external point of view, it is, as a whole, a process equally aimless, as crystals growing in an oversaturated salt solution, stones falling, or planets revolving around the sun caused by gravity (which does not contradict the fact that within living organisms there are purposeful, from an internal point of view, properties of structure and function and directed regulatory mechanisms serving survival and reproduction). It is undoubtedly astonishing that such a simple mechanism as natural selection can lead to great complexity and richness in living forms. This alone confirms the great importance and heuristic fertility of Darwin's theory. However, already simple physical processes can produce objects complex in their own way—it is enough to recall the icy flowers painted by frost on windows. The phenomenon of evolution and the appearance of icy flowers have no purpose or sense—these processes simply occur. Both constitute examples of self-organization of matter in the physical world. The main difference is that natural selection is a cumulative process: the information about the structure and function acquired in one generation is transferred to the next generation, which undergoes further verification, testing, modification, and (at least potentially) quantitative growth. For this reason, living organisms, resulting from billions of years of evolution, are much more complex than icy flowers or hypercycles of proteins and nucleic acids contained in coacervate-like bubbles.

The selfish gene

One could ask, *what,* in fact, evolves due to the action of natural selection: genes, whole genomes, organisms, their properties, populations, or maybe entire species? Different evolutionists give different answers to this question. Dawkins (1976), in his widely known book *The Selfish Gene,* promotes the thesis that particular genes take care exclusively of their own evolutionary interests, tending to multiply themselves in a possibly great number of copies. In this conceptualization, living organisms and humans become only transient carriers (containers) for the genes steering their development. Selfish genes realize their evolutionary success through possibly fruitful cooperation by creating an efficient phenotype of the whole organism. Their propagation occurs through the reproduction of organisms containing these genes. For an organism to survive and leave progeny in its environment, its genes must find themselves in the possibly best company of other genes. However, in the genomes of most organisms, DNA sequences exist that are even more egoistic than ordinary genes—a so-called parasitic DNA. They do not cooperate with other genes in the reproduction of the whole organism but tend to produce a possibly great number of copies of themselves within the genome of this organism. The discussed transposons, (in particular: retrotransposons) belong to them. This parasitic DNA implicated Dawkins's conception of the selfish gene.

This gene-centric point of view has been criticized by several authors (for example Oyama 1985; Lloyd 1988; Hoffmeyer 1997; Griesmer 2000). Some of the following arguments are similar to the arguments presented by these authors.

However, the claim that just "naked" genes alone constitute the unit of selection and the subject of evolution is undoubtedly a significant simplification, if not just a half-truth. As *self-copying systems* evolve, they are composed, first of all, of nucleic acids and proteins (see Chapter 3). It should be remembered that the sequence of nucleotides in genes has no sense or meaning without the genetic code, that is, without assigning an appropriate sequence of amino acids in the protein encoded by a given gene to the sequence of nucleotide triplets in the DNA of this gene. Therefore, genes need the genetic apparatus and, in the simplest case, the whole metabolism and structure of the cell in order to realize their propagation. They have to cooperate for the correct functioning of the whole organism. If, for instance, the gene encoding cytochrome oxidase ignores the rest of the genome and tries to be selfish, the whole individual would suffocate instead of executing its function, and the gene itself would die.

Additionally, the genetic information is not only genes encoding proteins (with the mediation of mRNA) and other RNA types (tRNA, rRNA, miRNA) participating in gene expression and its regulation, and the rest of the genetic apparatus (DNA and RNA polymerases, ribosomes), but also DNA fragments (enhancers and silencers) binding various transcription factors (proteins activating or inhibiting transcription, activators or repressors, respectively), and the transcription factors themselves, crucial for cell differentiation and functioning and morphogenesis. It should also be remembered that natural selection does not act directly on genes but on "ready" phenotypes. For all these reasons, this, which constitutes the unit of selection and subject of evolution, and which "really" changes in its course, is identity, that is, an individual's properties of a given self-copying system, emerging through the co-operation of various elements, among which, of course, genetic information is very important. Nevertheless, it is frequently forgotten that genetic information is not only recorded in the sequence of nucleotides in DNA (in genes and regulatory segments) but also information about how to read this information by the transcriptional and translational apparatus. Similarly, the information in this book is not only a certain sequence of letters but also requires knowledge of the English language and some of biology's fundaments from the reader. Dawkins' genes are slightly anthropomorphic—certain intentions are attributed to them (of course, evolutionary and not psychical intentions).

Nevertheless, such instrumental, operational "intentions" (purposefully directing on a certain goal—survival and self-replication) can be attributed only to whole self-proliferating systems equipped with numerous purposeful properties and mechanisms. It can be equally well said that genes encode appropriate proteins and, through them, form the whole organism and that proteins use DNA (genes) for recording their own identity—after all, just proteins, and not nucleic acids, are the active element of the entire system (machinery of the cell), while DNA constitutes only a passive recipe meaningless by itself. DNA needs not only proteins for its copying and propagation. An adult leopard will never develop spontaneously from naked leopard DNA. It needs the whole replication apparatus, reading genetic information and regulating this reading. It needs the whole fertilized egg cell and its metabolism (our "little robot"). Finally, it needs a female leopard, which bears the fetus in the uterus and feeds, protects, and teaches the newborn cub to live in the leopard's ecological niche (e.g., hunting strategies).

It is worth emphasizing that the relationship between the sequence of nucleotides in genes and the macroscopic properties of living individuals (their phenotypes) is terrifically complex. Every aspect of structure and function is conditioned by many (frequently thousands) genes and non-coding (regulatory) DNA sequences, and each gene affects numerous properties as a rule. Nucleic acids, proteins and other organic compounds create such an enormously intricate network of relations that it would be extremely difficult (in practice: impossible) to fully interconnect, understand, and analyze all the relationships between genes and phenotypes (features of structure and function) of the organisms carrying these genes.

Therefore, while the conception of the selfish gene seems interesting as an original formulation of the problem, it undoubtedly exposes the problem in a decidedly one-sided way. It has the advantage of simplicity, giving an untutored reader a deceptive impression of understanding. However, this does not constitute a gauge of its "correctness" and heuristic fruitfulness. Apart from this, the evolution of genetic information consists of modifying existing genes and the origination of new genes. Dawkins' conception says little about this topic, which is crucial for the macroscopic evolution of structure and function, construction plan change, and complexity increase. Finally, most DNA in higher organisms is the so-called "junk" DNA (containing, amongst others, pseudogenes, inactive transposons, and short and long repetitive sequences) with no known function, just transferred from generation to generation. It does not "care" about itself and does not act to increase the probability of its propagation or transfer to descendant genomes. Again, the "selfish gene" paradigm has nothing to say about this phenomenon.

Whether somebody likes this or not, in science, especially in its popular reception, there are frequently consolidated and broadly discussed conceptions that are very simplified, one-sided, or simply false but which instead can be easily and schematically formulated in a few words or sentences. They draw attention with attractive, frequently provocative vocabulary while at the same time do not require intellectual effort and knowledge. Examples of such conceptions are Dawkins' "selfish gene," "multi-universe," "universe as a hologram" (see Baggot 2010 for excellent discussion), or the already mentioned "Gaia hypothesis." On the other hand, subtler, more complex ideas much closer to the truth, which nevertheless require appropriate knowledge and intensity of thinking processes, for obvious reasons, have much greater problems with breaking through not only the social consciousness but also wider scientific circles. Unfortunately, the ability of a more general, integrative look is also not so frequent among scientists. It can be said that a scientist-expert differs from a philosopher in this way: the latter knows nothing about everything (because of the lack of any reliable, rational methodology), while the former knows everything about nothing (because of advancing specialization in narrow domains of knowledge).

Nevertheless, abandoning this digression, my opinion concerning this, which "really" constitutes the subject of biological evolution and unit of natural selection, is presented in Chapter 6 that is devoted to the cybernetic definition of life, particularly of the living individual.

Kin altruism/kin selection

Apart from natural selection, a few related mechanisms of evolution called "selection" can be distinguished, being, in principle, varieties of natural selection. The first is *kin selection* and related to it "*kin altruism*" (Williams 1997, 41-50; Futuyma 2013, 291-294), which in reality is not altruism at all (in evolution and biology in general, there is no place for "true" altruism). It consists of the fact that, in some cases living individuals, instead of possessing progeny by themselves, help their closest relatives increase the amount of progeny they produce. This "pays" in the genetic and evolutionary sense if such help causes the gain from "altruism" *g* of individuals with the degree of kinship to the "donor" of "altruism" *k* multiplied by this degree exceeds the costs of "altruism" *c*:

$$\mathbf{g \cdot k > c.}$$

For instance, the degree of kinship in sexually reproducing organisms where both sexes are diploid amounts to 0.5 between parents and progeny (each half of the genes of a descendant comes from one of two parents) and between siblings (they have half of the common genes coming from parents), while between grandparents and grandchildren, it equals 0.25. As an example of kin altruism, some birds in the first year of their independent life help their parents to feed and take care of their subsequent progeny (being younger siblings of our "altruistic" birds) instead of having progeny of their own. However, at first, the degree of kinship with their own progeny is the same (0.5) as with their siblings, and therefore for the set of genes (identity) of a given individual, it, in fact, does not matter whether they have progeny by themselves or they increase the amount of progeny brought up by their parents by the same number. Second, by helping their parents, young birds acquire experience in the care of progeny; because of this, they will be more efficient parents the following year.

Another example of kin selection (altruism) is associated with *group selection* (Futuyma 2013, 290-295) when a group of closely related individuals, for instance, monkeys, compete with other groups for available environmental resources.

Finally, a significant role is attributed to kin selection in the origination and functioning of animal societies: eusocial colonies of insects (wasps, bees,[4] ants, and termites) and mammals (naked mole rats, rodents devoid of fur living in branched systems of burrows in African semi-deserts). In the

---

[4] Many wasps and bees are solitary and do not form societies.

societies of these animals, a division into casts appears—reproductive casts (fertile females and males) (in hymenopterans: wasps, bees, and ants, the males die after copulation, and only the fertilized female-queen remains in the nest, while in termites and naked mole rats, the queen and king co-exist in the colony) and sterile casts which may include workers and possibly soldiers (both casts can appear in different varieties, differing, e.g., in size) (in hymenopterans, these are exclusively females, while in termites and naked mole rats there are representatives of both sexes) (Wilson 1971; Wilson 1975). The role of sexual casts is the production of a possibly great amount of progeny, while sterile casts perform all other tasks to which there belong, care of sexual casts and their progeny, feeding them, the acquisition of food, construction of the nest, regulation of the conditions prevailing in it (e.g., temperature, humidity), and defense against various threats, *etc*. The problem of what "really" constitutes the living individual: insets representing sexual casts, insects representing sterile casts, or the whole colony, is discussed in Chapter 6.

The origin and maintenance of such societies are usually attributed to kin selection. It is strongly expressed in hymenopterans: wasps, bees, and ants, where the haplodiploid system of sex determination appears: males are haploid (developed from unfertilized eggs), while females—are diploid (developed from fertilized eggs). Therefore, while parents and progeny are still related to each other to the degree of 0.5, the kinship degree of sisters equals 0.75: they possess half of the common genes coming from the mother (0.25) and all common genes coming from the father (0.5) (0.5 + 0.25 = 0.75). Therefore, the "altruistic" cooperation should pay off exceptionally strongly for them. In termites and naked mole rats, both sexes are diploid, but the degree of kinship between parents and progeny and siblings amounts to "as much as" 0.5. Is it, therefore, not profitable for them to cooperate and altruistically sacrifice themselves for the genetic-evolutionary good of the community, that is, the whole colony?

Unfortunately, the matter is not so simple. In many wasps, the lack of trials of reproduction of subordinate females participating in the initial phases of developing a colony (e.g., construction of the nest) has a behavioral background: it results from their suppression of the production of progeny by the dominating female founder of the colony (that can, although she does not have to be, their sister). When, for some reason, the dominant female disappears, her reproductive role will be taken over by the second female in the dominance hierarchy (Wilson 1971). Therefore, it is not so that subordinate females voluntarily "agree to" the role of sterile workers. The workers being the progeny of the dominant female, are already sterile physiologically—they have, for example, undeveloped ovaries. In socially advanced wasps, bees, ants, and termites, the sterile casts are conditioned by the rest of the colony (by appropriate feeding of the larvae or pheromones release) or genetically (Wilson 1971). Therefore, they have no "choice" concerning contingent reproduction: they are anatomically and physiologically sterile. Therefore, there is no place here for any "altruism." Even if the sisters in hymenopteran colonies are related to each other to the degree of 0.75, one sister cannot help another in reproduction.

Leaving aside the nutritional/pheromone/genetically conditioned sterility, the degree of kinship of 0.5 with the queen and 0.75 with other workers/soldiers in a colony is frequently elusive. First, in many cases, the queen is fertilized by many males. Second, in many ant species, many unrelated queens co-exist in one nest (Wilson 1971). As a result, the average degree of kinship between particular workers and between sterile workers and particular queens falls dramatically.

For all these reasons, kin selection does not seem to play any important role in maintaining current societies of insects and naked mole rats. It has been replaced by physiological-genetic-behavioral mechanisms "dooming" workers and soldiers to sterility. Similarly, physiological mechanisms stop or limit divisions of somatic cells in our bodies (cancers result from some cells "breaking free" from the control exerted by these mechanisms). Therefore, it is not "altruism" that hinders workers from reproduction. It is astonishing how many biologists still uncritically accept the thesis concerning kin selection in present societies of insects and naked mole rats. Nevertheless, one thing should be clearly stressed: while I cannot see an important role of kin selection/altruism in evolutionarily *advanced* insect societies, it certainly was very important during the *origination* of these societies (after all, most social insects are hymenopterans with a degree of kinship between sisters of 0.75). However afterward, during evolution, it was replaced by behavioral-physiological-genetic mechanisms determining the sterility of workers and soldiers.

Reciprocal altruism

Another type of apparent altruism, which is not altruism at all, is *reciprocal altruism* (Dunbar 1998; Futuyma 2013, 433). It appears in social organisms having a sufficiently developed neural system so that particular individuals can recognize each other and remember the result of previous meetings. The groups of individuals must be small enough for all their members to know each other. A purely egoistic calculation occurs: "If you help me, I will help you." At the same time, the costs of the cooperation must be lower than the profits drawn from it. A known example of reciprocal altruism is grooming in monkeys and apes, that is, the mutual removal of parasites and dirt from the fur (Dunbar 1998). The alternating reciprocal grooming of two individuals is much more efficient than doing this alone, particularly if it concerns the parts of the body that cannot be seen by a given individual, for instance, its back or head. At the same time, the following principle obliges: "If you previously groomed me, this time I will groom you." A consequent refusal of grooming by the other individual terminates the "co-operation"—hence the name "reciprocal." As both parties undoubtedly benefit here, it is difficult to call such behavior "true" altruism—it simply pays off (this is somewhat similar to symbiosis, where both partners profit from the cooperation). If you behave "egoistically" and will not groom another individual, it will remember this, and at the next meeting, it will refuse to groom you. Additionally, grooming plays an important social role.

Another example of reciprocal altruism is the food exchange between vampire bats (Futuyma 2013, 433). They live in small colonies and feed on the blood of big animals (also people). Finding a, preferably sleeping, blood donor is uncertain; it depends greatly on chance. An individual lucky on a given day could drink to satiety, even to excess. Another less fortunate individual is left with nothing. After returning to the colony, vampires can detect which individual is sated or hungry. The latter, through appropriate behavior, "asks" the former for a part of its meal and, as a rule, gets it mouth-to-mouth. Of course, during the following days, the situation might be reversed, and the first individual will return the favor to the second one insofar as it previously received help from it. Therefore, similar to the case of grooming in monkeys, this is no altruism but pure cold calculation.

The good of the species and the Gaia hypothesis

If we are already discussing altruism, I would like to mention another kind, which is neither altruism, nor does it actually exist. This is the cooperation of particular individuals and even scarification of their own good for "the good of the species" (Wynne-Edwards 1962). Why insist that such behavior cannot take place? Would it not be beautiful if all individuals of a given species would care for its good, survival, and success in competition with other species? Is it not so that people in various communities behave this way and cooperate for the common good?

First, animals (and plants, all the more) should not be compared to people. Humans are above the purely biological level, namely on the psychical, social, and cultural levels (and, nevertheless, egoism is common, although it is moderated by social control). Purely biological individuals only "care" about what is good for themselves from the evolutionary point of view, for spreading (proliferating) their genes and their identity (Williams 1997, 50-58; Futuyma 2013, 294). They act for purely egoistic purposes and simply do not care about the "good of the species." In principle, it is impossible to find properties or behaviors that would be good for the species but harmful for particular individuals. Let us make the following assumptions:

Assumption 1: the whole population has the allele *a* responsible for acting for the benefit of the species at the cost of their own interests;

Assumption 2: there appears (due to a mutation) a "selfish" allele *A* responsible for acting only and exclusively for its own benefit (for the benefit of the individual "carrying" it).

Soon, the unavoidable consequence will take place:

The consequence: allele *A* will very quickly oust allele *a* from the population, spreading as its "carriers" will win the competition with "good-of-species altruists" only because they will care exclusively about their own interests, thus increasing their chances of survival and propagation of identity (together with the egoism conditioned by allele *A*) in a possibly great number of copies, and will transfer this allele to progeny. Therefore, they will behave rationally from the biological-evolutionary point of view, at the cost of "idealists" with allele *a*.

All purely biological individuals are selfish, which does not exclude apparent altruism insofar as it brings profit.

Let us consider a subsequent example (Futuyma 2013, 294). Many birds that live in poor environments, for example, the tundra, where there is little food in some years, in the form of insects or seeds, will limit the number of eggs they lay, and therefore the amount of potential progeny. According to the conceptions of promoting the good of the species, the birds self-limit themselves not to exhaust the environmental resources accessible to the whole species. However, there is a much simpler, prosaic, and rational (and, first of all, true) explanation of such behavior, taking into account only the selfish genetic-evolutionary interest of the individual. Namely, birds estimate that in a given year, the accessibility of food is limited; because of this, they can feed only a limited amount of progeny. It is better to lay five eggs and bring up five chicks than to lay seven eggs and have seven chicks who die of starvation. This has nothing to do with the "good of the species."

Let us take another example (Williams 1997, 52-55; Futuyma 2013, 327-328). In most vertebrates, the ratio of males to females equals roughly 1:1. From the point of view of the good of the species, this seems to be pure wastefulness, as one male can fertilize hundreds of females. Females can produce only a limited amount of progeny, and just their number constitutes the "bottleneck," limiting the reproducibility of the population and species.

Therefore, a population caring about the good of the species should produce a hundred females for one male (at least in a species where males do not participate in feeding and taking care of progeny). However, from the point of view of egoistic individuals, the matter looks completely different. In a population where one male would fall per one hundred females, producing only male progeny by some individuals would give a huge evolutionary advantage to these individuals and enable effective propagation of their identity, as the males being their progeny, could procreate very intensively. Therefore, producing progeny with the male sex would be very favorable. This would last until the equalization of the proportion of sexes. If the pendulum swung in the opposite direction and males began to predominate in the population, an average male would have little chance to procreate and, therefore, to propagate his own identity in a large number of copies it would pay to produce females. The numerical force of both sexes would again undergo equalization. Therefore, it statistically pays to produce equal numbers of males and females. Alternatively, strictly speaking, it pays to dedicate the same amount of energy for the production of both sexes. Because if males are born much bigger and consume, for example, a much greater amount of milk, as it takes place in elephant seals, it would be optimal to produce a smaller number of them. Nevertheless, what matters here is only an individual's genetic-evolutionary interest and not the species' good.

A male lion (or a group of males), which subdues a pride of female lions and ousts its previous "owner" (owners), kills the existing cubs that still require feeding and care by their mothers. From the point of view of the good of the species, this is a pure wastage (I purposely leave aside here the ethical point of view, as the biological world is, in its essence, amoral) as the members of the species have already invested much energy and time in giving birth to, raising, and adapting this progeny to life. However, from the point of view of the new ruler of the pride, this is pure gain: a female deprived of her previous progeny quickly ovulates again and can give birth to his own progeny, spreading his identity (set of genes).

Maybe it is a good place to mention the amorality of living nature. Frequently, mawkish nature films are broadcast on television, in which a mother with commitment, and even love, looks after her progeny. Similarly, young individuals are often shown as aflame with vivid feelings for their birth mother and attendant. However, this is evident anthropomorphism. The care of progeny in animals constitutes only and exclusively an expression of the tendency toward the propagation of one's own identity in the world of living organisms, having nothing to do with "lofty" motivations, for instance, with parental love, present in humans.

Interpretations of this kind result from projecting the human psyche and wishful thinking into the world of animate nature. One can slobber over the lion female taking care of her young, but it is not easy to do the same in relation to the case mentioned previously, where a new lion male, after subduing the pride, kills the whole progeny of his predecessor. In some viviparous sharks, the progeny eats each other within their mother's body—by rule, only the two strongest individuals survive and are born. In a certain mite species (*Acarophenax tribolii*), young individuals, a male and several females, hatch from eggs in their mother's body. They feed by eating her from the inside. After some time, the male, not leaving the dead carcass of his mother, copulates with each of his immature sisters and soon dies. After reaching adulthood, the young females emerge from their mother's body only in order to be, after a few months, eaten by their progeny. From the point of view of human morality, we have cannibalism, incest, pedophilia, depravity, and even the rape of juveniles, masochism, and the whole thing seems to allude to necrophilia and pimping. It would be difficult to find a brighter example illustrating the thesis that the tendency to reproduce in nature is amoral and should not be subjected to ethical judgment, as this is pure anthropomorphism. Incidentally, I wonder how the described case corresponds, in their own opinion, to the beliefs of the protagonists of the "intelligent project." Indeed, it would have to be a perverse and obscene intellect… I will not continue the "ethical values" thread in the animate world; I think readers will find many examples analogous to those described above (see Williams 1997, 152-167, for a more general discussion).

However, more naïve and contrary to the factual state than the conception of the "good of the species" is the "Gaia hypothesis,"[5] according to which all organisms on the earth collaborate for the good of the entire terrestrial biosphere (Lovelock 1979). Nothing is more misleading—it is enough to realize the acute competition between individuals of the same or different species, the eating of organisms by one another, or epidemics frequently killing entire populations. Let us consider, however, a less drastic example—the height of trees. 20, 30, or 50-meter trees are an unnecessary waste of substance, energy, effort, and time to produce trunks. Of course, the point is not to be shaded by other trees or to shade competitors. However, it would suffice to agree to "make an arrangement" that all trees produce trunks of the height of one meter. Of course, there is no way to lead to such an arrangement, and even if it would happen by some miracle, every individual that broke the contract and produced a two-meter trunk would

[5] Gaia is the ancient Greek goddess being a personification of the earth, primeval mother of all life.

gain an instantaneous advantage over the "honest" individuals in the "evolutionary arms race" and its egoistic genes (identity) would propagate in the population. This evolutionary trend consisting of the increase in the trunk height would continue until the cost of producing a tall trunk would counterbalance the benefits drawn from it, and the high trunks would be unstable.

It seems very surprising that some people are still inclined to believe that living organisms' behavior is oriented towards the "good of the species" and the "good of the biosphere." This results from wishful thinking and a complete lack of understanding of the nature of biology, biological evolution, and living individuals.

Artificial selection

Another "variation on the theme" of natural selection is *artificial selection* carried out by humanity (Futuyma 2013, 356-357). As a result, there originated high-yielding varieties of cereals, different varieties of cows for meat and milk, pigs and dog species for utility (e.g., shepherd dogs) and decoration (Pekingese), cat and pigeon varieties, *etc*. There are two main differences between artificial and natural selection. First, the selection criteria are not determined by the natural environment but by humanity: this can be useful as a source of food or for aesthetics. This occurs in such a way that only the individuals that have most expressed the desired features are allowed to reproduce. Second, artificial selection proceeds much faster than natural selection. In natural conditions, organisms had millions of years to adapt to environmental conditions and therefore undergo only tiny, "cosmetic" changes, for instance, when adapting to slow climate changes. On the other hand, humanity's very strong selection pressure in an unprecedented direction leads to an unusually efficient selection of phenotypes and genotypes (combination of alleles), conditioning them.

Sexual selection

Another variety of natural selection is *sexual selection* (Futuyma 2013, 411-425). In most cases, it concerns males and is distinguished by some arbitrariness and accidentality of attractive properties for females of a given species and serves competition with other males for females. On the other hand, properties preferred by natural selection are beneficial "objectively," in a necessary way, as useful in survival and reproduction in the natural environment.

As a rule, these males are characterized by bright coloration, complicated mating rituals, long-lasting singing, and long tails to gain the consideration of females, or by large antlers or canines to fight other males directly for the right to reproduce. All these appearance/behavioral features require large energetic investments, disturb the functioning, expose them to predator attacks, and generally diminish survivability. However, it frequently pays for the identity of a given male individual to decrease the chances of survival if, at the same time, the chances of having much progeny are elevated even more. The fitness of a living individual is proportional to the probability of survival to the reproduction period multiplied by the expected number of offspring if it survives to this period (in reality, this is more complicated).

On the other hand, females are, as a rule, characterized by a modest appearance (the differences in coloration and structure between males and females are called *sexual dimorphism*). Where does this extreme dissimilarity of reproductive strategies between sexes come from—the behavior, protective coloration, and accepting (mating with) males that either present themselves most grandly (possess the most attractive and brightest coloration, sing most beautifully for the longest time, have the most impressive tails, perform most complicated mating dances, build the most complex and colorful gazebos of flowers, *etc*.), or win a direct fight for females (deer, elephant seals)?

First, a female bears incomparably greater expenses during reproduction than a male—comprising the production of large eggs equipped with nourishment stores, incubation, pregnancy, feeding, and care of progeny. In many species, the male participates in the incubation of eggs, feeding and care of the young—in which case, the differences between a male and a female are much less. On the other hand, a male's investments in reproduction are (in most cases) much smaller—this is frequently only the production of semen and copulation. Therefore, generally, different strategies "pay" for a male and a female. The strategy of a male is to fertilize the greatest number of females: a male can produce millions of spermatozoa, so in principle, there is no obstacle to having a great amount of progeny with hundreds of females. On the other hand, for purely physiological reasons (the cost of reproduction mentioned previously), a female can produce only a limited number of descendant individuals. Therefore, her strategy consists in being fertilized by the "best" male, with the most efficient combination of genes, that is, with a best-fitted identity—genes of the female accompanied by the "good" genes of the male will ensure her having strong, efficient and vital progeny (being, after all, carriers of her genes also, that is, her identity).

For this reason, it is just the males that exhibit bright coloration, impressive tails, parade around with long-lasting singing, or produce substantial antlers or canines and physically fight for females; while females accept males with the most beautiful coloration, longest tail, sing for the longest time or those that have won the direct fight.

In fact, it is not important which property of males consolidates in a given species as attractive for females—it is arbitrary and comes from chance—as long as it is possibly strongly expressed and requires a large energetic investment. In ancestors of a given species, a barely expressed male feature started, presumably due to an accidental mutation, to be preferred by females. Of course, in the following generations, males "struggled" to possess this feature expressed as intensely as possible, and females more willingly chose the males that were (because of appropriate mutations/recombination) successful in this. For obvious reasons, the increase in the intensity of such a feature could not advance endlessly: the functionality of a male in its environment and its chances of survival constituted an obstacle. A peacock male with a ten-meter tail could not move at all, while a male nightingale devoting all his time to intense singing would die of starvation and exhaustion.

However, why will a male with the brightest coloration, longest tail, most persistent and intense singing, with the greatest antlers and strongest in a direct fight with other males, in fact, be the best, ensure the greatest vitality of the common progeny? After all, none of these properties is useful in "everyday life"; just the opposite, they decidedly disturb it (e.g., make locomotion more difficult, expose to predator attacks, exhaust energetically). This is true. However, only a strong, healthy (free of parasites and wounds), well-fed, and efficient male can manage such a task and survive. He sends the message to the female: "Look, I have such good genes that I can afford to have very energetic and costly extravagances, and nothing happens to me." Apart from this, if a given male can produce attractive features for females, with a high probability, his male descendants will also be able to do so, which will also ensure the further transfer of the genes of the mother of these descendants.

Due to such a state of affairs, a specific equalization of the energetic investment of a female and a male ensues: the female invests in reproduction while the male in getting access to reproduction. In principle, it does not matter which "senseless" (not increasing survivability, but frequently decreasing it) way the male spends his energy, as long as he sufficiently spends it. The functioning of the sexual selection mechanism is supported by the situation prevailing in phalaropes. In these birds, while females lay the eggs (it is impossible to be otherwise), they leave the nest, and the male incubates the eggs and takes care of the progeny. It is he that bears *per saldo* greater energetic costs of reproduction. As a result, phalaropes females are more brightly colored and fight ritual battles for males, while protectively colored males wait for the winners of these duals to spawn progeny with them.

The coming into being in a given species of some accidental property of males that is attractive for females constitutes a one-way process, which cannot be reverted in principle. Unless the property becomes harmful regarding the chances of survival of males, that is, from the point of view of "classical" natural selection. In this case, such property will either regress or the species will become extinct. The first case was observed in sticklebacks (Futuyma 2013, 490). In these fish, an attractant for females is the red male underbelly. However, it attracts not only females but also predators. Therefore, in large water reservoirs (rivers, lakes), abundant in predatory fish, male sticklebacks are protectively colored, while in small pools, devoid of predators, they manifest their red underbelly in all their magnificence. In both cases, the selective pressure is oppositely directed—ultimately, the balance of gains and wastes counts. Giant deer represent a second case. They lived in a glacial period in tundra areas and possessed huge antlers spanning four meters—of course, a product of sexual selection. When the climate warmed up, taiga started to occupy the areas previously covered by tundra. The enormous antlers did not allow the giant male deer to move between trees, and most probably because of this, the species became extinct, as it did not evolve to have smaller antlers.

Reasons for the existence of the sexes

We already know more or less what sexual selection consists of. However, the principal question remains: what do the two sexes, male and female, and sexual reproduction exist for (Futuyma 2013, 399-404, 424-427)? At first look, it seems to do only harm and lower an individual's (its identity's) fitness. First, sexual reproduction is a waste of resources (energy and building material) for the production of males. As mentioned, in most cases, the proportion of sexes is usually about 1:1 (equal). The factor that limits the amount of progeny in each population is the number of females; because of energetic costs, a female can only produce a limited amount of progeny. Therefore, if the population/species were composed exclusively of parthenogenic females, it could yield twice as much progeny as a population/species in which females constitute only 50% (we assume that, because of the limited capacity of the environment, the total numerosity of the population is limited). Second, in the sexual type of reproduction, each of the parents transfers to progeny only 50% of its identity, its set of genes. Again, in the case of parthenogenic reproduction of females, this would be 100%.

There are additional costs to this: a waste of time and energy for courtship, mating rituals, bright coloration, exhaustive singing, fighting for females, searching for a partner, copulation, and so on. This time and energy could be allocated for additional feeding, enabling more better-equipped progeny to be produced. It seems, therefore, that sexual reproduction has only faults and evidently harms the species, lowering the fitness of its individuals.

Indeed, there are numerous species with an asexual type of reproduction, composed only of parthenogenically reproducing females (but not among the "highest" vertebrates: birds and mammals). However, these are relatively young species, which means that asexual reproduction originated in them recently. This suggests that organisms propagating asexually, in the long run, lose the competition with organisms "equipped" with two sexes. Why does this happen, then? How, despite the bringing of undisputable harm and apparent lowering the fitness, does sex supply such profit that it rules in the world of living organisms? Even bacteria have a substitute for sexual reproduction in the form of the conjugation and horizontal transfer of genes mediated, e.g., by bacteriophages, which was discussed above.

A few solutions to this problem have been proposed. Undoubtedly, sexual reproduction brings long-term benefits—it increases a population's diversity, allowing for a greater pace of evolution, greater elasticity in reacting to environmental conditions in space and time, and a higher rate of adaptation. However, evolution does not plan for the future and does not anticipate the variability of the environment in advance—natural selection acts online, that is, without planning for the future and profits that have beneficial effects "here and now" are necessary to explain the existence and maintenance of sexual reproduction.

The short-term benefits of sexual reproduction were postulated as resistance to diverse and quickly changing in genetic aspects parasites and avoidance of the so-called Muller's ratchet, that is, the cumulation of harmful mutations in clones[6] in small populations, which leads to their elimination from the population (Futuyma 2013, 406). Nevertheless, these explanations have not gained common acceptance among evolutionists.

In my opinion, however, the solution can be quite simple. In asexual reproduction, which lacks genetic recombination, only "moderately good" or "mean" descendant individuals are produced, as profitable mutations originated in various clones cannot meet in one individual. On the other hand, in the case of sexual reproduction, where intense recombination takes place (Fig. 4-4), various beneficial alleles of genes originated through mutations can meet in one genome, or even in one chromosome or gene, either due to

[6] A clone is a set of genetically identical individuals.

random distribution of chromosomes (chromatids) to gametes during meiosis, and the accidental meeting of chromosomes in one genome after fertilization or due to the process of crossing-over. Recombination can lead to forming "good" (profitable, giving high fitness), "mean" (moderately profitable, resulting in average fitness), and "bad" (disadvantageous, characterized by low fitness) combinations of particular alleles. At the same time, it should be added that the effect of meeting in one genome/chromosome of either beneficial or harmful alleles is frequently synergistic: the joined effect (positive or negative) of two (or more) alleles (profitable or harmful) is greater than the sum of these effects separately. One could argue that if asexual individuals only produce "mean" progeny, while sexual individuals: have "good," "mean," and "bad" progeny, the average effect is the same. However, this is an incorrect way of reasoning. If there is produced, say, ten times more progeny (in sexual organisms) and 20 times more progeny (in asexual organisms; let us remember that they have twice as many females in the same environment capacity) than parental individuals exist, then the limited capacity of environment will cause that only, respectively, 10% or 5% of the descendant individuals will survive and reproduce, while the remaining 90% or 95% will die without leaving offspring. Therefore, it does not matter that there are much fewer "good" allele combinations, and thus "good" descendant individuals than "mean" and "bad" allele combinations/descendant individuals, and that asexual parents produce twice as many "mean" individuals than sexual parents—individuals of different degrees of "goodness." Nevertheless, most probably, the minority "good" progeny of sexual individuals will enter into the composition of those 5% or 10% of descendant individuals that will survive and reproduce, as they are characterized by a significantly higher fitness than the rest of the progeny. In this case, quality counts more than quantity.

The shuffling of "good" and "bad" gene alleles during recombination due to crossing over and random segregation of chromosomes to gametes and their accidental joining after fertilization in sexual reproduction, giving as a result "good," "mean," and "bad" allele combinations (chromosomes, genomes), is illustrated in Fig. 4-9 and Fig. 4-10, respectively.

We can see in Fig. 4-9 that the exchange of parts of chromatids between homologous, "moderately good" chromosomes leads to forming a chromatid, one per four, that is decidedly "good." It will be passed to one of the gametes and, having some luck, can, after fertilization, meet another "good" chromatid coming from the other parent, giving, as a result, a "good" diploid descendant individual. Of course, in reality, a much greater number of genes and their "good," "bad," and "mean" alleles can participate in the described process.

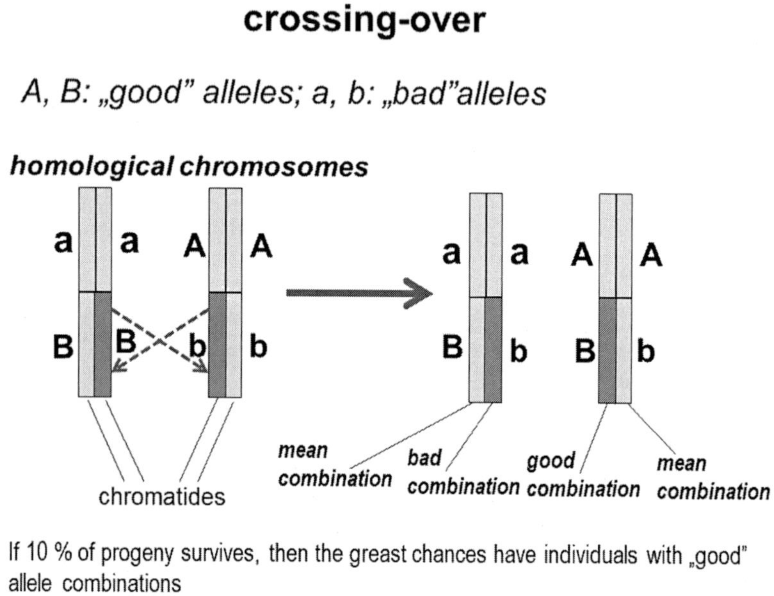

Fig. 4-9. Formation of "good" (AB), "mean" (aB, Ab), and "bad" (ab) (with high, moderate, and low fitness) combinations of "good" (A, B) and "bad" (a, b) alleles within chromosomes due to recombination in the crossing-over process in sexual reproduction. The frequency of these combinations is: "good"—25%, "mean"—50%, "bad"—25%. Assuming that ten times more descendant individuals are produced than can survive, although "good" combinations constitute a minority, most probably mainly just them will be among the 10% of individuals that have survived and reproduced. The fragments of chromatids that were exchanged between homologous chromosomes are marked in dark grey.

Fig. 4-10 shows that in a descendant of "mean" parents that have a "good" allele on one chromosome and a "bad" allele on another chromosome (only two non-homogeneous chromosomes are shown here; as a rule, there are many more of them and therefore the described process occurs with much greater intensity) there can appear, due to random segregation of chromosomes to gametes, a "good" gamete. If at fertilization, it meets another "good" gamete, a "good" zygote and a "good" descendant individual will originate. Of course, again, in the described process, many chromosomes, genes and alleles of a different degree of "goodness" can participate.

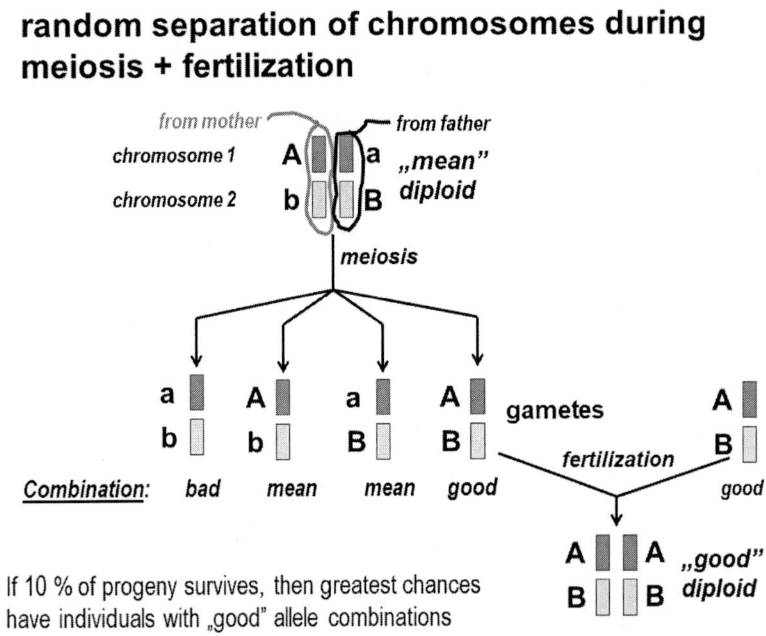

Fig. 4-10. Formation of "good" (AB), "mean" (aB, Ab), and "bad" (ab) (with high, moderate, and low fitness) combinations of "good" (A, B) and "bad" (a, b) alleles in sets of non-homological chromosomes (genomes) due to recombination through random segregation of chromosomes coming from two "mean" parents to gametes and fertilization in sexual reproduction. The frequency of these combinations is: "good"—25%, "mean"—50% "bad"—25%. At fertilization, a "good" gamete can meet another "good" gamete, due to which a "good" diploid descendant individual will originate. Assuming that ten times more descendant individuals are produced than can survive, although "good" combinations constitute a minority, most probably mainly just them will be among the 10% of individuals that have survived and reproduced. Chromosome 1 is marked in dark grey, while chromosome 2—in light grey.

Summing up, it pays to produce fewer descendant individuals, of which at least some are "decidedly good," than twice the number of "mean" progeny, even at the cost of the fact that a part of progeny will be carriers of "bad" allele combinations, and thus evolutionarily handicapped. Because if only 5 or 10% of progeny is to survive because of the limited environmental capacity, what matters is not the average statistical quality of progeny, nor its quantity, but the quality of the "best" descendant individuals. Similarly, if one country is represented in the marathon run by

24 "mean" "asexual" competitors, while another country—by only half (12) "sexual" competitors, of which three are excellent, six are mean, and three are poor, all the medals will be most probably won by the three "decidedly good" runners from the latter, "sexual" country.

Speciation and isolation

In sexually-reproducing organisms, the species is defined as a set of individuals capable of interbreeding with each other without any limitations in natural conditions (at least potentially) and able to produce fertile, fully efficient progeny. In organisms with an asexual type of reproduction, much more arbitrary, structural and biochemical criteria are used (the matter is complicated here by horizontal gene transfer, leading to the fact that a given living individual can be a hybrid of a few organisms of different origin). Of course, the cumulation and consolidation of mutations in a population of some individuals can eventually lead to *speciation*, that is, the actualizing of a new species (Futuyma 2013). One type of speciation is *phyletic evolution*, a gradual transformation of one species into another. It seems rather rare. Another possibility is constituted by divergent evolution, when two or more populations genetically isolated from each other evolve in a different direction, either due to living in slightly different environments or due to chance, and finally, from each of them, a new species originates.

Isolation is essential to prevent crossbreeding of individuals and the exchange of genes between populations—if such exchange occurred, both populations would undergo "averaging" and fusing into one population (Futuyma 2013, 461-504). The most important and, at the same time, most primeval kind of isolation, the "mother of all isolations," is *geographical isolation*, or, more broadly, *spatial isolation*. In this case, populations are spatially separated by a large-scale barrier: a sea, mountain range, desert, or river separates the areas inhabited by individuals of a given species (solid land and islands, valleys, forest areas, river banks). This can also take place at a smaller scale—then, we can speak about spatial isolation. For instance, tropical forests are composed of hundreds of tree species, and therefore particular individuals of a given tree species usually grow at a large distance from each other. Various insects inhabit just this and not another tree species; therefore, insects living on different trees of this species practically do not meet each other and therefore do not crossbreed. For these insects, each tree constitutes a kind of an "island", separated from other "islands" by the "sea" of trees of other species. This, among other reasons, is why insects have the greatest number of species, most of which live in the tropics. In another case, a mere creek can constitute a barrier impossible to cross for non-flying insects or other arthropods.

The two populations on both sides of the barrier will begin to differ more and more in genetic and phenotypic terms. They live in slightly different conditions and continuously adapt; some role can also be played by chance (genetic drift, see below). Finally, even when the barrier is removed, the individuals belonging to them will be different enough that they cannot effectively crossbreed with each other, and two new species will originate. New, secondary types of isolation responsible for this will develop. These are: *ecological isolation*, when neighboring populations occupy slightly different habitats, for instance birds feeding in a forest and an open area; *seasonal isolation*, when particular sub-populations reproduce in slightly different seasons; *mechanical isolation*, that is, incompatibility of male and female reproductive organs; *isolation of gametes*, when a spermatozoon and egg cell are unable, because of lack of compatibility, to form a functional zygote; *lethality of zygote*, when the zygote or early fetus dies soon after fertilization; *sterility of cross-breeds*, when due to interbreeding of individuals of two species, functional, but completely sterile progeny originate (e.g., a mule, a cross-breed of a donkey and a horse); *disability of progeny*, when cross-breeds are able to reproduce, but both they and their progeny are disabled in terms of fitness in relation to parental species—therefore they undergo elimination like the bearers of harmful mutations (this means that the two species have not completely separated yet); finally—*ethological (behavioral) isolation*, when each sex does not recognize individuals of the opposite sex of the other species as potential sexual partners because of different mating coloration, changes in the mating ritual, other singing in birds, or the chemical composition of pheromones in mammals or insects. This is the target, final, "highest" form of isolation, as individuals already belonging to different species do not waste their resources and time on senseless advances, copulation, bearing the fetus, giving birth to young or laying eggs, and or feeding and taking care of sterile or disabled progeny. All these actions are very burdening and decidedly negatively affect fitness, and therefore a very strong selective pressure on the evolution toward behavioral isolation will take place, which saves individuals from this sort of absurd and harmful action.

Nevertheless, all these secondary types of isolation could not originate without a certain kind of primeval spatial isolation, which initiated the entire process of separating species, preventing the free, panmictic (potentially "everyone with everyone") crossbreeding of individuals from two or more populations.

Genetic drift/the founder effect

Natural selection (and its different "varieties") is a directional mechanism of evolution aiming to increase the fitness of individuals in the current environmental conditions (ecological niche). However, there are non-directional mechanisms leading to accidental changes in the genotype and phenotype of individuals. For purely statistical reasons, chance can play an important role only in small populations. Because of random crossing-over during meiosis, the separation of the chromatids to gametes, and meeting of the gametes coming from parents, after a relatively small number of generations, one allele of a given gene can, because of pure chance, spread within the micro-population of all individuals, while its "rival" will be eliminated. Such a process is called *genetic drift* (Futuyma 2013, 257-281).

On the other hand, a "one-time episode" of such a drift, when a few individuals of random genetic composition (frequency of alleles) "buds off" from the parental population, frequently of statistically different genetic composition, migrates and populates a new area, is called the *founder effect* (Futuyma 2013, 255, 484-485). The new population originating from this small group's reproduction will differ from the initial parental population. The frequency of alleles has changed; therefore, evolution has occurred, although in a completely accidental direction.

For genetic drift and the founder effect to act, various gene alleles taking part in them must possess identical, or almost identical, fitness. Therefore, new alleles had to originate from old alleles due to neutral mutations. This directs our attention toward the latter. Not only can they enable genetic drift and the founder effect, but also, in new conditions, they can lose their neutrality and become beneficial or harmful. Therefore, they potentially prepare the ground for future evolution. In the face of this, the occurrence of neutral mutations can be regarded as a subsequent non-directional mechanism of evolution.

The limitations of natural selection

Not underestimating the role of natural selection in the evolution of the animate world, it should not be forgotten that it is not, as mentioned above, an all-mighty mechanism that explains and is responsible for everything in the process of transformation of one life form into another. First of all, the thorough change of their structure plan and dramatic increase in complexity in at least some developmental lines, being occasionally supported by a few of its "varieties" (kin selection, group selection, or sexual selection) and isolation, genetic drift, and the accumulation of neutral mutations. There is nothing more misleading. Natural selection is only a "censor," a selector of this that has already been created, and not a generator of new diversity and complexity. Generating a completely new structure (and functions based on it) occurs by modifying living organisms' existing structure (and function). This occurs in the case of single cells through modification of the gene expression regulation cascade inside the cell (see Chapter 5). In the case of complex multicellular organisms, especially those "equipped with" tissues and organs, the evolution of their structure plan and the possible increase in complexity takes place due to changes in developmental mechanisms steering the shaping of structure (morphogenesis) during embryonal development (embryogenesis) and individual development (ontogenesis) (see Chapter 5). All these changes and modifications are based on mutations and recombinations in regulatory genes supervising and directing these processes. These alone are responsible for generating diversity, offering to natural selection slightly or radically new construction plans, frequently characterized by a substantially higher degree of complexity than the hitherto existing structures, while natural selection only confronts them with current environmental conditions/properties of the ecological niche, and they are either accepted and passed to "mass production" in the process of reproduction, or not.

CHAPTER 5

WHAT ARE THE REGULARITIES AND WAYS OF THE COURSE OF EVOLUTION OF LIVING ORGANISMS?

The purposefulness of living individuals

When considering the purposefulness of living organisms or their lack thereof, it is worth distinguishing between "external" and "internal" purposefulness. From the external point of view of physics, the phenomenon of life has no purpose or sense—it is the same physical phenomenon as the falling of stones in a gravitational field or the "painting" of "flowers" by frost on window panes. Therefore, the "external" purposefulness of living individuals does not exist; this notion is devoid of any sense. However, from the internal point of view of biology, the features of living individuals are decidedly purposeful—they serve the survival and reproduction of these individuals. Therefore, biological individuals possess undeniable, well-determined and strongly expressed "internal" purposefulness (Williams 1997).

There is no sense in asking why the sun shines, why the earth revolves around the sun, why the terrestrial atmosphere has a particular penetrability for electromagnetic radiation of different wavelengths, why water has a particular dielectric constant, why a given chemical reaction has a determined equilibrium constant… on the other hand, it makes sense to ask why do animals have eyes formed in a given way, why do plants have leaves and photosynthetic pigments, why did the brain and the entire nervous system originate, why do peacocks have such long and colorful tails, and why do deer have antlers, why do enzymes have such specific kinetic properties… Therefore, while looking for purposefulness in purely physicochemical systems seems fruitless, this has a deep sense in biological systems seen from the inside. However, one should be careful not to confuse these two kinds of purposefulness and to determine which is being discussed at a given moment.

Every property, every element of structure and function of living organisms can and should be considered from the point of view of biological evolution and the purposefulness of the structure and function determined by it. For instance, thanks to the great progress in biology in the 20th and the beginning of the 21st century, we can speak reasonably about the purposefulness (and its kind, that is what a given purposeful solution serves) of such aspects of structure and function of living organisms, as:

⇒ the sequence of amino acids in an enzymatic protein, its three-dimensional structure and function, kinetic parameters;
⇒ the composition of lipid-protein membranes in the cell, transport of substances and ions through them;
⇒ the structure and function of organs: heart, brain, kidneys, *etc.*;
⇒ the behavior of living organisms—perception of various aspects of the environment and appropriate reaction to them;
⇒ the structure of the contractile apparatus and the mechanism of muscle contraction;
⇒ the size, shape and coloration of the body;
⇒ the course of the life cycle, amount of progeny, length of the maturation period;
⇒ structure and function of the eyes of vertebrates, cephalopods and insects.

The superior, primary purposefulness of a living individual expresses itself through the maximization of fitness—the chances for survival and reproduction, for the propagation of its identity in a large amount of possible similar copies. Individuals that are not "the best," most efficient in the realization of this task, whose particular aspects of structure and function are not the most optimal in given environmental conditions (ecological niche), will simply not transfer their properties (their identity) further on and will become extinct without leaving progeny. Biological evolution is based on a simple, absolute, indifferent and (from the human point of view) even repulsive directive: "be the best or die (childless)."

It is also worth remembering that while evolution perfects and optimizes, that does not mean it always finds the "best of all possible" and "ideal" solutions. Therefore, it is not all-mighty, does not plan for the future, acts "here and now" opportunistically, and does not predict far-going consequences of the changes that just occur (hence the so-called "blind alleys" or "dead ends" of evolution). A large role in evolution is played by chance (related, e.g., to directionless mutations). Evolution is cumulative to a certain extent—the earlier acquired properties condition (limit) the ways of further development of individuals being "carriers" of these properties. It builds the new based on the old and does not plan and create everything from the very beginning. Evolution must also consider the limitations of the physical world, which may seem trivial: living things lighter than hydrogen, exceeding the speed of light, storing and copying information with infinite accuracy, and so on, cannot exist.

Regardless, various limitations concern biological evolution and our evaluation of its results: the properties of living organisms known to us. Our knowledge is frequently insufficient to decide which solution is the most efficient, most optimal, and which is not—apart from evident cases of evolutionary "duds", burdened with obvious constructive and functional faults, such as the crossing of the respiratory and digestive tract in land vertebrates, complex eyes in arthropods, some details of the constructive plan of the eyes of vertebrates, or finally—defects in the structure and function our greatest glory, hopeless from the engineering point of view—the human brain (these structure-functional mistakes of evolution will be discussed in more detail below). It should also be remembered that this is the balance of gains and losses that ultimately matters: the best solution from the constructive point of view does not always lead to an elevation of fitness—this is true when the costs of production and maintenance of a given aspect/property of structure and or function and burdens resulting from its presence outweigh the profits brought by it. Frequently it is more beneficial for evolution to allocate the limited energy resources to developing other properties, more increasing fitness. Therefore, in the process of evolution, strict bookkeeping prevails over perfectionism, and the principle of "something for something" holds for optimizing various features.

Evolution is not omnipotent in the face of this; it does not always find the most optimal or the correct solutions from an engineering point of view. We often do not know whether it would be possible to find a better solution. An example is the sliding mechanism of the muscle cell and the entire muscle contraction. It is present in all muscle types: skeletal, heart, and smooth muscles, in all known organisms with muscle or muscle-like cells. Very generally, due to neural stimulation of the muscle cell (myocyte) and release (from sarcoplasmic reticulum cisterns) of calcium ions (Ca^{2+}) to its interior (cytosol), a sliding of myosin (thick) filaments past actin (thin) filaments, shortening of the myomere (unit of the contractile apparatus) and the whole myocyte, and consequently, muscle contraction occurs (Alberts et al. 1998, 536-551; Reece et al. 2011, 1104-1100). We do not know whether a different mechanism could come into being. The contractile apparatus in muscle cells, acting according to the sliding mechanism, originated in the course of evolution and the differentiation of cells and tissues in multi-cellular organisms from "thin" actin filaments, which until today enter into the composition of the cytoskeleton of unicellular eukaryotes (for instance, some amoebas), along which myosin heads transport different loads driven by ATP synthesis (Alberts et al. 1998, 536-537). The innovation during the evolution of muscle cells consisted of adding long tails to myosin heads, "splicing" them into a fibrous structure, which led to the forming of "thick" myosin filaments and resulted in their parallel arrangement between actin filaments. This is a subsequent case of the rule that evolution utilizes and modifies that which already exists instead of creating something completely anew. The sliding mechanism of muscle contraction seems astonishing in its "cleverness" and functionality. However, is it perfect and optimal, the best of all possible mechanisms? We may speculate to ourselves and imagine alternative mechanisms, for instance, consisting of a conformational change (shortening along the long axis of myocyte) of a string of globular (more or less spherical) proteins, joined with each other, arranged in a "chaplet" along this axis, or in the shortening of a protein "spring" (its coils becoming close to each other).[1] Our knowledge is decidedly insufficient, and will probably remain so forever, to decide whether such mechanisms could spontaneously evolve and, if so, whether they could be better than the known sliding mechanism. Did the sliding mechanism originate because it is the most efficient or because of chance? Or perhaps it was a consequence of pre-adaptation consisting of the fact that almost all elements necessary for its coming into being were already present in the unicellular ancestors of higher animals? We will probably never know this.

Summing up, purposefulness predominates among the properties of living organisms, but this does not mean that biological evolution always comes to the optimal or "best" solutions from the engineering point of view. Quite the contrary, it frequently "errs" and "goes astray." In its course, an important role

---

[1] One can also speculate whether there could be a possible mechanism of muscle (or muscle-like organ) contraction based not on myosin and actin, as in known animal organisms, but on microtubules and kinesin or dynein (Alberts et al. 1998, 525-530; Reece et al. 2011, 116-117). Like in non-muscle cells various loads are transported by myosin heads, for which the rails are actin filaments, so other loads, for instance "bubbles" "budding off" from the Golgi apparatus or mitochondria, are transported by kinesins and dyneins that travel along microtubules (all these types of transport are driven by ATP hydrolysis). Theoretically, a variety of the sliding mechanism of muscle contraction could be formed through adding long protein tails to kinesins or dyneins, joining them in "thick" filaments and parallel arrangement between microtubules.

is played by pure chance (and physicochemical limitations). Let us repeat: evolution is not omnipotent and does not plan anything for the future. For this reason, apart from astonishingly ingenious, even extremely clever solutions in the structure and function of living organisms, we find many incredibly stupid mistakes and constructive solutions among its creations.

Let us look now at a few examples of both the enormous "geniality" and "invention" of evolution, leading to incredibly "intelligent" solutions, and of its utter "stupidity," due to which every engineer respecting their job would feel deep awkwardness. Interestingly, the vertebrate (including the human) eye and human brain belong to both categories.

As the unbelievable "positive" achievements and creations of biological evolution are commonly known, I will mention only two of them (I am sure that every reader, even without a thorough biological education, will easily recall many other examples). The first case is a small fish called the pony fish (this example was taken from Williams 1997, 14-16). This fish lives in oceans with many natural enemies—in the first order, all predatory fish. Any object "suspended" in the depths of the water, when seen from below, against the background of the relucent sky, either in the full sun or at dusk, will be seen as a black shape, even if it is bright white. How can it become invisible to a predator in such conditions? Well, the belly of a pony fish emits a visible light (glow) of the same intensity as is necessary to compensate for the light coming from the water surface in various illumination conditions. Therefore, this fish regulates the intensity of its shine accordingly to the part of the day and the weather *etc.* (of course, it knows nothing about the part of the day and weather but only detects the intensity of light reaching it from the ocean's surface and reacts appropriately). This is a decidedly purposeful and surprisingly "clever" action, testifying to the great creative potential of biological evolution.

Another very "positive" example of the possibilities of evolution is the human eye (and of all vertebrates, which principally have the same construction plan). The human eye is a very complex organ, and at the same time, it is extremely well adapted to fulfilling its function, meaning the detection of dynamic pictures (spatial distribution of quanta of electrodynamic radiation, photons, in the visible range of various wavelengths and their changes in time) and relaying to the brain for further processing (preliminary processing, e.g., the extraction of spots from the "pixels" of the picture, which are stimulated photosensitive cells, already takes place within the retina; Banich and Compton 2018, 138-141). It is composed of many elements: the vitreous humor, filling most of the eyeball; the retina, a layer of photosensitive cells (rods, sensitive to even weak light, and cones, responsible for seeing colors)[2] lining the bottom of the eyeball; neural connections (neurons and their appendages: dendrites and axons), receiving signals from these cells, carrying out their preliminary processing, joining together into the optic nerve and sending information to the brain; the lens, focusing the picture on the retina; the cornea, protecting the eye; the iris, whose muscles, through contraction or relaxation, determine the size of the pupil, and thus the amount of light falling into the eye; the choroid, surrounding the retina and containing black pigment impenetrable for light; lying outside the sclera, protecting the eye and determining its shape, *etc.* This is illustrated in Fig. 5-1 (the upper panel). Particular "optical" parts of the eye, such as the lens, vitreous humor or cornea (and the anterior chamber situated between it and the lens, containing the aqueous humor) exhibit evident "knowledge of physical laws" that enable the possibility of clear sight: they are characterized by appropriate transparency, light refraction coefficients and curvature. The eye can respond by decreasing the pupil's diameter in reaction to a strong light to see sharply and not damage the photosensitive cells and by increasing the pupil's diameter when dealing with weak illumination. By contracting the muscles sustaining it, the lens can flatten, enabling it to see distant objects, while after their relaxation, it becomes more rounded, which enables short sight (this is called eye accommodation). Therefore, the human eye is undoubtedly a highly complex, purposefully shaped organ that performs well. As such, it constitutes a favorite example of the opponents of biological evolution (especially creationists and protagonists of the "Intelligent Design" concept) as an organ that could not evolve spontaneously from something significantly simpler through fully functional transitory stages ("after all, one cannot see with a half of the eye"—this is a sample of the intellectual quality of their argumentation). Nevertheless, even Darwin, of course, deeply convinced about the truth of his theory of evolution by means of natural selection, admitted that he had trouble imagining how such a complex, clever and purposeful organ as the human eye could spontaneously come into being in the process of evolution. Today we know that his concerns were not justified, as we know practically all the most important mediate (transitory) stages in the evolution of the eye between a light-sensitive spot on the body surface and the vertebrate eye (see below) and understand the developmental mechanisms leading to forming the eye during embryogenesis, which can undergo evolutionary modifications, and we are also able to carry out a simplified simulation of the eye's evolution using a computer.

---

[2] And some other cells, for instance ganglion cells and bipolar cells participating in first stages of picture processing (Banich and Compton 2018, 138-141).

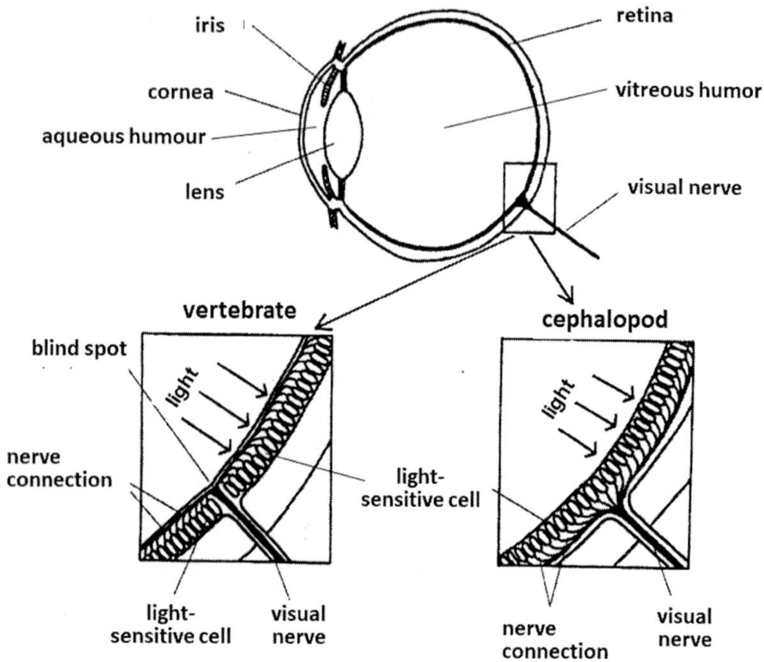

Fig. 5-1. Composition of the eye—common for vertebrates and cephalopods (top). The detailed structure of the retina (bottom) reveals clear differences. In vertebrates, neural connections go from photosensitive cells toward the interior of the eyeball, due to which they obscure the view and form the blind spot in the place where they join into the optic nerve that punctures the retina and leads to the brain. In cephalopods, the neural connections depart outside the eyeball, due to which they do not lie on the way of light rays and do not form a blind spot.

After these paeans to the glory of ingenuity, cleverness or even genius of biological evolution (after all, humanity still cannot repeat many of its achievements), it is time to pass to cases of not even stupidity but unbelievable imbecility (are we, therefore, dealing here with a Very Little Intelligent Project?). Its examples are less known, so I will discuss them a little more broadly. Furthermore, because we frequently regard ourselves—representatives of Homo sapiens—as the top achievement of evolution, even a "crown of creation," many of them will concern humanity.

Returning to the book by Williams (1997), I will quote the crossing of the digestive tract with the respiratory system, namely of the throat-gullet with nasal tracts and with the mouth cavity and the larynx-trachea, present in humans together with all land vertebrates (Williams 1997, 134-136). Due to this crossing, we can choke on liquid that accidentally gets into the trachea or even fatally choke on a solid food particle. Therefore, this is a botched anatomical-functional solution. After all, it would suffice to lead the respiratory tract from the nose to the trachea and lungs to omit the digestive tract so that the lumen of these tracts would not come into contact with each other (they would not cross). Where does such stupidity come from? Well, from the fact that land vertebrates evolved from crossopterygians, in which, as in all fish (and earlier—in primitive chordates, cephalochordates *Acrania*, such as the lancelet),[3] the gills (meaning their respiratory system) are (were) located in slits joining the digestive tract (throat) with the body surface. In order to breathe, fish intake water through the mouth and next pass it through gills, where oxygen is taken up. The primeval lungs of crossopterygians (and dipnoid fish) adopted the form of a sack, being only a diverticulum of the digestive tract. These fish continued to intake oxygen, swallowing it with the air through the mouth, which had become the evolutionary heritage of all their land descendants, including us. Again, evolution, when it can, uses and modifies what already exists instead of creating something anew.

A second example of an evolutionary bungle is also taken from Williams (2014) and concerns humans. The point is the course of testicular ducts in adult men (Williams 1997, 138-141). During individual development, the testicles descend from the abdominal cavity to the scrotal sac. The point is their cooling—in higher temperatures prevailing in the body interior, spermatozoa would not survive. However, past evolutionary occurrences resulted in the testicles, which, when descending to the scrotal sac and "drawing" the testicular ducts with them, "hook" them by the ureters, due to which the testicular ducts are ultimately much longer than they could be if they had traveled to the scrotal sack on the other side of the ureters. It might seem that this is a stupid mistake, although easy to correct. However, evolution does not think, does not analyze the ensuing situation and does not plan for the future.

Additionally, it is difficult to radically change the already-existing course of embryonic development. Apart from this, the evident constructive defect probably has a negligible impact on fitness. Therefore, the described state of affairs has not changed and probably will not change in the future.

---

[3] Indeed, gill slits "puncturing" the throat walls is a property characteristic, at least in some developmental stage (e.g., larval, embryonal), for all chordates, including tunicates, and most probably all deuterostomes, including hemichordates (*Enteropneusta* and *Pterobranchia*); it is supposed that echinoderms lost them secondarily in the course of evolution.

A third example of the silliness of evolution and the lack of long-term planning for the future, this time coming from me, is the human brain. It is the most complex system in the universe known to us. Thanks to it, we are who we are (in the psychical sense). It is the seat of our psyche, consciousness and self-consciousness—which emerge as an epiphenomenon, constituting an aspect or simply a "by-product" of the activity of the neural network in the human brain (the "ABC model of [self-]consciousness") (Korzeniewski 2010, 2017a, 2020). This alone constitutes a fundamental tool enabling the spectacular development of science, technology and medicine. Finally, as a result, we possess ethics and esthetics, including the fine arts, literature and music. The source of our humanity, with all its shines and shadows, lies within it.

However, it is enough to look a little at the structure and functioning of the human brain to consider many of the solutions appearing in it as fatal from the point of view of any rational logic (and engineering). First, neural connections between sense organs and the brain, and between different brain parts, are much longer and much more tangled than they could be if the brain were designed rationally by some engineer and therefore was constructively correct. This is illustrated in Fig. 5-2. It can be seen that neural connections relay visual signals from the eyes (situated at the front of the head) to the back of the head, to the occipital lobe. Afterward, these signals are integrated at many stages into still more and more complex sensory pictures in the occipital, parietal and temporal cortex. Signals from the receptors of touch, pain, warmth and cold located in the skin over the entire body surface, and signals from proprioceptors (receptors of "internal feeling") situated in the interior of the body, and kinesthetic receptors situated within muscles and tendons, reach the somatosensory cortex at the anterior (frontal) edge of the parietal lobe and also undergo integration, although not as complex as visual stimuli (Banich and Compton 2018, 136-154). Auditory signals from the ears reach the temporal cortex, not shown in the scheme.

All this represents the "ascending" (sensory-integrative) stream of neural signals. Somewhere (in fact, in several places) in the parietal cortex the signals (pictures) of different modalities (visual, auditory, olfactory, tactile) are associated with each other and relayed through long neural connections (occipitofrontal and longitudinal fasciculus; "horizontal" stream of neural signal transmission) (Banich and Compton 2018, 36-38) from the back (posterior) part of the brain to the most frontal (anterior) part of the brain—the prefrontal cortex, being the "headquarters" of the "operational memory" of the brain, the seat of the processes of thinking, planning and decision making (and also, in my opinion, being the "carrier" of the psyche and self-consciousness; Korzeniewski 2010, 2017a, 2020). The existing memory records are involved and exploited in forming sensory pictures in the parietal and temporal cortex and for thinking and decision-making. In turn, new sensory pictures, thinking processes, and the planning of future actions form new memory records. The undertaken decisions about a motor activity (this can be locomotion, manual operations, face mimics or speech generation) are transcribed in the frontal cortex into signals concerning the initiation of complex movements, activation of muscle groups and particular muscles, and next sent to particular (groups of) muscle fibers (motor units) by neural connections leaving the motor cortex, lying at the posterior (back) edge of the frontal lobe of the brain. This represents the "descending" (decisive-motor) stream of neural signal transmission (Banich and Compton 2018, 100-136). The connections between the receptors, effectors (muscles) and various parts of the human brain, firstly between these very parts, are characterized by large length and entanglement—they cross each other many times. This significantly worsens the speed and efficiency of communication between them. This is shown in the left scheme in Fig. 5-2.

Undoubtedly, neural connections in the human brain are long and tangled (they frequently cross each other). Nerves from the eyes, situated in the front of the head, go to the backmost part of the brain, meaning the occipital lobe, where the primary visual cortex is located. Signals coming from the eyes undergo multi-stage integration on an even higher level, more and more complex, first in the occipital lobe, and afterward in the parietal and temporal lobe, creating more and more organized sensory pictures. On the other hand, the connections from the body's surface lead to the somatosensory cortex situated on the anterior edge of the parietal lobe. Signals from them also undergo integration, although much less complex than visual signals. (Neural signals/connections from other sense organs, for instance, the ears, are not shown for simplicity in this general two-dimensional scheme). Finally, the integrated sensory pictures of different modalities somehow combine to form one united multi-modal sensory picture. Next, they are sent through the long "horizontal" neural connections to the prefrontal cortex, where they undergo their final unification and interpretation and are used as the "substrate" for planning and decision-making. Afterward, in subsequent steps occurring in the frontal cortex, decisions are transcribed over several stages into the activation of particular groups of (antagonistic) muscles, of particular muscles within these groups, and finally, the motor cortex, located at the posterior edge of the frontal lobe, sends dispositions concerning contractions to muscles in different body parts. One should pay attention to the long and neural connections crisscrossing each other in the human brain: eye—the occipital lobe, skin receptors—somatosensory cortex, the motor cortex—muscles, and the parietal lobe (and the temporal lobe, not visible in this scheme)—the prefrontal lobe (that is: the sensory-integrative—decisive-motor brain part).

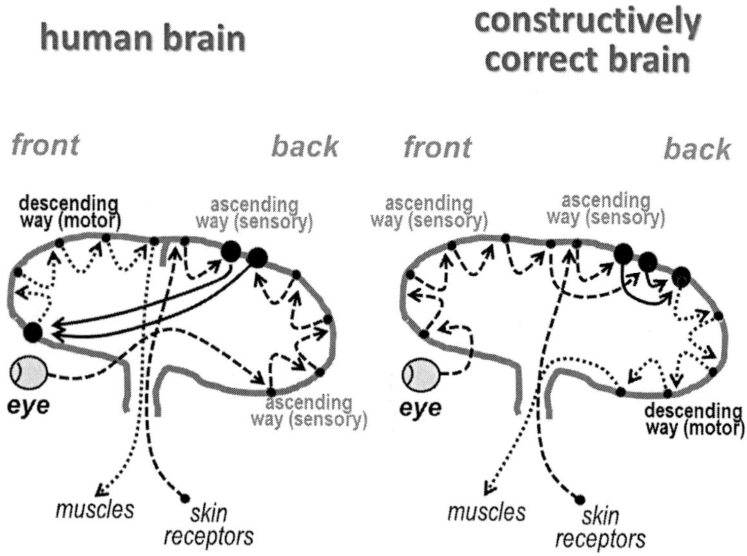

Fig. 5-2. Comparison of the neural connection between receptors (here: eyes and receptors in skin: touch, pain, warm, cold receptors), effectors (muscles) and various parts in the human brain and the "constructively correct" brain. Dashed arrows represent connections from sense organs to the brain and subsequent, still higher (forming more complex sensory pictures) stages of sensory data integration within the brain, that is the ascending (sensory-integrative) path. Dotted arrows indicate planning and decision making and their implementation, that is activation of particular muscle groups, its transcription into an appropriate spatio-temporal pattern of stimulation of particular muscles, and finally—sending signals to muscles, that is the descending (decisive-motor) path. Solid arrows correspond to the relay of signals, communication between the ascending and descending path, that is the "horizontal" path of neural signal transduction in the human brain. The neural connections in the "constructively correct" brain are overall much shorter and less tangled (over-crossed), than in the real human brain.

The situation looks completely different in the hypothetical constructively correct brain.[4] Here, the front and back of the brain are, to a large degree, interchanged. The visual cortex, supervising particular stages of visual stimuli integration, is located in the front part of the brain, near the eyes. Visual data are further integrated into complex visual pictures and associated with data of other modalities in the brain areas lying somewhat to the back but still in the anterior brain regions. Signals from the skin receptors flow to the somatosensory cortex situated on the back edge of the anterior part of the brain and undergo moderate integration. The "operational memory" and control of movement are situated in the posterior brain half. The most integrated, multi-modal sensory pictures are transferred from the integrative cortex in the anterior brain to the "highest" part of the decisive-motor cortex, that is, to the cognitive center responsible for thinking, planning and decision-making (a functional equivalent of the prefrontal cortex). Decisions are next made and executed through their transcription into the stimulation of the appropriate muscle groups and particular muscles, carried out by a sequence of neural connections heading the posterior part of the brain. Ultimately, instructions concerning muscle contractions are sent through neural connections leading to the muscles themselves. The total length of the connections between the receptors, effectors (muscles) and different brain parts is much shorter here than in the human brain.

Additionally, these connections cross each other only rarely and are therefore not very tangled as in the human brain (it is impossible to avoid this completely; of course, one should remember that all of this happens in three dimensions). Therefore, neural impulses communicating between the different brain areas must cover a much shorter distance here and require much less time. For this reason, the information transfer time is much smaller, while its efficiency is much higher. Thinking processes and consciousness related to them in the constructively correct brain are situated in the back part of the brain and not in the front part, as in the human brain. Generally, it seems that the human brain is constructed far worse than it potentially could be if some intelligent engineer endowed with consciousness planned it. Where does this stem come from?

The brain, to some extent, resembles a geological formation. Its new parts (layers) are added in the course of evolution to the old parts, "built on" them without any planning, here and now, step by step, without bothering about the final effect. This is, among others, why the neural connections

[4] I will probably receive many proposals as to how this brain could be further improved and made more efficient (constructively correct). They are most welcome—I do not claim that I have found a perfect solution.

reach from the eyes to the backmost part of the brain, the primary visual cortex in the occipital lobe. In our distant ancestors, whose brains were still very small, this did not matter. However, when the brain of *Homo sapiens* extended itself enormously to the present size, such a solution became a doubtless fault. In our ancestors, the somatosensory cortex, receiving signals from receptors, and the motor cortex, setting muscles and their groups in motion, were neighbors and exchanged signals (communicated) with each other, as other cortex types, responsible for higher-level integration and association, essentially did not yet exist. The fact that the sensory cortex was situated backward in relation to the motor cortex was most probably a matter of chance. Later during evolution, higher cortex regions appeared and started to grow, and they were deposited, probably again due to pure chance, *outside* the primary sensory and motor cortex and not *between* them. And so, the cortex carrying out the higher-level integration of signals coming from receptors into complex sensory pictures and related to the episodic memory (concerning single facts, objects and events) and semantic memory (related to general ideas), namely the parietal and temporal cortex, "accreted" *backward* of the somatosensory cortex (but forward of the primary visual cortex), while the cortex responsible for processes of thinking, action planning and decision making (and being, at least in my opinion, the "carrier" of self-consciousness), meaning the prefrontal cortex (also being the seat of a part of the semantic memory), and the cortex dealing with the planning of (sequences of) movements and its translation into the stimulation of particular (groups of) muscles, meaning the premotor frontal cortex (being, together with cerebellum, the carrier of procedural memory, concerning the execution of coordinated movements), were gradually added (first the premotor frontal, and next the prefrontal cortex) *forward* in relation to the primary motor cortex. As long as the brain was small, the excessive length and hopeless entanglement of the neural paths present in the brain did not cause many issues. They only appeared when the human brain reached the present size and degree of complication.

It should be added here that, for unknown reasons, the crossing of the sensory and motor neural connections leading out from and entering into the brain, leading from/to the entire body apart from the head, occurred: the left brain hemisphere operates the right side (half), of the body, while the right hemisphere—the left side of the body (the crossing of neural connections takes place in the medulla) (Banich and Compton 2018, 100-166). This undoubtedly lengthens neural connections and slows down body-brain and brain-body communication. As no rational explanation for this phenomenon has been found so far, the discussed constructive solution can again be regarded as an evolutionary defect/mistake.

Let us pass now to the vertebrate, in particular the human eye. Its cleverness, efficiency, complexity and purposefulness of structure and function were greatly praised. However, even this is not perfect from a constructional point of view. Interestingly, the higher cephalopod eye, while its general scheme is similar, in some respects represents a better engineering solution than the vertebrate eye. In vertebrates, the neural connections (neuron axons) depart from the photosensitive cells (in fact, directly from ganglion cells collecting and integrating neural signals from rods and cones) toward the interior of the eyeball, thus obscuring the view and creating a blind spot in the place where they join into the optic nerve, which relays neural signals to the brain (see the bottom-left scheme in Fig. 5-1) (Dawkins 1996, 159; Futuyma 2013, 57). It creates a "dead area" in the field of vision of the retina, which we are unaware of, as its existence is masked by special neurological mechanisms in the brain that extrapolate the blind area of the visual field neighboring it. This can be demonstrated using appropriate psychological tests. Therefore, despite all its astonishing cleverness and purposefulness, the eye of humans and vertebrates contains decidedly unfortunate solutions.

In this case, we do not have to speculate that this could be done better. Evidence that this is true is provided by higher cephalopods, e.g., octopuses. The general constructional plan of their eye is practically identical to ours—the same elements appear in it, arranged in relation to each other in the same way (Fig. 5-1, the upper scheme). The lens eye originated completely independently in vertebrates and cephalopods (eye evolution is described by Dawkins 1996, 126-179). First, these groups are unrelated and possess completely different constructional plans: vertebrates belong to deuterostome animals, while cephalopods to protostomes. Second, in both groups, eye formation during embryogenesis is different. While in the former, the eye (or at least the optic nerve and retina) separates from the brain and constitutes its "appendage", then in the latter, it is formed due to the herniation of the epidermis. Finally, what is most interesting for us in the discussed contexts is that some details of the retina's structure are different in both cases. Namely, in cephalopods, neural connections leave photosensitive cells from the outer side of the retina (under the layer of the retina looking from inside the eyeball). Therefore, these connections do not lie in the way of light rays, and as a result, do not obscure the view, and as they converge under the layer of photosensitive cells (and not above it, as in vertebrates), they do not form the blind spot. This is demonstrated in the bottom right scheme in Fig. 5-1. Therefore, the solution applied by cephalopods is undoubtedly better in the constructional respect. This example emphatically shows that humans, also vertebrates, although they overshadow other animals due to the size and complexity of their brains, are not the best constructed of all animals in every respect.

Let us finally move to the last example of an unfortunate evolutionary solution, namely the complex eye of arthropods, for instance, insects (this example is also taken from the Dawkins 1996, 126-179, but supplemented with elements coming from me). The relatively tiny constructional faults observed in the human eye are essentially nothing compared with the complete engineering disaster resulting from biological evolution in the case of the complex eye. Moreover, because such eyes are still possessed by most of the animal species presently living on earth (in particular crustaceans, chelicerates, including spiders and insects), the matter cannot be ignored.

To present the problem clearly, let us briefly discuss the process of the eye's evolution (see also Dawkins 1996, 126-179). Eyes able to create at least the simplest picture originated some 50-100 times independently in the history of the living world. Eyes first appeared about 540 million years ago during the Cambrian "explosion of life", when a huge number of various constructional plans of "higher" animals (*Eumetazoa*) appeared practically "from nothing." On the other hand, a large part of the cellular machinery used during embryonal development (for instance, the gene encoding protein opsin, responsible for photons absorption, or gene Pax6/eyeless[5] determining the location of the eyes) is common in all organisms possessing eyes (Futuyma 2013, 575-578). This suggests that their common ancestor already possessed some sensitivity to light (perhaps had light-sensitive spots on the body surface—see below). Practically all the main stages of the evolution of the eye can be found in presently-living organisms. Nevertheless, some unicellular eukaryotes, for instance, euglena, already possess the so-called "eyespot" (stigma) that distinguishes light from dark.

Analogically, completely flat "eye spots" in multi-cellular animals, located on the body surface and composed of a single layer of photosensitive cells, were most probably predecessors of today's eyes in different organisms. Slightly modified, moderately delved (concaved) into the body spots of this kind, in which a layer of cells containing black pigment impenetrable for light lies under the layer of photosensitive cells, are present in present plathelminths, for instance, triclads. Starting from the stage of the flat eye spot, the evolution of the eye followed two main, ultimately extremely different directions. This is illustrated in Fig. 5-3.

[5] An expression of the mouse Pax6 gene in e.g., the leg of a fruit fly leads to developing an eye (a complex insect eye) on this leg!

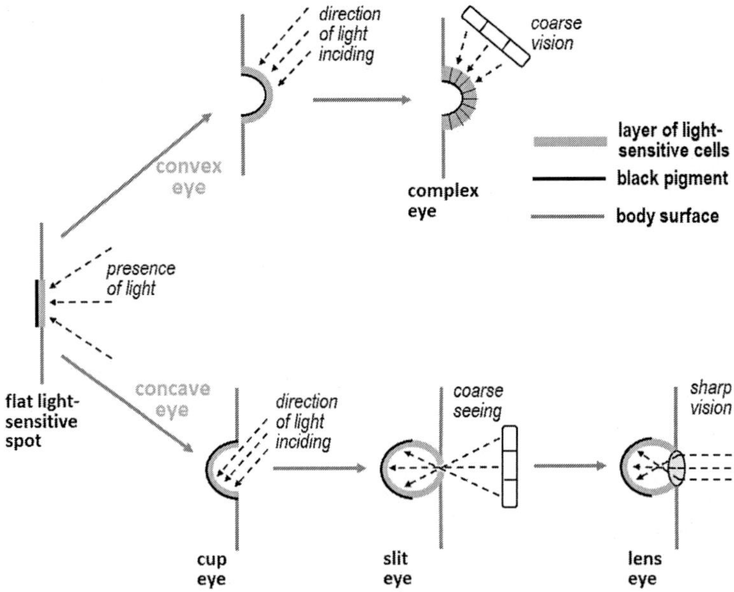

Fig. 5-3. Two main ways of eye evolution: convex eye and concave eye. The starting stage was a flat spot of photosensitive cells on the body surface, beneath which a layer of black pigment impenetrable for light lied (left scheme). From this phase, the evolution of the eye followed either the "convex" path or the "concave" path (see the text for details).

Let us start with the concave eye (the lower part of Fig. 5-3). The flat layer of light-sensitive cells, forming a spot on the surface of the body, constituting the starting phase (the left scheme in Fig. 5.3), can only detect the presence or absence of light (and, possibly, its intensity), and thinkably, if it possesses a lining of cells containing black pigment impermeable to light, decides whether it falls from outside or from the body's interior. However, the concavity (indentation) of this layer enables the relatively accurate determination of the direction of the falling of light, as light rays (photons) coming at a certain angle fall only on some cells of the photosensitive layer, while the rest of them are stopped by the layer of black pigment. This stage of eye evolution is called the cup eye—it has, more or less, the shape of a concave hemisphere, which is sometimes filled with a vitreous body of a light refraction coefficient higher than that of air or water (we will find it later in the lens eye). This is not a speculative phase—it appears in numerous contemporary groups, such as triclads, bivalves, polychaetes and snails. In the next stage of evolution, the hemisphere undergoes a closing in an almost complete sphere with a small opening in the front, and thus the slit eye originated. Through this opening, photons fall into the eye, while photons arriving from different directions fall onto different fragments of the retina, which enables forming of primitive pictures, still coarse-grained, blurry and devoid of details. This results from the fact that the opening in the eyeball cannot be too small, as too few photons would fall through it into the eye's interior, and the picture would be too "dark."

On the other hand, a large opening means that light rays from some point (detail) of the seen object undergo blurring on a relatively large area of the retina and therefore appear focusless. The split eye is present in primitive cephalopods alive now—nautiluses (it also appeared most probably in their close, extinct relatives—ammonites). This type of eye resembles, to some extent, a primitive camera called *camera obscura*, still devoid of the lens and having the form of a box with an opening in the front wall and a glass plate covered by a light-sensitive emulsion on the back wall.[6] Finally, the appearance of the lens focusing the light rays coming from (reflected by) some detail of the seen object on one determined point of the retina (together with the production of the cornea, protecting the eye, and iris, regulating the pupil diameter and amount of light falling into the eye) enabled forming sharp pictures with high resolution, that is seeing the details of the watched object. This stage, called the lens eye, currently appears in higher cephalopods and vertebrates. The lens eye resembles the modern camera: of course, the lens corresponds here to the lens, the iris—to the diaphragm, while the retina—to the photographic membrane with a light-sensitive emulsion or an electronic matrix with light-sensitive pixels in the digital camera. Only the mechanism of accommodation/focusing, ensuring the sharpness of the picture of once distant, once nearby objects, is different in the human eye and the camera. In the human eye, this occurs through a change of the lens curvature (shape), while in the camera—it occurs through a change of distance from the light-sensitive matrix. However, in some fish, the focusing mechanism is identical to the camera. Both in the lens eye and in the camera, the picture on the photosensitive layer (retina or membrane/matrix) is reversed "upside down" and "left-right" in relation to reality. We are unaware of this because of the brain's appropriate picture transformation.

As mentioned above, all main stages of the evolution of the eye (in particular, of the concave eye) are known in organisms living now, and all

---

[6] In order for the photograph to be light and sharp enough at the same time, the opening passing by the light had to be very small, and in the face of this the exposure time—very long. Indeed, it was necessary to pose motionless to a portrait for a couple of minutes!

of them are characterized by full functionality (although this does not mean that they create a perfect picture). However, it is also possible to simulate the concave eye evolution in a computer relatively simply. It was conducted by Nilsson and Pelger (1994) (see Dawkins 1996, 151-153). The computer program carrying out the *in silico* evolution of the eye contained some basic assumptions in its code. First, it took into account the elementary laws and parameters of optics, such as the law of light refraction, refraction coefficient and transparency. The criterion for selecting the generated eye structure plans was visual acuity, easily defined in purely physical terms. The properties of particular elements of the evolving eye (transparency, light refraction coefficient, shape, e.g., curvature) were encoded in the "genes" containing the values of particular parameters. These "genes" (parameter values) could mutate (change) in an accidental direction by a certain small value. Only those "eyes" characterized by the most acute vision were passed on to the next generation. They, in turn (the "genes" creating them), could further mutate and undergo selection in subsequent steps, that is, undergo normal Darwinian evolution.

The starting point for the virtual evolution was a flat light-sensitive spot on the body's surface (with a black layer impermeable for light beneath it). And what happened? After a few hundred generations, the light-sensitive layer began to curve and delve (herniate) into the body, adopting the shape of the hemisphere, which constituted an equivalent of the cup eye. Next, the hemisphere closed into a ball with an opening—the phase of the slit eye. In turn, the lens started to separate ("condense") in the opening, with an appropriately high light refraction coefficient. Finally, the cornea was formed outside the lens. Fewer than 1000 generations sufficed for this entire process, which is a ridiculously small period in the evolutionary (geological) scale of time (evolution had several million years to produce the cup eye and some tens of millions of years to produce the lens eye).

What is more important, the lens eye did not require any "intelligent projector"—the authors of the discussed computer model did not plan what the result of their *in silico* evolution would be in advance and only introduced elementary physical laws and the mechanism of natural selection into the computer program. They had no idea what would emerge from that. Nothing other than the commonly known lens emerged spontaneously and to some extent to the astonishment of the scientists who did not expect such a great success! Of course, it was much simpler than the real eye, but the computer program was also very simple. On the other hand, the time needed for the virtual evolution of the lens eye was incredibly short, much shorter than the millions of years it took to produce it in the real world. Let us pay attention to the fact that from the appearance of the first eyes in the early

Cambrian era to the origin of the lens eye, presumably in the late Cambrian, in the first Agnathan vertebrates, and maybe later, some tens of millions of years passed.

However, let us deal at last with the signalized "evolutionary mistake", which is the convex eye in arthropods (namely extinct trilobites, eurypterids and forms of unclear relations, such as *Anomalocaris*,[7] and crustaceans, arachnids, and insects living today; trilobites and anomalocarids are sometimes counted as crustaceans, while eurypterids belong to chelicerates). Its general evolution is presented in the upper part of Fig. 5-3. The starting point for the evolution of the convex eye, as in the case of the concave eye, was most probably a flat layer of light-sensitive cells forming a spot on the body surface. Supposedly, a simple convex eye originated, capable only of detecting the direction of the light falling. We are unaware of current or fossil organisms representing this hypothetical stage of evolution of the convex eye.[8] The next and final major phase was the complex eye, composed of many simple eyes (ommatidia), allowing only for forming blurry, coarse, "grainy" images devoid of detail. This mostly results from the size of the ommatidia (sing. ommatidium), which constitute the "perception unit," equivalent in the complex eye to the pixel of the picture, and which are complex structures composed of many cells. Their amount (corresponding to the number of "pixels" proportional to the resolution of the image) in general does not exceed several thousand. In dragonflies, the record holders for this, the eyes can number thirty thousand, but the eyes cover at least half of the dragonfly's head!

This remains in great contrast to the lens eye of humans (not to mention the eye of birds of prey, having even much better resolution), where the pixels of the picture correspond to single photo-sensitive cells, rods and cones, whose number in the retina is over one hundred million (the huge majority are rods). Therefore, the acuity and resolution of the picture created by the lens eye, the number of details recognized by it, is incomparably greater than in the complex eye. It has been calculated that in a human, the complex eye would have a diameter of one meter to possess such picture resolution (Dawkins 1996, 167) rather than the diameter of the lens eyes (of the eyeball) of around 3 cm, which evolution has equipped us with. Arthropods simply had big evolutionary bad luck that their ancestors "chose" due to completely random events (mutations), the "convex" way of eye evolution and could no longer turn back from it. In the beginning, this way brought significant benefits (it is better to see something than nothing),

---

[7] A two-meter *Anomalocaris* was the apex predator of the Cambrian seas.

[8] Perhaps the complex eye evolved through multiplication of the simple convex eye—one ommatidium would correspond to a single simple eye.

but in the end, it constituted a constructional-functional trap, a "blind" or, in fact, "dim-sighted" alley of evolution. This did not prevent arthropods from achieving great evolutionary success, as with their usually small dimensions, eyes with high resolution are not crucial, and their lack was compensated by other features useful in the occupation of many ecological niches.

Both the complex eye and lens eye are characterized by undoubted purposefulness and represent a result of biological evolution directed at the maximization of fitness, in this case, navigating in the environment more efficiently. Nevertheless, the lens eye is certainly a more fortunate, optimal and efficient solution for performing its function according to any rational and "objective" estimation. This results from the fact that ancestors of contemporary, say, insects and mammals entered, through pure chance, this particular path of eye evolution (convex or concave eye), and afterward, they had to follow it through. They could not predict what the ultimate result of their "choice" could be. This constitutes a subsequent example of the thesis that evolution is cumulative. Though, of course, insects could not, because of their size, possess lens eyes of the same resolution as we have, even if the evolution of the eye in their ancestors took another (accidental) course.

We come here to the last problem dealt with in this subchapter, perfectionism, which is possibly the most efficient functioning of a given organ. For instance, is the highest visual acuity always desired regarding evolutionary success, which is the maximization of fitness, in cases where it could be achieved? Of course, the answer is: no. Only the balance of gains and losses, which involves functionality, energetic investments needed for producing and maintaining a given organ and profits brought by it, counts. In optimizing the structure and function of living organisms, the rule "something for something" is in force. Instead of allocating energy and building material for producing super-efficient eyes, it is possible to use them to produce, e.g., a greater number of better-equipped progeny or improve the functioning of other organs, more needed in a given ecological niche. Let us consider a snail (see Dawkins 1996, 129-130). If it were lucky enough to receive evolutionary heritage as advanced lens eyes, then in order to see equally well as we do, it would have to have eyes comparable to human eyes (here, the minimal size of the eyes with a given resolution of the picture, that is the number of pixels, is determined to a large extent by the size of photo-sensitive elements = cells). With its dimensions, the snail would have to draw them behind it on the eye antennae. Furthermore, even if it were possible to miniaturize them in some (miraculous) way, the snail would have no profit from this—seeing a predator from a hundred meters away, it could only become stressed, as at its speed, it would not be able to escape from the danger. Therefore, instead of having the hawk's sight, it pays to invest energy and building material into developing other properties (e.g., a shell with thicker walls), increasing its chances of survival and reproduction in its ecological niche.

General evolution rules

It is frequently impossible to derive in a trivial and obvious way the causal relationship between the mechanisms underlying evolution, chiefly natural selection, on the one hand, and its general rules, regularities and course, on the other. Therefore, below I will discuss several basic rules of the process of evolution, emerging in a sense on a "higher" level from the mechanisms responsible for it at a "lower" level. I will start by enumerating them. Thus, here are some of the important "macroscopic" properties of the phenomenon of biological evolution:

- cumulativeness
- irreversibility
- directionality
- possibility of complex degree increase
- a variable pace
- the diverging (branching) of evolutionary lines
- great periodical extinctions.

Cumulativeness was mentioned a few times on different occasions (and will be mentioned again in the discussion of subsequent problems, as it is a property whose role is difficult to overestimate for understanding evolution and interpreting its products). Evolutionary changes occur mostly due to the gradual accumulation of usually small genetic modifications and corresponding phenotypic changes. Each subsequent change is based on the organism's existing structure and function. Evolution cannot proceed in any direction—the existing construction and functioning of the living being in the biochemical, physiological, embryonic development, and behavioral aspects determine a certain (still huge) range of the possibilities of the future development routes. The evolution of a given living organism cannot reject its hitherto history, start itself once again from the very beginning, and plan its further stages from the foundation.

On the contrary, it can act only through the modification and development of what already exists—the present structure and function of an individual fixes a certain limited, although still very broad, spectrum of the potential ways of further transforming its identity from generation to generation. The cumulativeness is related to the *continuity* of evolution, the occurrence of gradual, step-by-step changes of quantitative properties, a determined sequence of which can ultimately lead, due to cumulation, to a large qualitative change (in the construction scheme and or degree of complexity). On the other hand, we do not know any documented large sudden, saltatory changes in the structure and function of living individuals, although at the genetic level, chromosomal and genomic mutations, which can instantaneously lead to speciation (mutated and unmutated individuals possessing a different number or structure of chromosomes, cannot efficiently crossbreed), can be regarded as such discrete changes.

We met the property of cumulativeness when discussing the evolution of the eye, where the organisms that were unlucky to enter the "convex" way of evolution had to continue with it and exploit it to the end, or the genesis of the structure and function of the human brain, in particular the cerebral cortex, where new areas were superimposed on old parts, with no rhyme or reason, in an accidental direction. Let us consider a few subsequent, rather simple examples. A mutation or a change in the color of a flower can occur as the flower has already evolved. The limbs of land vertebrates could originate because their ancestors—crossopterygians—possessed a skeleton of the pectoral and ventral fins shaped in a such-and-not-another way (the skeleton of paired fins of those fish already possessed analogs of humerus/femur, radius/tibia, ulna/fibula, hand/foot bones of land vertebrates). Bats could not produce a pair of wings in addition to two pairs of legs for walking, as their ancestors—ground-based mammals—had only two pairs of limbs. They also cannot efficiently walk on two limbs, like birds, as their ancestors were not bipedal, like bird-like dinosaurs, but quadrupedal and glided with the aid of skin lappets outstretched between limbs, similarly to the present gliders (*Dermaptera*).

The cumulativeness of evolution is frequently reflected in the course of embryonic development. The presence in mammalian fetuses of gill slits and arches that undergo later transformation into the middle ear and auditory ossicles can serve as one of the numerous examples. This constitutes a derivative of the fact that the gill arches of agnathan vertebrates (*Agnatha*)[9] were adapted in the course of biological evolution for various purposes and transformed into the bones of the mandible and jaw, hyoid cartilage, larynx and trachea cartilages and three auditory ossicles of the middle ear: malleus, incus and stapes (Kardong 2002, 236-237; Reece et al.

---

[9] Today lampreys and hagfish, and conodonts, kin to them, existing from the Cambrian to the late Triassic and armored jawless fish (*Ostracodermi*) common in the Silurian and Devonian belong to them.

2011, 705, 721; DeLaurier 2018). The general conclusion is that evolution, when it can, uses and modifies what already exists instead of creating something completely new.

The *irreversibility* of evolution consists of the fact that evolutionary transformations never return (apart from the simplest cases) to the starting state. This results, among others, from cumulativeness—it is very difficult to repeat hundreds and thousands of evolutionary changes in exactly the same order, but in the opposite direction. It is possible to reverse, e.g., a single point mutation (so-called back mutations). It is a little more difficult to reverse simple adaptive changes (e.g., reduction of eyes in the mole due to the underground living style or shortening the fur in some mammals due to the warming of the climate, although here the "backward" evolution will not pass the same stages in the same reversed order, which were previously passed by the "forward" evolution). On the other hand, it does not pay at all to get rid of "overall-improving" properties that are useful and elevate the efficiency of functioning in very different conditions/environments/ ecological niches, such as the big brain, four-chamber heart, viviparity and homeothermy in mammals.

The *directionality* of evolution expresses itself in the fact that in different developmental lines, a certain invariable evolutionary trend consisting of a constant and gradual increase (or decrease) of the intensity of a given property (set of properties) can persist for a long time (millions or tens of millions of years). For instance, the evolutionary history of modern horses (genus *Equus*) is characterized by a constant increase in body size (the Eocene ancestor of horses was the size of a medium dog) and a reduction of the number of fingers and toes from five to one (retaining the medium, third one) and an increase of the size of the hoof, with which this finger/toe is terminated. This was very well documented based on fossils from the developmental line of horses from different eras (in fact, it was not a line, but rather a tree with numerous, although rather short, branches). Supposedly, a gradual lengthening of the neck took place among subsequent ancestors of giraffes, although here, the fossil record is much poorer. However, the only living relative of a giraffe, namely the okapi, seems to represent an intermediate stage—it has a relatively long neck but is much shorter than a giraffe's. The constant trend toward increasing body dimensions also occurred in many dinosaur groups. However, while directionality occurs quite commonly, it does not constitute a universal rule of evolution—it is not observed in many lines.

A subsequent property of the process of biological evolution is the possibility of increasing the complexity of the structure and function of living organisms. Namely, in many evolutionary lines living individuals (their identities) were becoming more and more complicated. After all, humans are more complex than an archaeon, from which they originated over several billions of years of evolutionary transformations. Here are a couple of examples of evolutionary phases (many more can be quoted, in particular among higher plants and animals) of a degree of complexity much higher than the phases preceding them:

⇒ hypercycles in coacervates;
⇒ first cells;
⇒ LUCA;
⇒ prokaryotes;
⇒ eukaryotes (endosymbiosis);
⇒ multicellular organisms;
⇒ organisms composed of tissues and organs;
⇒ the human brain.

However, it should be mentioned that "possibility" does not mean "necessity". A common opinion is that a clear trend toward increased complexity is present in biological evolution. This is completely untrue. Evolution leans toward the maximization of fitness and not complexity. More complex organisms do not have to be "better" from the point of view of biological efficiency. Judging from the morphology of fossils from the Archaic eon (Archaeozoic, 4.0-2.5 billion years ago), prokaryotic organisms (bacteria and archaea) did not change regarding their general appearance for at least the last 3.5 billion years. Even today, the greatest circulation of matter and energy in the biosphere is carried out by photosynthetic bacteria living in the oceans (and descendants of some of them—chloroplasts). They also supply most of the oxygen on our planet. The most common (most numerous) organism on the earth is an oceanic bacterium, *Pelagibacter,* feeding on organic matter suspended in water. If a nuclear war started or a great asteroid collided with our planet, perhaps all "higher" forms of life, including humans, would die out, and only bacteria and, especially, archaea "accustomed" to various extreme conditions would survive. The fact that *Homo sapiens* situate themselves on the top of the "ladder of life," celebrating themselves as the culmination of evolution or "crowning of creation," constitutes a manifestation of extreme anthropocentrism. According to biological-evolutionary criteria, the "best" is not the most complex (possibly intelligent and self-conscious) but what will survive and leave progeny.

Additionally, a decrease in the degree of complexity can occur in the evolution process, provided that the ecological niche of a given individual allows this or even promotes it (the maintenance of unnecessary complexity means unwanted energetic and material expenses). For instance, regress occurs in parasites, with a decrease in the complexity of structure and function, which is associated with ceding many vital functions to the host. In the extreme case of viruses and viroids (see Chapter 6), all life functions have been transferred to the host cell. In moles, the reduction of eyes occurred because of the subterranean lifestyle, while in axolotl—not only the decline of the eyes but also skin pigment due to living in water reservoirs in caves, where sunlight does not reach, took place.

Generally, an increase in the complexity of organisms in the course of biological evolution can occur (when they "discover" an appropriate, still unoccupied ecological niche), but it does not have to—this is no common nor unavoidable trend in biological evolution toward an increase of complexity. "More complex" does not automatically mean "better fitted." In many lines, increased complexity results from the fact that life "started" from simple forms (which seems obvious) and "struggled" to fill, through random evolutionary "wandering", all accessible "ways of existence" or *modi vivendi*, namely, the ecological niches. Some of them are niches for complex multi-cellular tissue organisms "equipped with" systems of organs, however, decidedly not all, nor even most of them. Taking a look at the dazzling diversity of (construction planes of) unicellular organisms, such as prokaryotes, various kinds of flagellates and amoebas, ciliates, radiolarians, cryptomonads, green algae, yeasts, diatoms and so on (frequently having quite complex structure), in order to see that "searching" for new ecological niches frequently did not occur in the direction of multicellularity and tissue structure is sufficient.

A variable pace also characterizes evolution. An example of a slow evolution pace is the morphology (shape, size) of bacteria (and archaea), which has evidently not changed significantly for over three billion years, or the most known "living fossil": coelacanth (*Latimeria*)—a crossopterygian fish that does not differ significantly from its Devonian ancestors from before about 400 million years ago. The reason for evolutionary constancy in the first case can be from achieving the simplest structurally and functionally evolutionarily stable phase of free-living (not parasitic) organisms in the real physical-biological environment (Korzeniewski 2005). In the second case, on the other hand, it is the extreme permanence and monotony of the environment in space and time (oceanic depths) associated with a very low frequency of mutations and no directional selection pressure.

There are also known examples of the unusually high pace of evolution. In order not to look too far, the human evolved from a form slightly differing from today's chimpanzee (the last common ancestor of humans and chimpanzees probably resembled the fossil genus *Sahelanthopus*) very quickly in the geological scale of time, over six-seven million years, although these organisms differ from each other fundamentally. Presumably, this was favored by fast environmental changes caused by climate cooling, self-driving systems: brain + prehensile upper limb + stereoscopic sight and the evolution of social behavior, language, culture and self-consciousness (Korzeniewski 2020) (social relations had a particular significance for the size and structure of the brain, and the shape of the mind, the "social mind", Dunbar 1998, 2005, 2016). Another example is the "explosion of life" in the Cambrian era, during which very many various construction plans of "higher" animals (*Eumetazoa*), including chordates, suddenly appeared "out of nowhere". The reason for this event is not exactly known, although it is supposed that it was caused, at least partly, by passing, due to random mutations, through a kind of evolutionary "bottleneck."

Evolution can proceed at a changeable pace even in the same developmental lines, sometimes accelerating and sometimes slowing down for a long time, essentially to zero (Futuyma 2013, 605-611). An extreme expression of this fact is the concept called *punctualism* or *punctuated equilibrium,* proposed by Eldredge and Gould (1972). Accordingly, episodes of violent, qualitative *punctualistic* changes in the structure and function of organisms are separated by long periods of *stasis*, meaning practically the lack of any changes. Today it is commonly thought that this concept contains a germ of truth, although it seems a rather exaggerated and simplified description of reality (Futuyma 2013, 605-611).

If we investigate the course of evolutionary changes of various species or larger systematic groups in time, their history will be arranged into an *evolutionary tree*. A simple schematic example of such a tree is demonstrated in Fig. 5-4.

The evolutionary development of some group of organisms (both the whole living world and, e.g., arthropods, chordates, placental mammals) always starts from the common ancestor of a given group. Due to *adaptive radiation,* many developmental lines (species or higher systematic units) originate from this ancestor. Some of these lines undergo gradual changes in time, not dividing into "descendant" lines. Others, on the contrary, undergo branching into many lines. Sometimes the simultaneous differentiation into particular developmental lines is so intense that a subsequent "secondary" adaptive radiation (such radiations are marked by small circles in Fig. 5-4) takes place. For example, after the "primary" adaptive radiation of placental mammals, a sequence of "secondary" adaptive radiations within predatory mammals, hoofed mammals, bats and so on occurred. Some developmental lines become extinct with time. This happens for different reasons.

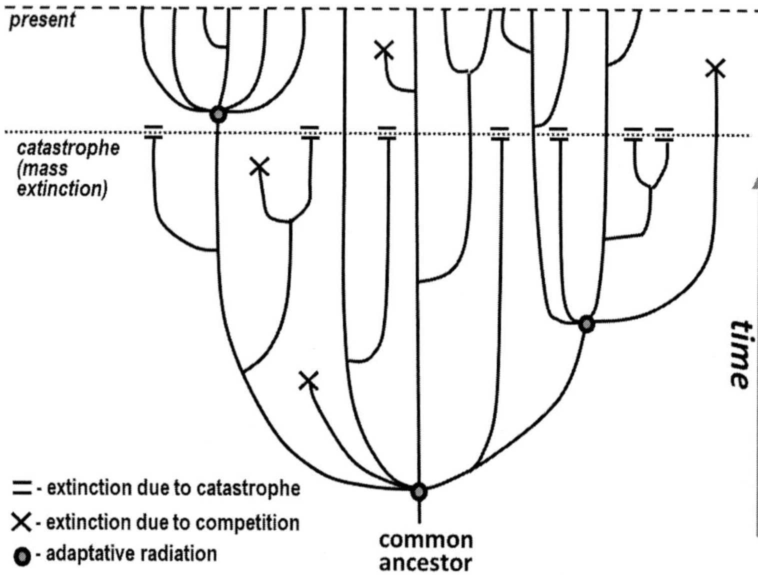

Fig. 5-4. Schematic evolutionary tree of living organisms. Particular developmental lines correspond to either species or greater systematic units.

Cataclysms, having their source outside the biosphere, for instance, in the cosmic space or the earth's crust plates tectonics, belong to the most spectacular of them. We know the five greatest documented extinctions of living organisms in the history of our planet (humans are carrying out the sixth): at the end of the Ordovician, Devonian, Permian, Triassic and Cretaceous. The largest of them, in which over 90% of living forms vanished from the surface of our planet, when, for example, the last trilobites (already reduced in number by earlier extinctions), many other sea forms (e.g., eurypterids, acanthodians) and landforms (many insect groups, predatory mammal-like reptilians) became extinct. The best known, all non-avian dinosaurs became extinct then (but also flying reptiles—pterosaurs, sea reptiles—plesiosaurs and mosasaurs and ammonites), is the extinction from the break of the Cretaceous and Tertiary (Paleogene) eras. The main reason for the extinction on the Cretaceous-Tertiary border was an impact on the area of the present Mexican Gulf (near the Yucatan peninsula) of an asteroid with a diameter of about 10 km (it left a crater of about 180 km in diameter, now covered with oceanic sediments, whose age, ~65.5 million years, coincides well with the time of this extinction) (Fastowsky and Weishampel 2016, 414-446). In most of the remaining cases, much activity in the earth's crust tectonics was the leading reason for the mass extinction, especially large-scale lava outflows. During the Permian-Triassic extinction, the so-called Siberian traps (composed of basalt, that is, consolidated magma) were formed, at a depth of several kilometers, covering huge areas of present Siberia. Great lava outflows—present now as deposits of basalts in various parts of the world—were also temporarily correlated with the remaining mass extinctions. Even in the hecatomb of living beings at the Cretaceous-Tertiary boundary, these lava outflows in the present Indian sub-continent (then a separate continent) are preserved as Deccan traps. It is unclear whether these outflows resulted from the asteroid impact or, by pure chance, coincided temporarily with that event (Macdougall 1996; Futuyma 2013, 103-134; Knoll 2021).

On the other hand, a huge impact crater (about 480 km in diameter) was recently discovered under the ice cover on Wilkes Land in Antarctica, whose age (~251 million years) correlates very well with the time of the Permian-Triassic extinction. It is likely that in both cases (the Permian-Triassic extinction and the Cretaceous-Tertiary extinction), the shock waves caused by the asteroid impact were propagated along the earth's crust and met in the point on the surface of our planet opposite the impact, cumulated and caused huge lava eruptions. In fact, in both cases, the places of lava outflows were located at that time (remember the plate tectonics and continent drift) on the opposite "end" of the terrestrial globe than the impact point. On the other hand, the lava outflows (basalt deposits) around the present Atlantic ocean in Africa, South America and North America (the so-called Central Atlantic Magmatic Province, CAMP) were associated with the Pangea super-continent breakdown (leading to the Atlantic Ocean opening) and the end-Triassic mass extinction some 200 million years ago. So far, there is no evidence that this event was related to some asteroid impact.

Catastrophes of this kind usually cause the extinction of many developmental lines that are frequently unrelated. The mentioned cataclysm on the Cretaceous/Tertiary boundary killed not only (non-avian) dinosaurs but also pterosaurs, plesiosaurs, mosasaurs and ammonites (belonging to cephalopods), and many other groups. The reason for this group's extinction was most probably excessive specialization that did not allow for a quick adaptation to the new environmental conditions after the dramatic change of the latter (cutting off sunlight by huge dust clouds and temporary blocking of photosynthesis leading to plant extinction, acidic rains, climate cooling and so on). One such catastrophe causing the simultaneous extinction of many lines is presented in Fig. 5-4 (marked by a double short horizontal line).

Even relatively small changes in the physical environment, for instance, climate, can significantly affect the course of evolution. We remember the case of the giant deer, which became extinct due to climate warming during an inter-glacial period. On the other hand, glacial periods favored the origination and existence of such species adapted to cold climates as the wooly mammoth or wooly rhinoceros. Our ancestors came down from trees to the ground due to climate change in Eastern Africa from a wetter and warmer climate to a cooler and drier one, which was associated with the replacement of forest areas by savannah. The Sahara Desert did not exist a few thousand years ago—the sudden drying of the land in its present area led to the extinction of the species living there and, consequently, the entire ecosystem.

The extinction of other developmental lines can take place due to competition with other organisms. Crosses mark the cases of this kind in Fig. 5-4. They are characterized by the fact that the dying out of different lines is not correlated with time. Huge, flightless predatory birds (somewhat similar to ostriches, but with a much greater head and beak), phorusrhacids ("terror birds") occupied the ecological niche of large land predators in the area of South America for a long time (from about 60 to two million years ago). Only the emergence from the ocean of the Panama Straight, connecting South America with North America, resulted in an invasion of modern predatory mammals (cat- and dog-like) from the north, which, as more efficient, ousted phorusrhacids, causing their complete extinction (Macdougall 1996; Futuyma 2013, 103-134; Knoll 2021).

The present variety of living organisms, representing those developmental lines that reached the "present" in Fig. 5-4, therefore constitute the result of the separation of particular lines into descendant lines, gradual, frequently directional transformations within particular lines or the extinction of some lines because of violent cataclysms, slower changes of the environment or competition with other organisms.

Evo-devo, that is, morphogenesis vs. evolution

In "evo-devo", "evo" stands for evolution, while "devo" stands for development. Namely, this is embryonic development (embryogenesis) and individual development (ontogenesis). Therefore, "evo-devo" is a popular abbreviation for the branch of science (biology) dealing with the relations between biological evolution and forming the structure of the organism (morphogenesis) during embryo- and ontogenesis (Maynard Smith and Szathmáry 1995, 241-254; Hall 1999; Futuyma 2013, 565-604). This means that the matter is the influence of changes in the morphogenetic cascade (see below) caused by mutations in regulatory genes (see below) on modifying the structure of an adult individual, changes in its construction plan and a possible increase in complexity.

We already understand that the structure of biological systems, in particular of complex (tissue) multi-cellular organisms, underwent modifications and (in some cases) increased its degree of complexity due to biological evolution. Nevertheless, what did this consist of? Chapter 4 seems to give a simple answer to the question of where the entire diversity and complexity of the structure of living individuals comes from—the basic mechanism responsible for this is natural selection. Is this explanation completely correct? Yes, and no. It is certainly not exhaustive. As mentioned, the problem is that Darwinian natural selection can only choose from what already exists; it cannot generate (new versions of) structure and complexity.

Therefore, the question is what causes the modification and possible complexity increase in multi-cellular organisms. If, in fact, natural selection can only select from what is proffered but cannot generate any structure or new construction plans, and in particular, an increase of complexity, then what mechanism can do so? Thanks to huge achievements in the field of embryology (and, more generally, developmental biology), we know that the complicated structure of complex multicellular organisms originates during embryonal (and individual) development due to the action of a hierarchically-organized cascade of morphogenetic (structure-creating) processes (Hall 1999, 94-96; 102-103; Twyman 2001, 38-44). The answer to our question is the statement that these are mutations in regulatory genes (encoding transcription factors) (Alberts et al. 1998; Twyman 2001, 69-80 Reece et al. 2011, 1280-1283) controlling appropriate elements of this cascade that can lead to a modification and or increase in the degree of complexity of structure.

Nevertheless, the problem is broader. In order to understand the mechanisms underlying the modification and increase in structural complexity of living organisms, one should first consider the structure and function of a single cell and also its genesis during its growth, division and cell differentiation. This will help us understand how various cell types evolved, both in uni- and multicellular organisms. It should be realized how forming the complex (first of all polymeric) organic compounds is possible, how they get organized, after being synthesized in a proper place, time and amount, into various complexes of macromolecules (ribosomes, chromosomes, elements of cytoskeleton, lipid-protein membranes), into larger cellular structures (organelles, such as the nucleus, mitochondria, chloroplasts, lysosomes, Golgi apparatus or endoplasmic reticulum), and eventually into whole cells, and first how the entire adult organism in multi-cellular and tissue forms, structurally and functionally complicated, develops from the fertilized egg cell. There are mutations in genes encoding various proteins, especially regulatory proteins (and so-called miRNA—micro RNA) (Maynard Smith and Szathmáry 1995, 219-223; Twyman 2001, 26-36, 69-80; Reece et al. 2011, 1280-1283; Futuyma 2013, 575-578), involved in the regulation of gene expression,[10] responsible for growth, divisions, organizing and differentiation of living cells and whole multi-cellular (tissue, containing organs and their systems) organisms, that can either modify the existing structure of both a single cell of a given type and of whole multicellular organisms or possibly lead to an increase of the degree of complexity of this structure (e.g., increase the number of the levels of organization hierarchy).

On the other hand, the sieve of natural selection can only pass the new version of the structure (and function) of an organism or not. Very frequently, it is forgotten that what has already originated due to evolution, namely a particular sequence (cascade) of morphogenetic procedures (leading to forming the structure of an adult organism), constitutes the starting point for further evolution, determining potential ways it could follow. Mutations in regulatory genes, able to disturb the course of morphogenesis (its particular stages), constitute an at least equally important mechanism of macroscopic evolution of structure and function, as the famous natural selection.

The problem can be presented in the following short way. *Morphogenesis* is based on the *morphogenetic cascade*, a complex of sequentially-initiated developmental procedures, with instructions contained in the regulatory genes. The cascade is organized hierarchically—

---

[10] Segments of DNA binding regulatory proteins (transcription factors) that regulate a given gene expression are also important. *Enhancers* bind *activators* that stimulate gene expression, while *silencers* bind *repressors* that inhibit gene expression. The same regulatory protein (transcription factor) can serve as an activator and or repressor of various genes. Enhancers and silencers can be situated outside the gene, whose activity they regulated, or within the introns of this gene (Murray et al. 2012, 384-387, 422-426). Mutations in enhancers and silencers, affecting their efficiency of binding of transcription factors and strength of affecting of gene expression, can also significantly affect the gene regulation cascade during cell and full organism development.

morphogenetic centers on a higher level initiate forming morphogenetic centers and body structures on a lower level (Wolpert 1981; Maynard Smith and Szathmáry 1995, 225-240; Twyman 2001, 30-32). The morphogenetic centers produce chemical substances (usually proteins)—*morphogens*—that supervise forming structures of a given hierarchy level and lower-level morphogenetic centers (Wolpert 1981; Maynard Smith and Szathmáry 1995, 219-223; Slack 2001, 67-71; Twyman 2001, 30-32). Due to diffusion, gradients of morphogen concentrations originate—their concentration is lower, the greater the distance from the centers producing them (this concentration dependence on the distance is exponential) (Wolpert 1981; Maynard Smith and Szathmáry 1995, 234-240; Slack 2001, 69-71; Twyman 2001, 30-32). Forming secondary morphogenetic centers and body structures in an appropriate place and time is determined by the combination of concentrations of morphogens produced by primary (in relation to them) morphogenetic centers. This is the so-called *positional information* (Wolpert 1981; Maynard Smith and Szathmáry 1995, 234-240; Twyman 2001, 30). The centers being secondary in relation to certain centers can be primary (situated higher in the hierarchy) in relation to other centers (determine their formation through the positional information resulting from the ratios of the concentrations of secondary morphogens released by them). All this will be discussed in detail below.

It is relatively easy (which does not mean that it is simple) to answer the question concerning the genesis of the structure of the cell during the life cycle in the case of unicellular, especially prokaryotic, organisms. The adaptation to the current environment conditions consists in the improvement (due to mutations and selections) of the existing genes, both those encoding enzymes and structural proteins, and those coding for regulatory proteins (the transcription factors) responsible for the control of expression of other genes (miRNA also participates in this process), that is determining the intensity of the synthesis in a given place and time of the proteins encoded by them. Because of their amino acid sequence, the newly produced polypeptides can sometimes, with the help of other previously synthesized proteins (chaperons, proteases), fold their unidimensional chain into a three-dimensional native structure and possibly join with other proteins. They can incorporate themselves into the two-layer cellular or organelle membrane and possibly pass during transcription or just after its termination through such a membrane on its other side—special carriers (yet other proteins) are responsible for the transport of newly synthesized proteins through appropriate membranes within the cell[11]—and only then adopt the native three-dimensional conformation. The self-formation of proteins' spatial structure (associated with the appropriate function) constitutes the crucial transition from a unidimensional—sequential—record of information (sequence of amino acids) to a three-dimensional structure and function. Other organic compounds (including DNA and RNA, lipids and sugars) are synthesized, taken up from outside, transformed and incorporated into appropriate cellular structures or transported to appropriate compartments with the participation of proteins, especially enzymes and membrane transporters.

In unicellular organisms, the origin of new genes (especially through copying and differentiation of the existing genes), but also the "horizontal gene transfer" between individuals of the same, or even different species, with the mediation of plasmids and viruses sometimes takes place. However, the modifications of structures related to the last process are rather small. Indeed, as fossil data indicate, bacteria and archaea's morphology and degree of complexity did not change significantly from the moment of their origin, over three and a half billion years ago. More pronounced changes—in the form of the origination of many new genes, the appearance of complex cellular structures, and also the re-construction of the entire genetic apparatus responsible for copying and reading the genetic information and the regulation of this reading, and for separation of DNA to descendants during cell divisions—had to take place in the case of creating nucleus-containing organisms (eukaryotes) from devoid-of-nucleus organisms (prokaryotes), to which bacteria and archaea belong (it is supposed that eukaryotes evolved from the latter). Here, a great increase in both the degree of complexity and diversity of construction plans of the cell (not to mention its size) already took place. In some unicellular eukaryotes, a "promise" of higher organization levels, such as organs and

[11] The transport of newly synthesized proteins to an appropriate compartment in the cell (incorporation into a proper lipid-protein membrane or transport to another compartment, for instance endoplasmic reticulum or proper organelle, e.g., mitochondria) is enabled by the so-called signal sequence (Murray et al. 2012, 555-558) on one of the ends of the polypeptide chain, which is cut off after fulfilling its function. Proteins that just after the termination of translation get to the cisterns of endoplasmic reticulum can be transported by "bubbles" constituting "budded off" fragments of its lipid-protein membrane to the Golgi apparatus cisterns, to which these bubbles are incorporated, releasing their content to its interior. The Golgi apparatus, in turn, produces its "own" bubbles, which can be incorporated into the cellular or organelle membrane, thus compensating the losses of its substance or increasing the size of its surface (material) (Alberts et al. 1998, 104-124, 462-466).

their systems, appearing in complex multicellular animals can already be seen in the structure and function of a single cell. For instance, a paramecium, belonging to ciliates, possesses such structures as a "throat" (peristome), "mouth" (cytostome), "digestive tract" (food vacuoles, an analog of the stomach and guts), "anus" (cytoproct), "kidney" (contractile vacuoles, to which radiating canals lead; an analog of the excretory system), the "nervous system" (a network of proteins called neurofibrils lying beneath the surface of the cell and coordinating the work of cilia, which cover the body of the paramecium and propel it using synchronized movements) (Reece et al. 2011, 584-585). Although the cellular level is undoubtedly situated much higher on the ladder of structural complexity and function than the molecular level, we know quite a lot about how the former emerges from the latter, from the cooperation of its elements. Even the entering into endosymbiosis of the ancestors of mitochondria and chloroplasts with the eukaryotes-to-be and the origin of the simplest multicellular organisms could be initiated by one of a few mutations, leading, respectively, to a lack of digestion of the incepted bacteria and the lack of separation of cells after divisions.

We understand quite well how the complex elements of eukaryotic cells, and consequently—entire cells, are formed. For instance, the proteins localized in the basic cytosol are synthesized in ribosomes located in this cytosol. The proteins entering into the composition of various organelles are produced by ribosomes attached to the rough endoplasmic reticulum, they pass to the interior (cisterns) of this reticulum before adopting a native three-dimensional conformation by the unidimensional protein chain, they are transferred within bubbles surrounded by double-layer lipid-protein membrane to Golgi apparatus (whose function is modifying substances produced by the cell, e.g., proteins or lipids, their sorting and distribution within the cell), and are next transported to their destination (appropriate organelles or outside the cell in the case of exocytosis) in small bubbles (vesicles) that separate them from the dictyosomes (cisterns) of this apparatus (Alberts et al. 1998, 104-124, 462-4660. We know the mechanisms responsible for forming such bubbles and their separation from the "maternal" membrane and fusion with the "destination" membrane. (If we deal with releasing the content of a bubble outside the cell, we speak about *exocytosis*; the incepting of substances or objects, e.g., microorganisms constituting food, from the outside and closing them within a bubble is called *endocytosis* (Alberts et al. 1998, 472-477). The proteins entering the composition of mitochondria and chloroplasts are either synthesized within them (on the matrix of their own DNA, using their own transcription-translation apparatus) or in the cytosol and then transported to the interior of these organelles by appropriate protein pores. We also know much about forming such cellular structures as various lipid-protein membranes (the Golgi apparatus plays a key role here), chromosomes or cytoskeleton elements (Alberts et al. 1998; Reece et al. 2011). The principles of building and maintaining the structure of the cells of complex tissue organisms do not differ basically from those met in unicellular eukaryotes. Appropriate enzymes synthesize lipids, sugars and other organic compounds. The creation of the structure and function of the cell, both in uni- and multicellular organisms, should therefore be regarded, to a large extent, as explained by science. The same concerns the mechanisms underlying cell division (*cytokinesis*), nucleus division (*karyokinesis*) and the separation of chromosomes to descendant nuclei (Alberts et al. 1998, 549-571; Reece et al. 2011, 228-255).

Generally speaking, both the maintenance and production of the structure and function of the cells, chiefly during the differentiation of cells into different cell types during embryonic development in "higher" plants and animals are steered by regulatory genes. They encode regulatory proteins (also called *transcription factors*, activators or repressors), which, binding to DNA (its segments called enhancers or silencers) near various, strictly specified genes (frequently within introns of these genes; Futuyma 2013, 578-584), activate or inhibit their transcription into mRNA, frequently co-operating with other proteins (factors) of this type. In multicellular eukaryotic organisms, in particular, higher plants and animals, cascades of regulatory genes are ubiquitous (or: *cascades of the regulation of gene expression*), in which one such gene on a higher level of hierarchy influences the expression of many regulatory genes on a lower level of the hierarchy, switching it on or off, or modulating its intensity (Maynard Smith and Szathmáry 1995, 219-240; Alberts et al., 1998, 257-273; Twyman 2001, 8-12). Such cascades are multi-level, hierarchically-organized—genes on a higher level of hierarchy influence the expression of genes on a lower level. It is worth adding here that the expression of particular genes is regulated (activated or inhibited) by many different regulatory proteins (transcription factors). Therefore, the cascade of the regulation of gene expression resembles a network. On the lowest level, enhanced or silenced genes encoding "structural"[12] proteins, responsible for the structure and function of the cell, are present. In this way, an activation/inhibition of one (or a few) gene(s) on the highest level of the hierarchy can direct (switch) the differentiation of the embryo cell within a complex organism in the

[12] In fact, they should be called "structural-functional" (as opposed to "regulatory") proteins, especially since a large part of them are enzymes, carriers *etc*.

direction of an, e.g., epidermal, nervous or muscle cell, as such activation/inhibition affects the intensity of the synthesis of many proteins, and thus can determine a huge amount of cell properties. Many regulatory DNA sequences encode miRNA, which does not serve as a matrix for protein synthesis. miRNA (micro-RNA, small RNA strand containing 22 nucleotides) is partly (or completely) complementary to short fragments of mRNA of many genes. By binding them, miRNA disenables or limits (slows down) the translation process blocking the synthesis of appropriate proteins on the matrix of those mRNA strands (Maynard Smith and Szathmáry 1995, 219-223; Reece et al. 2011, 1280). Of course, we can suppose with a large dose of probability that the modification and complication of the structure and function of cells (especially eukaryotic) in the course of biological evolution was based on a modification and development of the system of steering gene expression based on regulatory proteins (transcription factors) and miRNA.

Therefore, the understanding of both the construction of complex cellular structures during the life cycle of particular cells and entire uni- and multicellular organisms and the gradual origination and complication of these structures during biological evolution does not meet any significant conceptual problems, which does not mean, however, that we know all the details of these processes.

On the other hand, it is more difficult to explain the evolution of structure (and function) of more complex (multi-cellular, containing tissues, organs and systems of organs) living organisms constituting macroscopic objects that reach far above the molecular and cellular level. For this purpose, we must discuss the most general principles of these organisms' embryonic development. The transition from the fertilized egg cell to an adult individual does not resemble either the building of a house from fundaments to the roof or the construction of a car, where previously produced sub-elements are assembled into elements; these are combined into bigger parts into sub-systems, and finally into complete systems that ultimately make the full car. As mentioned, the embryonic development of a biological individual consists in the self-construction of a little micro-robot—the zygote—which self-develops, frequently with the significant help of its mother (which, however, does not participate in the determination of the course of the entire process, but only supports it), building its body from substances available in the surroundings (e.g., deposited in the egg), leading ultimately, through the phase of the fetus, to the adult individual.

The advancing self-construction of the embryo, that is, embryogenesis, proceeds in stages. Its essence comprises two parallel processes: growth (an increase of the dimensions of the whole embryo and its particular structures) and differentiation (the formation within the initially homogeneous mass of the embryo and its parts, fragments differing from each other, areas and elements). First, the most general structures are formed, e.g., the front and back of the body and its dorsal and ventral sides become distinguished, the division into particular segments (in segmented animals such as insects) or sections of the body, separation of the head, thorax and terminal (tail or abdomen) section takes place; in plants, this can be a separation of the stalk and root (Hall 1999; Slack 2001; Twyman 2001) Next, germs (anlages) of particular systems, e.g., the digestive tract or nervous system (in vertebrates formed through bulging inside the embryo walls) and the limbs, originate. These systems, in turn, differentiate into particular segments and organs, e.g., the digestive tract is comprised of the gullet, stomach, liver, different kinds of guts, *etc.*; the brain, together with its main parts (forebrain, midbrain…), the spinal cord and peripheral nerves are separate entities which differentiate within the nervous system. While, in addition, the upper limb becomes "divides" into the arm, forearm and hand. Afterward, still smaller and smaller structures, such as the detailed architecture of the liver, folded brain cortex and the construction of the whole brain in general, the eye with all its elements (lens, iris, retina) or fingers in the hand composed of a few phalanxes are formed (differentiated). In parallel, forming particular tissues from non-differentiated cells of the embryo occurs—e.g., a finger is built of connective tissue (bone, cartilage, tendons, skin lining), muscle tissue (muscles flexing and extending particular finger segments), epithelial tissue (epithelium), secretory tissue (sweat and grease glands), and nerves and blood vessels enter. Generally, most organs and body structures are built of cells belonging to different tissues, appropriately structurally and functionally organized. It should once again be stressed that the embryo grows and differentiates simultaneously (Hall 1999; Slack 2001, 67-71; Twyman 2001).

What is the mechanism of the above-described formation of structures in the "top-down" manner, from the most general to the most detailed structures? It occurs through a *hierarchically-organized cascade of sequentially initiated developmental (morphogenetic) instructions (procedures)* (Wolpert 1981; Slack 2001; Twyman 2001, 28-43). In this cascade, an instruction (procedure) of a higher order initiates the execution of many procedures of a lower order, and each of them stimulates procedures of yet a lower order, and finally, it comes to the production of the entire structure of the organism with its tiniest details. This has something in common with the generation of fractals, where a certain simple mathematical procedure applied many times produces incredibly (although only apparently) complex and beautiful structures as a result.

Embryogenesis differs from fractal generation because particular procedures and the elements of the structure produced by them are frequently very different on the different levels of the hierarchy. Apart from this, fractals serve nothing (apart from pleasing our eyes and exciting the enthusiasts of deterministic chaos), while the procedures generating the structure of living organisms must be such that the structure and function of an adult individual (that is, its biological identity) realize their basic biological purpose—the maximization of the chances of survival and the multiplication of a given structure (identity) (together with the entire cascade of morphogenetic procedures underlying it), its propagation in a possibly large number of copies. In principle, it can be said that the subject of biological evolution, which competes with other subjects for the accessible resources of the environment, is not the "finished" adult individual, but its whole developmental cycle, and first—the discussed hierarchical cascade of developmental (constructional) procedures contributing to the entire process of embryogenesis and the cascade of the regulation of gene expression (in unicellular organisms only the latter is present). It should also be mentioned that the correctness of the course of embryonic development of living individuals is supervised by a whole set of regulatory mechanisms, especially negative feedbacks, that aims to counteract any deviations from the "assigned" course of the morphogenetic cascade, the multi-dimensional "trajectory" of morphogenetic processes leading to the ultimate construction plan of an adult individual. Of course, this plan is indirectly recorded genetically in the form of the entire cascade of procedures (executive instructions) and the regulatory mechanisms controlling it (regulatory genes "participate" in it, meaning genes encoding transcriptional factors and miRNA, and enhancers and silencers).

On the other hand, the reading of the genetic record is controlled by these procedures/instructions/mechanisms. Therefore, we again have, so typical for living organisms, a vicious circle, contributing to the causal closure of living individuals. Within this closure, genes are not an especially distinguished or privileged element.

Additionally, it is worth mentioning that in the case where a developing embryo/larva must simultaneously and independently gain food for its growth and development, e.g., in most sea invertebrates and insects, such an embryo must be functional in the surrounding environment at each stage, which makes the development and differentiation more difficult (it is not easy to perform two completely different tasks at the same time: development and re-construction of its structure and function, on the one hand, and acquiring food, avoiding predators and, generally, managing oneself in the surroundings, on the other hand). Such a situation can significantly limit the ultimate complexity of structure and function of an adult organism. Therefore, the most complex living organisms either deposit a great number of nutrients in the egg (birds) or give birth to young individuals very advanced in development, which as an embryo, must have ensured nutrients and protection within the mother's body (mammals, especially placentals), and afterward, the mother or both parents frequently take care of and feed the young individual until it reaches independence. In such a case, the embryo can be completely functionally not self-reliant and dealing exclusively with its development and differentiation, as it does not have to manage itself in the external world. This frequently concerns young individuals after hatching/birth. As a result, the free-living developmental stadia, e.g., the larvae of sea invertebrates, tadpoles and caterpillars, seem categorially different than, e.g., the early human embryo.

Let us come back, however, to the morphogenetic cascade responsible for developing the structure (and function) in complex "higher" plants and animals (*Eumetazoa*) during embryogenesis (and ontogenesis). How does forming various structures on particular levels of hierarchy proceed? The general rule is quite simple. Certain morphogenetic (structure-creating) centers produce a substance—*morphogen*—which diffuses within the embryo, due to which its concentration decreases (exponentially) as a function of the distance from such a center (the so-called *morphogenetic gradient* is formed). Here, where an appropriate concentration of some morphogen or (more frequently) the appropriate combination/proportion of the concentrations of two or more morphogens (positional information) is present, a new, secondary morphogenetic center originates. Various secondary centers are produced in different places in different combinations of morphogen concentrations. These secondary centers produce their own secondary morphogens, acting at relatively shorter distances (because of slower or shorter-lasting diffusion), determining forming subsequent centers situated lower in the hierarchy, and a morphogenetic cascade, already known to us, is formed. Morphogenetic gradients not only initiate forming secondary morphogenetic centers but first control forming structures on different levels of hierarchy. This occurs through the determination of various patterns of gene expression in particular cells (with the mediation of regulatory proteins and miRNA) in particular places of the embryo. This can result in certain parts of the embryo growing (due to intensive cell divisions and or growth) in determined directions (e.g., limb germs). Other parts necrose and undergo atrophy (e.g., the spaces between fingers or toes being formed), yet other parts undergo differentiation in a tissue of a certain type (see below). The gradients of morphogens also constitute a guidepost for different types of cells migrating from some parts of the embryo to others. For instance, proto-neural cells migrate from the neural tube in the embryo of chordates to different body parts, forming peripheral nerves. As mentioned, the specific combination of concentrations of appropriate morphogens in a given place of the embryo, determining the production of a specific structure and formation of new centers, is called positional information. It alone decides that, e.g., a thorax, left lung, eye, or the second lower incisor are to develop in a given place. Morphogenetic gradients, variously spatially oriented, frequently crossing each other, are responsible for the differentiation of the parts of the embryo, for instance, the segments of the body along the front-back axis, segments of the hands from the shoulder to the tips of the fingers, or the fingers of the hand from the thumb to the little finger. The primeval front-back axis is determined either externally, by the orientation in the mother's body (e.g., in the oviduct, whose walls/cells supporting the embryo excrete mRNA constituting the matrix for the synthesis of appropriate proteins-morphogens in the front and back ends of the embryo), or spontaneously, through the self-generated heterogeneity of the zygote, in which a small pellet of cytoplasm is concentrated at one end (pole) of the cell, while the rest is occupied by fat fulfilling the function of a reserve substance (this is a kind of self-organization discussed during of hydrophobic interactions) (Maynard Smith and Szathmáry 1995, 227-229; Twyman 2001; Slack 2001). The embryonic and individual development in plants is also based on morphogen gradients—e.g., probably three morphogens A, B and C (encoded by three genes a, b and c), produced in a greater and greater distance from the center of the flower germ, are responsible for the concentric generation of particular parts of the flower—starting from the middle: the style, filaments, petals and sepals (these structures are all modified leaves) (the generation of subsequent kinds of structures is determined by the activation of expression of, respectively, c, b+c, a+b, a genes) (Twyman 2001, 430-433; Futuyma 2013, 572-573). These problems will be discussed in more detail below.

Let us start with the developmental cascade leading to the production of the structure of adult individuals. Fig. 5-5 presents a very general and simplified scheme of the developmental cascade of higher organisms or a fragment of such a cascade. The black/gray/white circles represent morphogenetic centers, while the arrows represent the generation of new, secondary centers on lower hierarchy levels. The horizontal lines under the morphogenetic centers mean that just these centers participate in generating secondary centers, to which the arrows from these lines lead. Only four hierarchy levels are shown in the scheme, corresponding to the polarization of the entire body along the front-back axis and next to large, medium and small structures. Only some morphogenetic centers are shown in the third and fourth-level cases not to confuse the scheme.

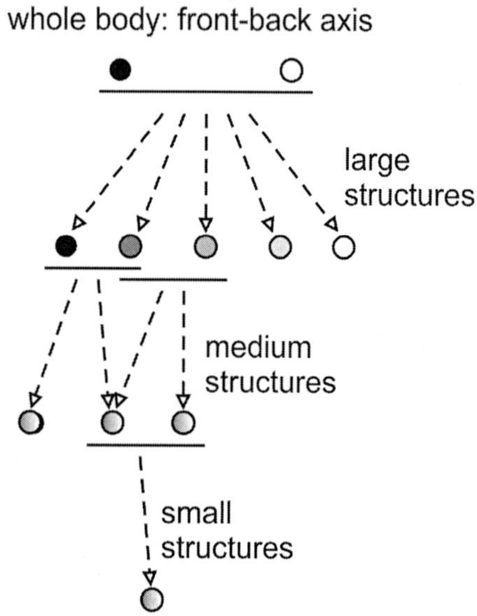

Fig. 5-5. General scheme of the morphogenetic cascade of complex living organisms. Morphogenetic centers (circles) on higher levels of hierarchy determine forming morphogenetic centers on lower levels of hierarchy.

The above mechanism is presented in more detail, considering morphogenetic gradients and positional information, in Fig. 5-6. Two morphogenetic centers that produce two morphogens, A and B, of which one determines the front, while the other, the back of the body, are present on the highest hierarchy level (1). As mentioned above, this primeval polarization can result from the self-organization of the substance in the egg cell or from the arrangement of the embryo in the mother's body. The concentration of morphogens falls with the distance from the centers producing them, forming a gradient. An appropriate combination (ratio) of concentrations A/B (five such ratios are arbitrarily chosen in the figure: 10, 3, 0.7, 0.3, 0.1) determines forming morphogenetic centers and large structures on the second level of hierarchy (2)—this ratio corresponds to the positional information mentioned above. These large structures can correspond, e.g., to body sections: the head, neck, thorax (chest, belly) and tail in vertebrates or particular body segments (or their groups) in insects. The combinations of concentrations of second-level morphogenss (*C*, *D*, *E*,

F, *G*) determine the origination of morphogenetic centers and medium-size structures on the third level (3). Only three centers and morphogens produced by them (*H*, *I*, *J*) formed in the places of appropriate ratios of concentrations of morphogens *C* and *D* and *D* and *E* are shown as an example. They, in turn, induce, through the formed morphogen gradients, morphogenetic centers and small structures on the fourth level (4) (only one center and morphogen—*K*—is shown as an example). Although the presented scheme represents the idea of the mechanisms of morphogenesis relatively well, it possesses a few important limitations and simplifications. First, as mentioned, the number of hierarchy levels in the developmental cascade is significantly greater than four in more complex organisms. Second, importantly, Fig. 5-6 shows morphogenesis only in one dimension because of the purely technical presentation possibilities. However, the real development of the embryo occurs, of course, in three dimensions, which creates an incomparably greater number of possible combinations of spatial arrangements of morphogen gradients, thus enabling the generation of much more complicated structures. Third, finally, the distance (x-axis) in subsequent stages of development (1 to 4) is identical in Fig. 5-6, while in reality, this distance significantly increases with time as the embryo grows during development (at least at some of its stages).

Summing up, the entire structure of complex, multi-cellular living organisms is formed during embryonic development based on the mechanism of the hierarchical cascade of developmental instructions, in which a sequential induction of morphogenetic procedures takes place. This occurs through morphogen gradients formed due to the production of various morphogens by appropriate morphogenetic centers and, next, the diffusion of these morphogens in the embryo's body. Various combinations of morphogen concentrations in a given concrete place (determining the positional information) lead, in a given set time, not only to the creation of subsequent secondary morphogenetic centers but also to the activation/inhibition of appropriate cascades of the regulation of gene expression within cells, and therefore to their proper differentiation. In other words, a determined combination of (concentrations of) morphogens within the morphogenetic cascade in the entire organism, in turn, initiates a cascade of the regulation of gene expression within the cell that is also organized hierarchically. This cascade, through the agency of activation or inhibition of the expression of various genes by appropriate regulatory proteins and miRNA, leads to a differentiation of this cell (or its possible migration in an appropriate direction).

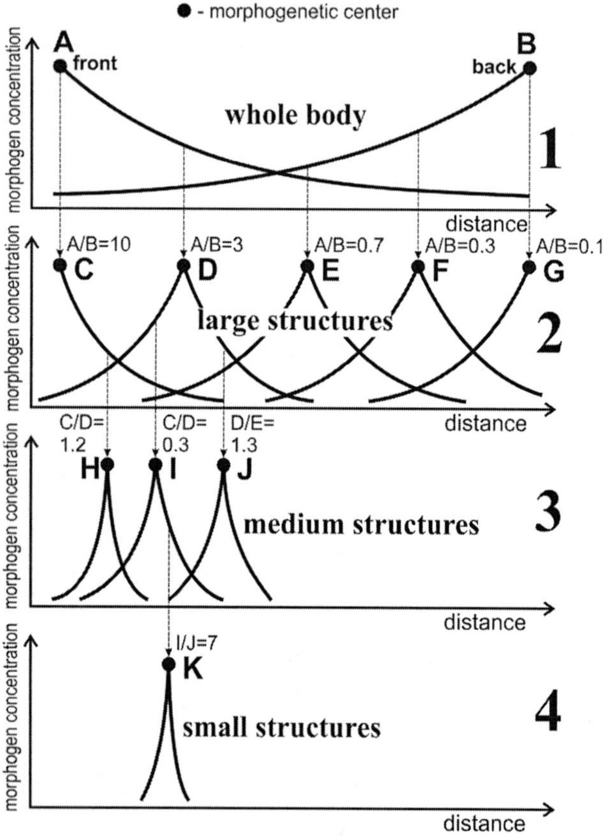

Fig. 5-6. Extended scheme of the morphogenetic cascade from Fig. 5-5, taking morphogenetic gradients and positional information into account. 1-4, distinguished levels of hierarchy; *A-K*, morphogenetic centers and morphogens produced by them; *A*/*B* = 10, *C*/*D* = 1 *etc*., ratios of morphogen concentrations corresponding to positional information. Subsequent diagrams 1-4, corresponding to subsequent levels of the hierarchy of the morphogenetic cascade, present at the same time, occurring one after another in time events within the same developing and growing embryo.

Let us have a closer look at these processes. The hierarchically-organized cascade of the regulation of gene expression within the cell is illustrated in Fig. 5-7 (Maynard Smith and Szathmáry 1995, 219-223; Alberts 1998, 257-273; Twyman 2001, 71-81; Futuyma 2013, 539, 583). A "superior" regulatory protein I, or a set of a few such proteins, constitutes the highest level in the cell's gene regulation cascade hierarchy. This can be a receptor protein (set of proteins) in the cell membrane or nucleus detecting the presence and concentration of (a) determined morphogen(s) in the neighborhood of the cell. In differentiated cells, these can be receptor proteins binding various hormones or neurotransmitters (that not only modify the current functioning of the cell but can also influence the temporal pattern of gene expression). Regulatory proteins of the first order (I) can activate or inhibit the expression of genes encoding regulatory proteins of the second order (II) by binding themselves to appropriate DNA fragments (enhancers or silencers) nearby these genes (or even within them, located in their introns). Generally speaking, the proteins influencing gene expression (transcriptional factors) have a very characteristic domain binding to DNA, with a very similar structure in different transcription factors, the so-called homeodomain, while the gene encoding them contains a segment called homeobox, corresponding to this homeodomain (Alberts et al. 1998, 260; Twyman 2001; Reece et al. 2011, 655). It is supposed that the mentioned proteins, or at least the homeobox segments encoding the homeodomain of regulatory proteins binding to DNA, originated due to the multiplication in many copies and later the differentiation due to the cumulation of mutations and evolution of a certain ancestor gene/homeobox.

Returning to the cascade of the regulation of the gene expression, the genes activated by (a) protein(s) of the I order serve as the matrix for the synthesis of the regulatory proteins of the II order, which, in turn, can activate or suppress the expression of genes, that is the synthesis of regulatory proteins encoded by them, of yet lower orders (III, IV…). Several such levels in the gene expression regulation cascade can be present. On the lowest level of the hierarchy, the production of structural and functional proteins is ultimately stimulated or inhibited, according to an appropriate spatial-temporal-quantitative pattern, which leads to shaping the ultimate form of a given cell.

Morphogens, and, strictly speaking, positional information equivalent to the combination of their concentrations in a given location in a determined time,[13] determine not only which macro- and microscopic structures are produced here and now but first which types of cells differentiate in a given place of the embryo (or undergo migration from/to other parts of the embryo). In other words, they direct the differentiation of initially omnipotent embryonic cells ("capable of everything") onto specific routes

[13] And therefore, in principle, it should be called the positional-temporal information.

and determine their target shape, size, internal structure and function. The process of cell differentiation into epidermal, muscle, nervous, excretory cells, *etc*., is based on the cascade of the regulation of gene expression discussed above (Twyman 2001, 8-12; 187-204; Slack 2001). Morphogens and their assemblies fulfill the role of the railway switches that re-direct developing a given cell (train) onto the appropriate rails. Through the activation of the synthesis of regulatory proteins (e.g., through the agency of receptors in the cell membrane or nucleus), morphogens determine the differentiation of a given embryo cell in the direction of one of several hundred types of cells appearing in most complex organisms, for instance in a human. This process occurs in several stages, through the subsequent activation of the appropriate genes/synthesis of the appropriate transcription factors, and ultimately of structural/functional proteins. For instance, in the first stage, the "turning on" of regulatory protein *A* can direct the development of a given cell on the route leading to a muscle cell (myocyte) in general. Next, activating the gene/regulatory protein B decides this will be a skeletal muscle (not the heart or smooth muscle) cell (skeletal muscles are responsible for locomotion, manual operations, speech generation and facial mimicry). Finally, an expression of gene/synthesis of transcription factor C will cause this to become a muscle cell (fiber) of an oxidative nature (producing ATP mostly in the process of oxidative phosphorylation, with little contribution of anaerobic glycolysis). Therefore, our cell can be specified as A(+)B(+)C(+). At the same time, it should be emphasized that the fact of "turning on" some gene/regulatory protein is equally important as is "turning off" (lack of turning on). The combinations A(-)B(+)C(-) or A(-)B(-)C(-) can determine the origination of other, equally well-specified kinds of cells. This is illustrated in Fig. 5-8. Of course, in real systems, a much greater number of (combinations of) "turned on" and "turned off" genes/transcription factors can determine the differentiation of a much greater number of various types of cells. For instance, several sub-types of oxidative skeletal muscle fibers exist.

However, the expression of one or a few regulatory genes, meaning the synthesis of regulatory proteins encoded by them, can determine not only the differentiation of cells but also of whole smaller or greater body fragments. As mentioned above, in animals, the body's structure on different levels, from the front-back axis, through sections/segments of the body, up to the tiniest details, are determined by morphogen gradients delivering appropriate positional information. Morphogens are either proteins themselves or are produced by proteins, and therefore an appropriate regulation of gene expression in morphogenetic centers is of crucial importance for the form of the morphogenetic cascade and, therefore, also for the appearance and function of the "target" organism.

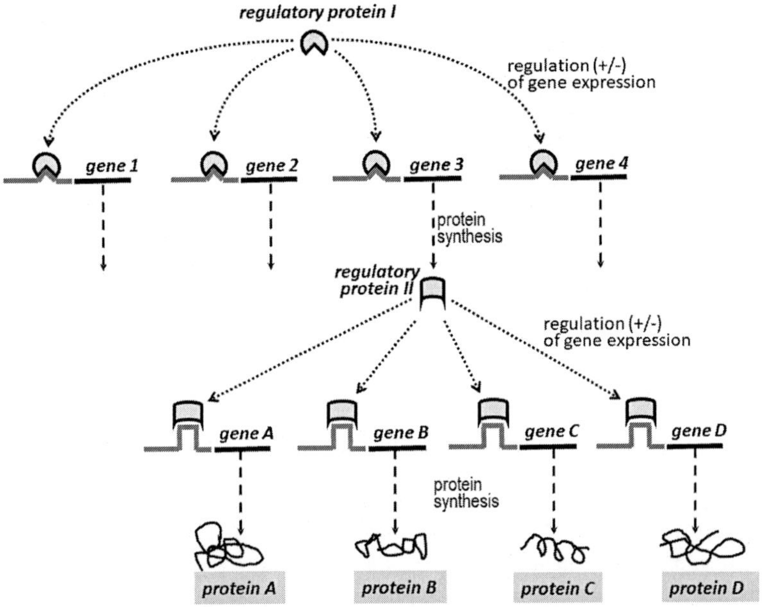

Fig. 5-7. Gene expression regulation cascade in the cell. Superior regulatory protein(s) I (of the first order) control(s) (turns on/off) the expression of genes (1, 2, 3, 4) encoding appropriate regulatory proteins II (of the second order). They, in turn, steer the expression of genes on a still lower level of hierarchy. As an example, gene/protein 3 is shown, which influences the activity of genes A, B, C, D, and therefore also the intensity of production, that is ultimately the concentration, of the proteins encoded by them. These proteins on the lowest level of hierarchy of the regulation of gene expression (depicted in the figure, in reality there can be much more of such levels) are structural/functional proteins, which are responsible for the shaping of the target architecture and complex of functions of the cell.

Let us look at a few examples of the determination of the structure due to the expression of appropriate regulatory genes and forming morphogen gradients in higher organisms. Let us start with plants. The flowers in flowering plants (belonging to angiosperms) are composed of four concentric (complexes of) structures, starting from the center: the style (the female reproductive organ, which evolved from connate sporangium-bearing leaves), filaments (the male reproductive organs), petals (fabulously colorful, whose function is to lure insects) and sepals (green, similar to "normal" leaves). All of them evolved from leaves: purely photosynthetic leaves or sporangium-bearing leaves (sporophylls), and these, in turn, from complexes of telomes (see Subchapter 5.4). Their formation in the germ of the flower (having roughly the shape of a hemisphere) is determined by three genes expressed in concentric rings at the top of the germ. Gene C is activated in a central circle of the germ, gene B—is in a somewhat more external ring, and gene A—is in the most external ring.[14] The areas of the germ on which genes A and B and B and C are expressed partly overlap. This is presented in Fig. 5-9. Due to this, four concentric zones are formed, in which the following genes are expressed: only gene C (circular central zone), genes C and B (internal ring zone), genes B and A (medium ring zone) or only gene A (external ring zone). These combinations of activated genes A, B and C determine forming subsequent (assemblies of) structures of the flower: the style (C), filaments (CB), petals (BA) and sepals (A) (Fig. 5-9). This system seems alluringly simple but only starts certain cascades of morphogenetic processes. For a style to be formed, for instance, the cooperation of hundreds, if not thousands, of genes is needed, both regulatory genes of lower orders and structure/function genes encoding appropriate proteins. Nevertheless, if only one of the superior "switches" is lacking, the entire construction collapses like a house of cards. For instance, when efficient gene *C* is absent due to a mutation, in the place of the flower only concentric rings made of petals and sepals are formed, while the lack of functional gene *A* results in the production of only the style and filaments (Maynard Smith and Szathmáry 1995, 227-229; Twyman 2001, 430-433; Futuyma 2013, 572-573).

In all animals having bilateral symmetry (*Bilateralia*),[15] in principle, the same complex of genes conditions forming subsequent body sections (e.g., segments) along the front-back axis. This complex, called the HOM/Hox[16] complex, is so characteristic and universal that it has been proposed as a principal element defining bilaterally symmetrical animals. Nevertheless, two genes of this complex are expressed even in cnidarians (e.g., hydra, sea anemones, not belonging to *Bilateralia*), where they determine the anterior (mouth) and posterior (opposite to mouth) end of the body.

---

[14] Strictly speaking, we are dealing here not with single genes, but with groups (classes) of genes A, B and C encoding proteins that have the nature of transcription factors. Sometimes this simple model is supplemented with gene classes D and E (Twyman 2001 430-433; .Futuyma 3013, 572-573)

[15] Some groups, for instance echinoderms, secondarily lost bilateral symmetry in favor of radial symmetry.

[16] It was discovered independently in fruit flies, and called HOM, and in mice, where it was named Hox.

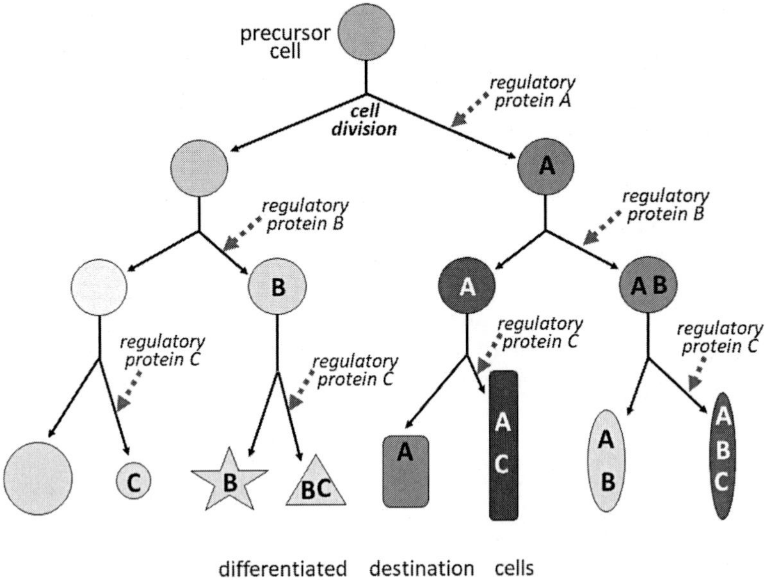

Fig. 5-8. The differentiation of cells into different types during embryogenesis due to activation or its lack of genes encoding superior regulatory proteins A, B and C. This activation can consist in the binding of a given morphogen to a specific receptor, which leads to the expression of the gene encoding a given regulatory protein. In subsequent stages the synthesis of subsequent regulatory proteins can be activated or not, which causes a gradual "detailing" of the fate of a given cell. The target combination of "switching on" and "switching off" of the synthesis of three regulatory proteins A, B and C leads to maximally eight combinations and thus eight cell types. In real complex living organisms, the number of "superior" regulatory genes, and, in particular, the combinations of their "switching on" and "switching off" can be much greater, which enables the differentiation of a much greater amount of various cell types.

Before particular body sections undergo differentiation along the front-back axis, the embryo's front and back ends must become fixed. In the *Drosophila* fruit fly, this occurs through the orientation of the embryo in the mother's oviduct. At one end of the embryo (being in the phase of the syncytium, that is, a multi-nuclear cell), mRNA for protein (morphogen) Bicoid (Bcd), determining the frontal end of the body, produced by the mother's organism, is concentrated, while at the other end—mRNA for protein (morphogen) Nanos (Nos) determining its rear end, is concentrated (Twyman 2001, 211-229; Reece et al. 2011, 1280). Appropriate morphogens are synthesized on the matrix of those mRNAs in the front and back part of the embryo, where they diffuse toward the opposite end of its body, creating morphogenetic gradients (of morphogen concentrations), and the ratio of the concentrations of these two morphogens constitutes positional information for the genes determining the differentiation and development of particular parts and segments of the body. In other animals, particular details of this process can be different, but regardless, two morphogenetic centers are localized at the front and back end of the body and produce morphogens diffusing toward the opposite end, thus forming gradients of concentrations and positional information. This is presented in the left scheme in Fig. 5-10. The differentiation of the dorsal (top) and ventral (bottom) sides of the embryo occurs analogously—they are assigned through gradients of appropriate morphogens.

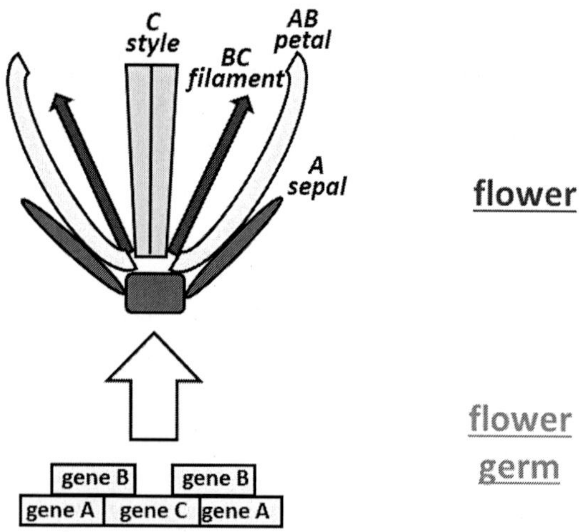

Fig. 5-9. Determination of forming the parts of the flower by superior regulatory genes *A*, *B* and *C*. In the germ of the flower (lower scheme) these genes are expressed in concentrically situated areas: central circle (gene *C*), internal ring (gene *B*) and external ring (gene *A*). These areas partly overlap with each other, due to which four concentric zones originate, in which genes are subsequently activated: *C* (zone I), *C* and *B* (zone II), *B* and *A* (zone III) and *A* (zone IV). Due to this, the style is produced in zone I, filaments—in zone II, petals—in zone III and sepals—in zone IV (upper scheme).

Let us return, however, to the front-back axis. Now the aforementioned HOM/Hox gene complex enters the scene. These are homeotic genes, containing the homeobox sequence, which, as we remember, encodes the homeodomain of regulatory genes (transcription factors) binding to (enhancer and silencer sequences in) DNA. These genes are activated (undergo expression) sequentially along the front-back axis, according to the positional information determined by morphogens fixing the front and back end of the body. Their expression pattern is principally the same in fruit flies, mice, humans and all other studied, in this aspect, "bilateral" animals (Maynard Smith and Szathmáry 1995, 225-240; Twyman 2001, 235-243; Futuyma 2013, 566-571).

Interestingly, these genes are arranged side by side on a chromosome in the same sequence, in which they undergo expression along the long body axis. This is called *collinearity*. Proteins encoded by them are synthesized in subsequent body sections starting from the head and finishing in the tail or abdomen. There are 10 (in fruit flies) or slightly more (13) (in mammals) of such genes, and they determine the developing subsequent body sections. This is illustrated in the right scheme in Fig. 5-10. Of course, these proteins only constitute the tops of appropriate, hierarchically organized morphogenetic cascades leading to the formation in subsequent body sections of proper structure. Various morphogenetic centers, morphogens and regulatory proteins "participate" in these cascades on subsequent levels of hierarchy.

The activation of subsequent genes of the HOM/Hox complex is determined (in fruit flies) by the positional information resulting from the gradients of morphogens Bicoid and Nanos. However, the matter is more complicated, as an intermediate stage between the expression of Bicoid and Nanos genes and the expression of HOM/Hox complex genes is the activation/production of the so-called "gap genes/proteins" and "segment-polarity genes/proteins" (Twyman 2001, 207-208; Futuyma 2013, 584), however for the sake of popular-science I will allow myself to simplify the problem and not go into detail.

All evidence indicates that the complexes of homeotic genes HOM/Hox originated in the process of biological evolution from a single regulatory gene "equipped" with a homeobox (homeotic gene), through its multiplication and the following differentiation of particular copies. Furthermore, single homeotic genes, kin to the genes from the HOM/Hox complex, participating in the cell structure formation occur in unicellular eukaryotes. As if all that were not enough, while in most animals, one HOM/Hox gene complex occurs, in vertebrates as many as four such complexes can be found (Futuyma 2013, 569). They originated through the multiplication and differentiation of one ancestor complex. These complexes are expressed and participate in the differentiation (determination) of body sections along its long axis (front-back), but also of limb segments along the proximal-distal axis (from the base to the tips of the fingers/toes) and along the transverse axis of the hand/foot (thumb/big toe—small finger/small toe).

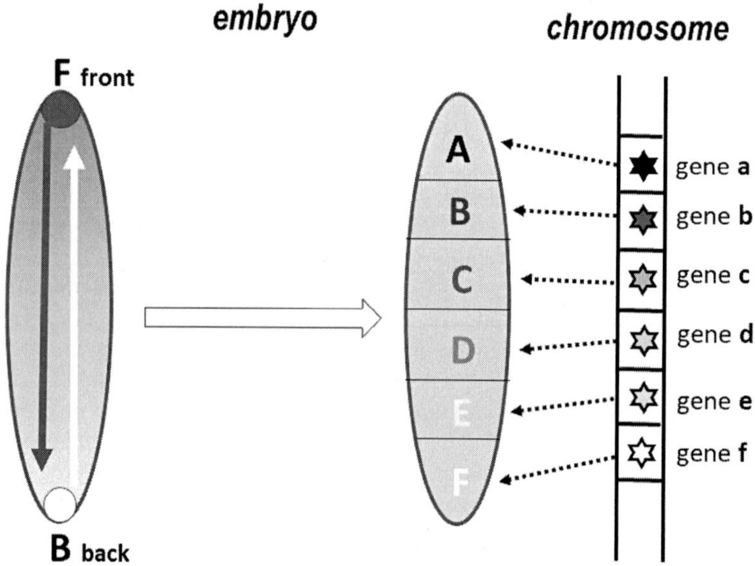

Fig. 5-10. The first stages of the determination of particular sections of the animal embryo. Two proteins-morphogens determining the front (F) and back (B) are synthesized, in morphogenetic centers located at the front and back end of the embryo, on the matrix of two mother genes and mRNAs corresponding to them. These morphogens diffuse, form concentration gradients, and specify the front-back axis (left scheme). Positional information equivalent to the ratio of concentrations of these morphogens in a given place of the embryo determines the expression along the long body axis of particular homeotic genes from the HOM/Hox complex (marked here conventionally from a to f, lying on a chromosome in the same sequence, in which they are expressed along the body axis). This leads to the production of morphogens characteristic for subsequent body sections (from A to F), which boot morphogenetic cascades causing the production of structures proper for given parts of the body (right scheme).

Mutations in homeotic genes/transcription factors acting on the top levels of the morphogenetic cascade or their expression in non-standard parts of the body leads to macroscopic "monstrous" effects. A mutation in gene eyeless/Pax6[17] causes the absence of eyes in an animal, while its expression in body parts other than the head leads to developing an "additional" eye in place of the expression. For instance, an artificial expression of the human gene Pax6 in a *Drosophila* leg stimulates the formation of an eye (insect, complex eye) on this leg (this means that the human gene efficiently stimulates eye development in insects, but the kind of eye that will develop depends on the "local" morphogenetic cascade) (Futuyma 2013, 575-577). A mutation in the ultrabithorax gene, belonging to the HOM/Hox complex, leads to the development of a second pair of wings in fruit flies (flies normally have only one pair of wings) (Futuyma 2013, 566). An expression of gene antennapedia, also belonging to this complex, leads to developing legs on the head of fruit flies, where antennae normally develop (Futuyma 2013. 566-567). It is also worth mentioning that the distal-less gene stimulates the development of different appendages in various, frequently very evolutionarily distant groups of animals: limbs in vertebrates, antennae, mandibles and legs in arthropods, legs in polychaetes, ambulacral legs in echinoderms, appendages in caterpillars and onychophorans, and so on. A mutation in this gene can lead to a lack of these appendages (Futuyma 2013, 586-588).

It should also be added that homeotic genes, although not from the HOM /Hox complex, supervise sex determination (Twyman 2001, 151-160; Reece et al. 2011, 289-290).

Knowing the genesis of structure in complex multi-cellular organisms, we can analyze how this structure undergoes modification and complication and adapts to environmental conditions during biological evolution.

As one can guess, the changes in the macroscopic structure of adult multi-cellular living organisms and the possible increase in their degree of complexity in the process of evolution are the results of mutations in regulatory genes encoding transcription factors or miRNA, or in enhancer/silencer DNA segments, underlying the cascade of morphogenetic procedures. A mutation can concern, e.g., modification of the concentrations of produced morphogens, changes in the structure and properties of the morphogen (morphogens are mostly proteins), sensitivity (reaction threshold) of particular cells to the concentration of a given morphogen, the kind of response of the cell to the morphogen, temporal synchronization of forming particular morphogenetic centers, the addition/removal of such centers, creation of new levels in the morphogenetic cascade (see Fig. 5-6) or modifications of regulatory mechanisms compensating deviations of the course of the embryonic

[17] This gene was named "eyeless" in insects and "Pax6" in mammals.

development from the correct/assigned course—these mechanisms begin to regard another course as correct. The mentioned mutations can lead to a significant modification of structure, changes to the construction plan and or an increase in the degree of complexity of living organisms (individuals). Generally speaking, the higher (in the hierarchy) the morphogenetic procedure that concerns a given disturbance (mutation), the stronger its effect on the entirety of the course of embryonic development, and the bigger the structures it concerns. Mutations on the highest level can form an undifferentiated spherical mass of cells instead of a normal newborn. Mutations on a slightly lower level lead, e.g., to the lack of a (normal) head or thorax in insects or developing legs over the eyes where antennae are normally produced (see above). On a lower level, an additional sixth finger in the hand can be formed—this case sometimes occurs in humans (it is not always certain whether such an anomaly results from a mutation or an environmental disturbance). Finally, mutations on the lowest levels can change the amount of tiny elements, such as chaetae or liver cells, and the proportion, spatial orientation and shape of various body parts. For instance, the gradual (although not necessarily synchronous and occurring at the same pace) enlargement of the brain (especially of the prefrontal cortex) and skull, reduction of the jaw size, leg elongation, arm shortening, standing erect, the disappearance of the opposition of the big toe, vanishing, thinning and shortening of hair, *etc.*, took place in our evolution. However, it should be added that a huge majority of large, qualitative, macroscopic structural changes (e.g., the origination of the four-chambered heart from the two-chambered heart in mammals and "supreme reptiles," including crocodiles and birds, or jaws from gill arches in vertebrates) are due to the gradual accumulation (stacking up) of rather small quantitative changes. Only a tiny chance exists that an accidental large, saltatory change in the structure and function of an organism will not only not handicap its abilities for survival and reproduction but even appear to be beneficial. This results from the fact that each property of an organism is related to (associated with) thousands or millions of other properties (conditioned by an assembly of hundreds of co-operating genes and the spatial-temporal-quantitative pattern of their expression), which co-evolved together and underwent optimization for millions of years. Every large accidental change of such a property could disadvantageously influence many features correlated with it, disturbing them in a random, uncoordinated direction.

On the other hand, small, gradual changes in a living organism's structure can appear beneficial. This automatically opens the door for modification of this structure using natural selection—mutations causing accidental changes in the structure are sifted through the sieve of natural selection, and those few that appear to be profitable in the given conditions will be consolidated in the population. Mutations can also cause certain quantitative changes in particular procedures in the cascade of morphogenetic instructions. This leads to a moderate quantitative modification of the structure of the organism. However, over long periods, new procedures can appear, existing procedures can disappear, and even new levels in the hierarchy of procedures can be created. As a result, this can lead to large qualitative changes in the structure of a living organism in its construction plan and a significant increase in complexity. In extreme cases—the emergence of entirely new levels in the hierarchy of structures. According to this principle, we can easily imagine, at least theoretically, the evolution of structures from the stage of a hydra, a sack composed of two layers of cells, to humans. The fundamental rules of this process are generally known—modifying the existing procedures in the morphogenetic cascade and adding new procedures to this cascade; that is all.[18]

Here we come to an important rule of the course of the evolution of the structure of complex living systems: the changes concerning macroscopic structural properties are, to a large extent, a derivative of mutations in the genes (encoding regulatory proteins and miRNA) steering embryonic development[19] (although, e.g., some metabolism defects can elicit a non-specific dwarfness resulting from ineffective metabolism, and not from instructions contained in the morphogenetic cascades). These mutations modify and possibly extend the discussed cascade of morphogenetic procedures. It was already mentioned that very (in a sense: infinitely) complex fractals can be generated using a very simple mathematical procedure applied many times. Therefore, despite phenomenological complications, fractals are, at a deeper level, extremely simple creatures. The ice "flowers" "painted" by frost on windows or snowflakes constitute

---

[18] However, this that we believe we know general principles of the embryonic development does not contradict the fact that there still remains a long way to the complete quantitative description of the entire developmental cascade leading from the zygote to an adult human in terms of particular procedures, and that most probably numerous surprises await us.

[19] For example, a human differs significantly by structure and appearance from a chimpanzee, despite the fact that these species exhibit huge genetic similarity. This apparent contradiction can be explained by the fact that the discussed structural differences are a derivative of the differences in the genes steering the morphogenetic development of the embryo, and not in the genes encoding structural proteins or enzymes (e.g., cytochrome oxidase) in the cell. Therefore, in this case, not a general genetic similarity is important, but the similarities/differences in the assembly of regulatory genes (encoding transcription factors) supervising the morphogenetic cascade.

other examples, this time taken from the real world, of apparently very complex structures generated by simple rules/processes.

What about living organisms? Of course, instead of one very simple procedure or a set of a few such procedures, we are dealing here with a very complex cascade of various developmental instructions, different from one another. However, in my opinion, for instance, the human organism, despite its huge complexity, is generated during individual development due to the actions of as few as several hundred (I hesitate to say iconoclastically: some dozens of) morphogenetic procedures (with the reservation that the same procedure, e.g., the instruction "differentiate into a muscle cell" or "migrate upstream a given morphogen gradient," can appear in many places of the entire cascade). I also believe that in the future, biology will lead to a quantitative explicit "decomposition" of all these cascades into particular procedures.

Fundamental events in the evolution of structure

Generally speaking, the evolution of living organisms can be somewhat arbitrarily divided into *microevolution* and *macroevolution* (Futuyma 2013, 605-629). The changes in the frequency of the occurrence of alleles in a population, on which the synthetic theory of evolution and population genetics are focused, lead to a gradual transformation of various features of organisms. These organisms can become larger or smaller, darker or lighter in color, while the proportions of various body parts can undergo modification. Gradual, quantitative, relatively small changes of this kind in the population's genetic composition and, which follows, in the phenotypic properties of its individuals are called *microevolution*. A characteristic property of microevolution is that it does not lead to significant changes in the construction plan of the entire organism or its parts. As a result, lower systematic units: varieties, species, or at most, genera originate. The idea of microevolution is illustrated in Fig. 5-11.

The schematic organism increases its length and changes color to a darker one, but these are quantitative changes, not leading to the origination of qualitatively new properties. In the case of microevolution, an increase in the degree of complexity of an organism does not follow.

On the other hand, *macroevolution* can lead to important, qualitative changes in the construction plan of the whole organism or its parts. Additionally, it can be associated with increased degrees of complexity. While microevolution is firstly related to the origination of new alleles of the existing genes and changes in their frequency in the population, then the origination of new genes, and also mutations in regulatory genes steering the process of embryogenesis, also called *macromutations*, frequently lies at the basis of macroevolution. Many macromutations are harmful and lead to non-functional monsters or chimeras forming. Nevertheless, some less dramatic macromutations may appear advantageous for the organisms carrying them and be consolidated by natural selection.

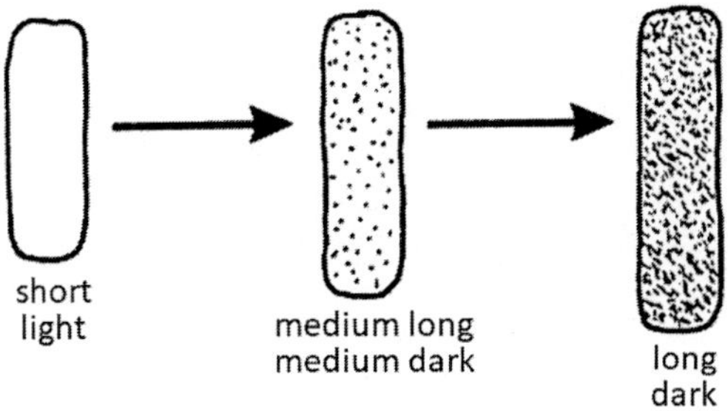

Fig. 5-11. Schematic example of microevolution. Changes in the course of evolution from left to right concern particular features of the presented abstract organism: its length increases, while the color becomes darker. On the other hand, the general construction plan of the organism does not change.

The influence of many genes on the phenotype is pleiotropic (Futuyma 2013, 204-205, 214), meaning these genes condition many different properties of an organism (the reverse statement is also true—a given property is conditioned by the cooperation of many genes) and is very important for macroevolution. An example of such a pleiotropic effect is the mutation in one of the genes responsible for the production of the pigment melanin, causing albinism. Albinos are characterized by pale-pink skin, white hair, red irises, and greater sensitivity to sun rays, but also an atypical course of neural routes in the brain. In turn, the "agouti" gene responsible for certain fur coloration in mice is lethal in homozygotes, leading to excessive obesity in heterozygotes (Futuyma 2013, 582-583). Therefore, single genes can influence various properties of an organism, frequently apparently unrelated to each other. This firstly concerns homeotic genes regulating the expression of other genes. This fact suggests that modifying a relatively large number of phenotypic properties, among others, associated with the construction plan can result from relatively few changes in the genetic record.

As mentioned above, macroevolution consists first of a change in the construction plan and can additionally lead to increased complexity. As a result, new higher taxonomic groups, such as families, orders, classes and types, originate. The creation of multi-cellular organisms from unicellular organisms, developing three germ layers (ectoderm, endoderm, mesoderm) and the body cavity (coelom) in bilateral animals, the appearance of body segmentation in annelids, arthropods and vertebrates, forming jaws in fish (placoderms and acanthodians) (and three ear ossicles, the hyoid bone, and cartilages of larynx and trachea in mammals) from gill arches in agnathans, the transformation of fins into five-finger limbs in the ancestors of amphibians, developing a four-chambered heart in mammals and birds (and most probably in the remaining, non-avian dinosaurs), the appearance of the egg resistant to drying out and fetal membrane—amnion—in first reptiles or development of the placenta in placental mammals (Reece et al. 2011, 534-735; Futuyma 2013, 77-133), can serve as examples of macroevolution.

There are still controversies concerning the question as to whether the same mechanisms govern macroevolution as microevolution (and therefore whether macroevolution constitutes the "sum" of many microevolutions) or whether other specific factors should be found in order to explain the former. In the first case, macroevolution can be treated as the cumulation of very many tiny changes, while in the second case, the possibility of sudden, qualitative, saltatory changes in the organism structure should be accepted (Futuyma 2013, 96-101). Such saltatory changes took place in some cases, for instance, during the origination of eukaryotes due to endosymbiosis of prokaryotes with proto-eukaryotes, or in the case of the transformation of unicellular organisms into multicellular ones. Regardless, there is no sharp borderline between these possibilities (gradual and saltatory changes). Whatever the ultimate answer to the above problem would be, microevolution seems to be more associated with mutations of "genes of structure" (and function, e.g., encoding enzymes) and with changes in allele frequencies, while macroevolution is more associated with the cumulation of numerous small changes, mutations in genes regulating embryogenesis and the origination of new genes.

Below, we will analyze several examples of the macroevolution of plants and animals' structures and construction plans (the course of evolution is presented in: Reece et al. 2011, 534-735; Futuyma 2013, 77-133).

Plant eukaryotes appeared in the oceans about one and a half to two billion years ago in the Precambrian era. Until the end of Silurian, they mostly had the form of unicellular benthonic (settled on the sea floor) and planktonic (floating in the water) algae. There are known Precambrian fossils of multicellular algae, but these organisms probably played a marginal role in the ecosystems of that time, and their body constituted a thallus that was not differentiated into tissues and organs. They did not need the epidermis to protect them from drying out or vascular bundles to transport water and minerals.

Developing plant tissues and organs, opposite to animals, occurred only after plants populated the land. This took place at the end of the Silurian era. The first "true" land plants (I am leaving unicellular algae aside here) were *Psilophytales* (this is an umbrella taxon containing not closely related forms like *Rhynia*, *Cooksonia* or *Psilophyton*) belonging to primitive vascular plants. They did not yet possess roots and leaves, while their stems were covered with epidermis and contained the core, whose important components were vascular bundles. Psilophytales originated most probably from green algae (chlorophytes) (their ancestors were closely related to the group of green algae—charophytes), to which the chemical similarity (among others, the same set of photosynthetic pigments) between these algae and contemporary vascular plants, being descendants of *Psilophytales*, testifies. Mosses and liverworts also evolved from chlorophytes, and have never fully become land plants, limiting their occurrence to wetlands and aquatic habitats.

The shoots of primitive *Psilophytales* had a simple construction plan being the starting point for the evolution of all overground organs of vascular plants. This structured scheme's basic, repeating element was the *telome*, a symmetrical branching of the cylindric stem (together with the core containing vascular bundles) into two identical (of the same diameter) shoots (Zimmermann 1952). Such a telome is shown in the left scheme in Fig. 5-12.

The first Psilophytales were, in fact, assemblies of several telomes joined together, some of which grew up from the endings of others, and most had sporangia on their tops (omitted for simplicity in the middle scheme in Fig. 5-12). The complexity of the telomes constituted the starting point for developing all overground organs of vascular plants, such as the stem, branches, leaves, and sporangium-bearing leaves or parts of the flower. The origin of the stem with branches and leaves is shown schematically in Fig. 5-12 (lower schemes). The stem originated through deprivation of the symmetry (equality) of the telome bifurcations, where one (superior) branch became thicker and entered into the composition of the main axis of the stem, while the other, thinner branch became an inferior side offshoot. The leaf, in turn, was formed through the branching of vascular bundles (that became the "veins" of the leaf) on one plane and the flattening of the pith (parenchyma) that formed the leaf blade filling the space between the bundles. Petals and sepals of the flower are simply transformed leaves, while the style originated through the fusion of several sporophylls (sporangium-carrying leaves) with macrospores (female spores, which afterward transformed into germs)[20]-producing sporangia facing inward. Similarly, the filaments evolved from sporophylls carrying microspores (male spores)-producing sporangia on their tops. Therefore, developing the whole diversity of vascular plants, including the dominant angiosperms, constitutes the result of various modifications in the process of evolution of a certain very simple structure element. The principle: "first multiply, then differentiate" (as I call it) finds a ubiquitous application in the increase of complexity of living organisms during evolution. Previously, we came across it during the discussion of the origin of new genes, where particular copies of the same gene evolved in divergent directions, and in the case of the origination of multicellular organisms, whose initially identical cells next differentiated and entered into the composition of tissues fulfilling various functions. In Subchapter 5.5, various cases of this principle will be summarized, as it is difficult to overestimate its significance for biological evolution.

Psilophytales constitute the main group (as mentioned, this is an artificial taxon grouping loosely related to each other primitive vascular plants) of land plants at the end of the Silurian and the start of the Devonian eras. Cryptogamic vascular plants possessing developed stems, leaves and roots (or at least rhizomes), namely *lycopods*, *horsetails* and *ferns* appeared at the end of the Devonian. Their arboreal forms (e.g., huge lycopods—*lepidodendrons* or *scale trees* and tree-like horsetails—*calamites*) formed wet Carboniferous forests. Most of the coal deposits have been formed from the dead remains of these plants. The first *gymnosperms*, *seed ferns,* also commonly occurred in the Carboniferous forests. Other groups of gymnosperms loosely related to each other, namely *cordaites*, *coniferous plants* (*Pinopsida*), *cycads* and *ginkgos*, appeared in the Permian and Triassic eras. They dominated among land plants through most of the Mesozoic era to the middle of the Cretaceous. Then, a violent differentiation (adaptive radiation) of angiosperms (dicotyledons and monocotyledons), wholly dominating most land environments until today,[21] took place.

---

[20] Strictly speaking—into gametophytes that, after fertilization of the egg cell, produced germs.

[21] Although gymnosperms—mostly spruce trees—still dominate in taiga and many high mountains.

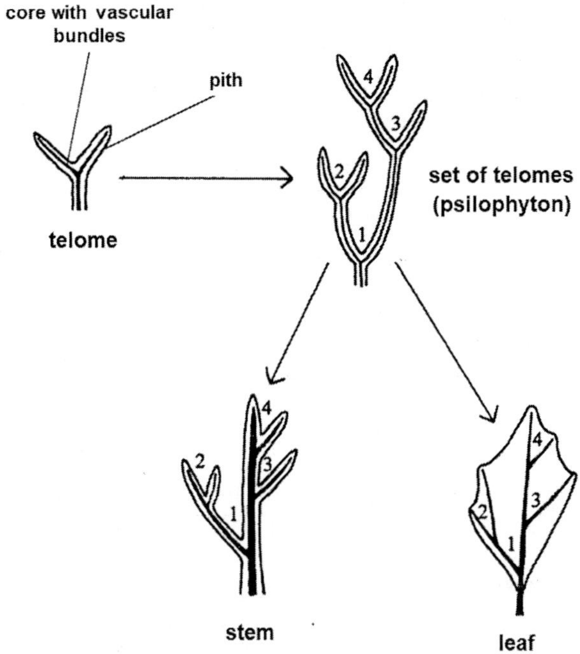

Fig. 5-12. Origination of the stem and leaf of vascular plants according to the "telome theory". These organs (and also all other organs, e.g., parts of the flower) evolved through appropriate transformations of stems of primitive *Psilophytales*, constituting assemblies of telomes, of which many were terminated with sporangia (not shown).

The development from *Psilophytales* through higher-organized spore plants, and then gymno- and angiosperms consisted of, among others, the differentiation and perfection of vegetative organs, such as leaves, roots and stems, toward a possibly most efficient adaptation to a water-deficient land environment. The roots anchored the plant in the ground, simultaneously taking up water and mineral salts. Flattened leaves ensured a large surface for photosynthesis, while stems equipped with vascular bundles delivered water to the leaves. The overground parts of the plants became covered by the epidermis, while the trees formed bark, preventing excessive water deprivation by evaporation. In developing land plants, the general tendency to improve all the above systems prevailed; for instance, the vascular bundles transporting water and mineral salts (xylem) in angiosperms (made of vessel elements, i.e., trachea) work more efficiently than analogous bundles in gymnosperms or spore plants (containing no trachea).

The evolution of generative (reproductive) organs and the reproduction process of land plants, tending towards making the process of reproduction of land plants independent of water, were occurring in parallel. This was associated with the reduction in the developmental cycle of the *gametophyte* (haploid phase, producing gametes) in relation to *sporophyte* (diploid phase, producing spores), up to the extreme simplification of gametophyte and its complete dependence on sporophyte (developing the residual gametophyte, no longer being a separate plant, occurred within the sporophyte). In spore (cryptogamic) plants living on the land—pteridophytes (lycopods, horsetails, and ferns), gametophyte already constitutes a small, although still independently growing plant, adopting the form of a thallus undifferentiated into tissue. Nevertheless, spermatozoa still need shallow water, for instance, that covers the surface of plants and ground after rain, to reach egg cells—therefore, spore plants cannot live in truly dry environments. On the other hand, in seed plants (beginning from seed ferns), the reduced female gametophyte develops from a spore within the sporophyte, remaining in the ovule all the time, while the male gametophyte in the form of a few-cellular pollen tube grows from a drying-resistant pollen grain transferred onto the ovule by wind or insects. Thanks to this, fertilization in seed plants does not require water. In angiosperms, the female gametophyte is simplified to only seven cells (eight cell nuclei), while the ovule and seed are protected from drying out and other environmental influences by, respectively, the walls of the style ovary and seed coat. As a result (with the parallel adaptation of vegetative organs), these plants can exist in very different environments, including desert areas. The so-called flowering plants (not constituting a formal systematic unit nowadays) developed the commonly known flowers in addition to leaves and stems (the role of flowers is to attract insects and other pollen-transferring animals).[22] Nevertheless, all of these are only assemblies of multiplied and modified telomes.

Animal eukaryotes originated, like plant eukaryotes, roughly 1.5 billion years ago. Taking into account that it was only necessary to acquire (due to endosymbiosis) ancestors of mitochondria for the origin of the first, while the plants-to-be had to additionally "domesticate" cyanobacteria, ancestors of future chloroplasts, it seems likely that animal eukaryotes had to have existed a little earlier, some 1.5-2 billion years ago, which is supported by fossil data.

---

[22] The co-evolution of flowers and animals that mediated in effective fertilization (insects, birds or bats) took place there.

They have existed exclusively as unicellular organisms for a long time (more or less half their history). The first multicellular animals, probably possessing tissues, appeared in the late Precambrian, over 600 million years ago (the so-called fauna of Ediacara, an excavation position in Australia). They were devoid of a hard external skeleton. Principally, they are dissimilar to any later living animals. Their general body scheme resembled a "quilted pillow." The most common form was *Charnia*, settled on the sea floor form having a shape similar to a feather, fern leaf or some colonies of today's coelenterates (sea feathers).

A true "explosion of life," meaning the adaptive radiation of sea invertebrates succeeded at the beginning of the first period of the Paleozoic Era, that is, the Cambrian period. Practically all known types (systematic units) of animals, representing enormous diversity of construction plans, essentially originated "at once" then (in the geological scale of time). The Cambrian fauna was characterized in most cases by possessing a hard external armor (carapace, shell), and its representatives are present as fossils in a large abundance in the rocks of the Cambrian from almost the very beginning. In this period, *trilobites* (belonging to arthropods), *penis worms* (priapulids, rooting in the sea bed), *brachiopods*, *sponge*, *coelenterates* (e.g., corals), many groups of *mollusks*, *crustaceans* and other *arthropods*, *echinoderms* (e.g., *crinoids*, but also several now extinct sedentary groups), *graptolites* (belonging to hemichordates), relatives of *velvet worms* (onychophorans, e.g., *Xenusion, Hallucigenia*), *annelids* and *chordates* existed. Many groups of animals then did not resemble any of the presently existing groups.

The diversity of Cambrian tissue animals, and also of their descendants in later geological periods until the present times, should not obscure the fact that the general construction plan of these animals (apart from the sponges, not being "higher" animals, *Eumetazoa*) constitutes more or less a modification of a certain fundamental construction scheme. All of them possess (and most probably possessed from the very beginning) very similar first stages of embryonic development, most probably roughly mirroring their evolutionary history. Free-living forms representing subsequent stages of the embryonic development of eumetazoans exist until today. Therefore, the course of embryogenesis of the latter constitutes, to a large extent, a recapitulation of the course of eumetazoan phylogenesis (evolutionary development) in them. This results from the cumulativeness property—evolution modifies and develops what already exists, improves the old and "builds on" it instead of creating something entirely anew. Apart from this, as mentioned above, the change of the construction plan and possible increase in complexity are realized through modifying the morphogenetic cascade during embryonic development. Therefore, the change of the construction plan and increase of complexity in higher animals occurred through adding new elements of this cascade to the assembly of old, already existing morphogenetic procedures and through modifying the latter.

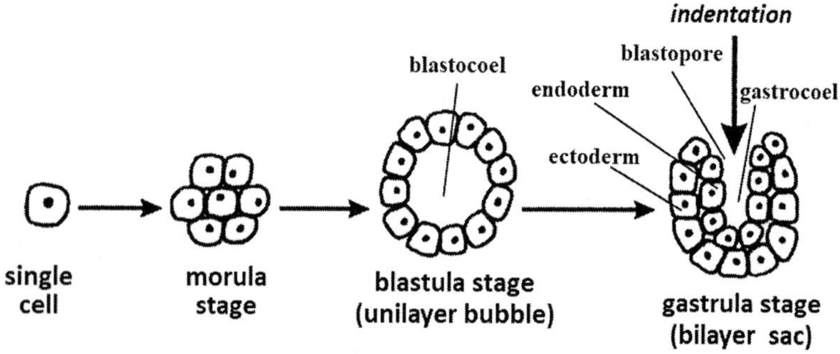

Fig. 5-13. First stadia of the embryonic development of higher animals (*Eumetazoa*) corresponding to the first stages of evolution of the spatial organization of the ancestors of these animals. The morula is a solid pellet of several to several tens of cells. The blastula is a single spherical layer of cells surrounding an empty cavity—blastocoel. The gastrula is a bi-layer sac surrounding a cavity, to which a single opening—the blastopore—leads.

This "concurrency" of the first stages of embryogenesis and phylogenesis of eumetazoans is illustrated in Fig. 5-13. It shows the embryonal development of animals proceeding from a unicellular *zygote* (fertilized egg cell) through the phase of *morula* (compact pellet of several tens of cells), *blastula* (an empty bubble surrounded by a single layer of cells) and *gastrula* (a bi-layer sack; it originates through bulging [herniating] of one side of the blastula into the other, analogous to squashing one side of a football, from which the air was removed, into the other). These stadia are represented by contemporarily living species whose individuals clearly "stopped" their development at particular developmental-evolutionary stages. For instance, unicellular eukaryotes (e.g., flagellate *Chlamydomonas*) correspond to the phase of a zygote, flagellates *Pandorina* (composed of 8, 16 or 32 cells) and *Eudorina* (composed of 16, 32 or 64 cells) represent the morula stage, volvox (*Volvox*) (an empty uni-layer sphere of cells)—the blastula stage, while hydra (*Hydra*) (built according to the scheme of a sac composed of two layers: external exoderm and internal endoderm)—the gastrula stage (Reece 2011, 655; Hall 1999, 93-124; Twyman 2001, 161-177; Slack 2001, 94-97, 110-114).

The gastrula stage is represented by the so-called bi-layered animals (*Diblastica*), whose individuals possess only two body layers, while embryos have two germ layers: exoderm and endoderm. They are also equipped with one opening leading to their gastrovascular cavity (commonly called mouth, but also fulfilling the role of the anus) (in embryos, this cavity is called a gastrocoel, while the primeval mouth opening is the blastopore). Cnidarians and ctenophores (the latter already possess the anus) belong to them. These two similar layers, not "folded" into a sack but flattened, are present in *Placozoa* of an unclear systematic position (maybe they represent the simplest known higher animals, *Eumetazoa*, closely related to their common ancestor).

Further (although still initial) stages of embryogenesis and (most probably) phylogenesis are common in "bilateral" (bilaterally symmetrical, *Bilateralia*) animals with primarily bilaterally symmetrical bodies (having identical left and right sides) (in some lines the bilateral symmetry could be secondarily disturbed or disappear completely, at least in the adult stage, but persist in the embryonal/larval stage, for instance in echinoderms). These animals possess three body and germ layers: the exoderm, endoderm and mesoderm.[23] Therefore, these are "tri-layer" animals. The so-called body cavity (coelom) develops within the mesoderm. Bilaterally symmetrical animals (in most cases) possess both the mouth and anus and are therefore "equipped" with a pervious digestive system "passing through" their body from the front to back. Forming these properties in the process of embryonic and phylogenetic evolution is presented in Fig. 5-14.

The presented process undoubtedly constitutes an example of macroevolution, a significant structure modification, a dramatic change of the construction plan, and increased complexity. Their developmental background was discussed in Subchapter 5.3.

We do not know in which order particular characteristics for bilaterally symmetrical animals appeared. It is unclear whether a separate mouth and anus or mesoderm and, therefore, three-layer organisms appeared earlier. Ctenophorans, having only two layers (they lack the mesoderm), already possess a separate mouth and anus. On the other hand, flatworms

[23] For instance, in humans, the epidermis and its derivatives (nails, claws, horns *etc*.), sweat and grease glands, external elements of the eye, enamel of teeth, the epithelium of the mouth cavity and the nervous system are formed from the exoderm, the ependyma of the digestive tract and respiratory system, liver, pancreas and urinary bladder—are formed from the endoderm, while the remaining parts, including muscles, connective tissues (bones, cartilages, tendons, blood, lymph), ependyma of the circulatory and lymphatic system and kidneys—are formed from the mesoderm (Twyman 2001).

(*Platyhelmintes*), belonging to three-layer and bilaterally symmetrical animals (*Bilateralia*) , are not yet endowed with a pervious digestive system with a separate mouth and anus. Therefore, it seems more likely that the mesoderm appeared in the developmental line of bilateral animals before the pervious digestive tract. The matter is still not fully explained.

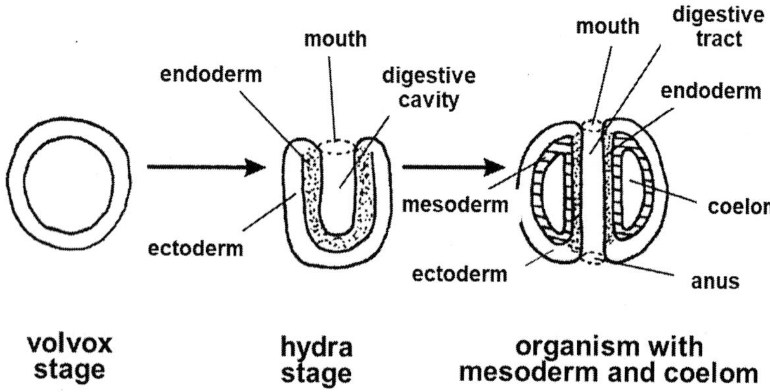

Fig. 5-14. The first stadia of embryogenesis and phylogenesis of bilaterally symmetrical animals (*Bilateralia*). The construction plan of an organisms (shown in cross-section) changes from a single layer of weakly differentiated cells forming a sphere (empty inside), through a two-layer sack (composed of exoderm and endoderm) surrounding the gastrovascular (digestive) cavity (in the embryo—gastrocoel), coming out by one opening (mouth opening, being the anus at the same time, and called the blastopore in the embryo), to a three-layer pipe, through which a pervious digestive tract passes, beginning with the mouth and terminating with the anus. The body cavity (coelom) is formed within the third layer (mesoderm).

The origin of the second, apart from the mouth-anus (the equivalent of the blastopore in the embryo at the gastrula stage), opening in the two- or three-layer sack led to forming a pervious digestive system. At the same time, two alternative (opposite) possibilities concerning the direction of the flow of food through this system existed, presented in Fig. 5-15. Either the previous "mouth-anus" became the definite mouth opening, while the new opening took over the role of the anus, or the new opening became responsible for taking up food (the mouth), while the existing opening started to exclusively serve the excretion of undigested remains (the anus).

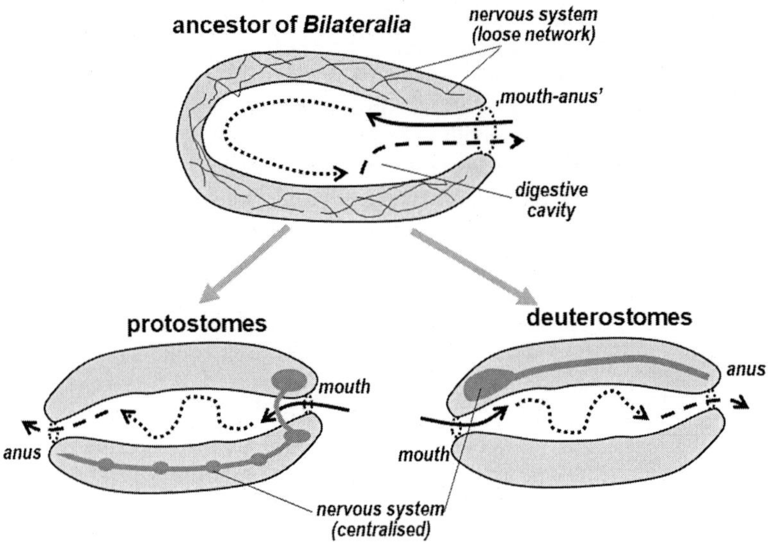

Fig. 5-15. Two alternative ways of origin of the pervious digestive tract in bilaterally symmetrical animals (*Bilateralia*). Their ancestor only possessed an opening in the body wall leading to the gastrovascular cavity, where taken up food was digested. It fulfilled the role of both the mouth and anus, as it served both food intake and removal of undigested food particles. During evolution, a second opening situated on the opposite body side to the first opening appeared (was perforated). In protostomes, the primary opening was ultimately transformed into the mouth, while the secondary opening became the anus (during their embryonic development the blastopore in the gastrula is transformed into the mouth, while the anus is formed on the opposite side of the gastrula). In deuterostomes, the primary opening ultimately underwent a transformation into the anus, while the new opening becomes the mouth (in the course of embryogenesis, the blastopore in the gastrula becomes the anus, while the mouth is ultimately formed on the other end of gastrula). While the nervous system of the ancestors of *Bilateralia* had the form of a loose network of neuron-like cells, then in protostomes neurons underwent concentration into ganglions and trunks of the central nervous system on the ventral (bottom) side of the body, while in deuterostomes it became condensed (ultimately into the brain and spinal cord in vertebrates) on the dorsal (upper) side of the body. Black arrows: solid—food uptake, dotted—digestion, dashed—elimination of undigested remnants.

Two main groups of animals of primary bilateral symmetry "chose" either the first or the second of the above possibilities. These animals are called protostomes, in which the already existing "mouth-anus" became the definite mouth, while the newly formed opening adopted the role of the anus. The remaining animals are called deuterostomes, for which the former "mouth-anus" served next exclusively as the anus, while the ultimate mouth originated *de novo* in the body wall. Again, the course of the early evolution of both these groups is documented (recapitulated) by their embryonic development: in protostomes, the blastopore transforms during embryogenesis into the mouth opening, while the new opening—into the anus, while in deuterostomes an exactly opposite situation takes place. A great majority of invertebrates belongs to protostomes, while echinoderms, hemichordates and chordates (together with vertebrates) are deuterostomes.

Of course, the perforation of a second opening in the body wall—the anus or mouth—is advantageous. Taking up food through the "mouth-anus," digest it in the gastrovascular cavity and assimilate through its walls, and finally remove the undigested remnants through the same "mouth-anus," as it takes place in animals devoid of the pervious digestive tract; all these activities can be executed simultaneously if an animal has a pervious digestive system at its disposal. This is illustrated in Fig. 5-15. Additionally, the continuous flow of nutritious substances through the "open" digestive tract enables the specialization of its particular sections in subsequent phases of digestion and the assimilation of these substances.

This end of the pipe (body), on which the definite mouth opening was formed, automatically became the anterior (front) end (it is obvious that it is more convenient to have the mouth at the front than in the back).[24] Also, most sense organs were developed near the anterior end, so the concentration of neural cells into ganglions occurred, which was associated with forming the head (encephalization). While the common ancestor of protostomes and deuterostomes probably had a nervous system in the form of a loose network of nervous (neuron-like) cells devoid of any significant concentrations of nervous cells, then the neurons underwent condensation into nervous ganglions and trunks in proto- and deuterostomes, and in the most advanced cases: into the brain and spinal cord. However, while in the former, the "central nervous system" originated on the ventral body side (apart from the super-esophageal ganglia—the "brain", directly "driving" the eyes and antennae), then in the latter, it is localized on the dorsal side. For instance, a concept once existed that deuterostomes originated from protostomes by exchanging the front and back body end and top and bottom body side. However, it seems difficult to imagine, e.g., a reversal of the flow of food and (functional) exchange of the mouth and anus (or relocation to the back of the body of neural ganglions formed at the front end).

---

[24] Although in some situations, e.g., during fast movement, some cephalopods have their mouth on their back side.

The majority of animals, including flatworms (*Platyhelminthes*) (that do not possess the anus yet), roundworms (*Nemathelminthes*), ringed worms (annelids), arthropods (crustaceans, insects, arachnids), mollusks (snails, bivalves, cephalopods), echinoderms (crinoids, starfish, sea urchins) and chordates, including vertebrates and ourselves belong to bilaterally symmetrical "three-layer" animals with a pervious digestive tract (*Bilateralia*). The further stages of their phylogenesis and embryogenesis already differ fundamentally (although there is a significant similarity in, e.g., larvae structure in some of them). Their evolutionary development occurred, while embryonic development occurs in diverse directions, resulting in adult individuals' having completely different structures (construction plans).

As we remember, all overground organs of vascular plants, in fact, represent a scheme of telome transformed in various ways, meaning a symmetrical branching of the cylindric shoot. In animals, such a fundamental construction scheme is the stage of a two-layer sack, corresponding to the gastrula, which contacts with the surroundings by a single opening: the "mouth-anus" (blastopore), and an oblong three-layer pipe derived from this sack that was formed at the moment of the creation of the second opening in the wall of the sack. Presently, the type of structure closest to the gastrula stage is represented by cnidarians, which digest their food in the gastrovascular cavity, take up nutrients, and remove undigested remnants through the mouth, corresponding to the blastopore. Afterward, a new opening appeared in the sack corresponding to the gastrula. In this way, the sack transformed into a pipe, the interior of which corresponds to the digestive tract (along with the origin of the mesoderm, the number of layers building the pipe walls increased from two to three). Generally speaking, all tissue animals starting from flatworms, roundworms and ringed warms, meaning both triclads and humans, are in different degrees transformed and developed versions of the same fundamental construction plan. Therefore, the most elementary structural property differentiating vascular plants from bilateral animals is that the construction of the former is based on the scheme of the telome, while the latter is on the scheme of a pipe (or, earlier, a sack). Of course, this difference results from the fact that animals, as heterotrophic organisms, need some interior to digest food, while land plants require richly branched structures (equipped with vascular bundles), ensuring a possibly large surface for photosynthesis.

While the most fundamental construction plan of all higher animals (*Eumetazoa*) is the same, their secondary construction plans exhibit huge diversity, being various modifications of this primeval scheme. Several main types of such modifications can be distinguished. Two of them are presented in Fig. 5-16. The starting scheme of the structure is presented in a simplified way as a cylindrical pipe, which is perforated along its long axis by the digestive tract that begins with the mouth (at the front of the animal) and terminates in the anus (at the back of the animal). This pipe is differentiated into the dorsal and ventral sides due to their different structures (e.g., the presence of different organs).

On the other hand, the left and right side of the pipe constitute their mutual mirror image, and therefore the animal has *bilateral symmetry*. The pipe is also equipped with organs of different sorts, both internal (e.g., neural ganglions, muscles, excretory organs) and external (e.g., jaws, limbs, gills). They are presented schematically in the figure as just one sample organ.

The first type of modification mentioned above of this basic construction plan consists of *segmentation*, which is the division of the body along its long axis into sections (segments) that are similar to each other (bottom left scheme in Fig. 5-16). The segmentation is associated with the multiplication of different organs (nephridia, transverse blood vessels, nerves and ganglions, muscles, legs, gills) with the number of copies corresponding to the number of segments (or fewer), and these copies are arranged one after another *along* the digestive tract, like the segments themselves. Additionally, septa can appear in the body cavity dividing it into separate compartments and separating particular segments (of course, they are perforated by the digestive tract, blood vessels and nerves). At the same time, the animal preserves its bilateral symmetry. An advantage of segmentation can be the increase in the number of the various organs (e.g., excretory organs) and improvement in the efficiency of movement through the division of longitudinal muscles into sections. This kind of segmentation has developed to a greater or lesser extent in many various groups of animals, for instance, annelids, velvet worms and their extinct Cambrian relatives, arthropods (trilobites, crustaceans, myriapods and insects, arachnids), primitive mollusks (*Monoplacophora* and chitons) and chordates.[25] This construction plan is associated with mobile rather than sedentary organisms.

---

[25] For instance, in lancelet, a primitive chordate, segmented are muscles (divided into so-called myomeres separated by myosepta), gill slits and arches, blood vessels and nerves. In man we can notice traces of segmentation in the form of vertebrae in the vertebral column, nerves coming out from the spinal cord or segmented abs (belly muscles).

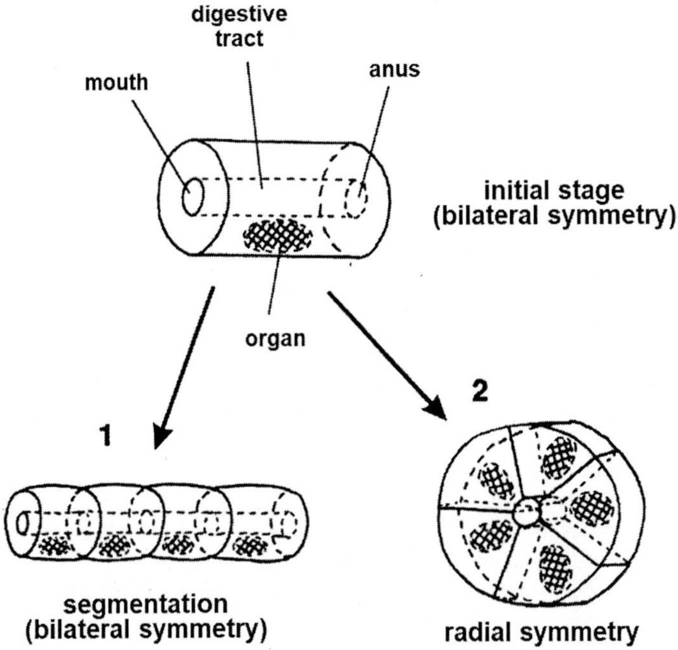

Fig. 5-16. Two types of secondary modification of the fundamental construction plan in the form of a simple pipe with bilateral symmetry: 1. segmentation – elements (organs) of the wall of the pipe copied along the axis fixed by the digestive tract with bilateral symmetry preserved; 2. radial symmetry – elements (organs) of the wall of the pipe copied radially around the digestive tract.

Modifying the primeval pipe can also consist of various organs or their complexes being multiplied not along but *around* the digestive tract. In such cases, similar body sectors are dislocated radially in different directions around the digestive tract, and the animal acquires radial symmetry (bottom right scheme in Fig. 5-16). Such symmetry is n-fold if the same set of organs is multiplied *n* times. The mouth opening and anus end here to fix the front-back axis, and they begin to lie on the top-bottom axis (this means that the animal has its mouth oriented upwards and anus oriented downwards, or inversely). Radial symmetry is typically associated with either sedentary organisms or those with limited mobility. Echinoderms with five-fold symmetry are typical examples of radial symmetry; these organisms have dislocated elements, e.g., ambulacral or nervous systems and arms and armor elements radially. Admittedly, four out of five currently living groups of echinoderms (starfish, sea urchins, brittlestars and sea cucumbers) contain organisms that can move (slowly and clumsily) on the sea bed, although their ancestors and extinct echinoderms were sedentary, attached to the ground by a long stalk, like present-day crinoids, or simply anchored in the mud.

Let us return, however, to animals segmented along the body axis. Examples of identical segments are not found anywhere among currently living and fossilized segmented animals. This is physically impossible—the first and last segments cannot have the same structures they would have if they were to connect to another segment—their free ends must be spatially separated from the external environment. Therefore, even in the hypothetical starting phase of segmented animals (of a given group), the first and last segments must differ slightly from the remaining segments. This phase is shown in Fig. 5-17a, where legs are an example of a multiplied organ due to segmentation. The fact that this phase is not encountered in nature (provided it appears at all) (although extinct Cambrian relatives of velvet worms, *Xenusia*, were not far from it) suggests that it underwent fast evolutionary transformations.

On the other hand, a slightly more advanced phase in which the first and last body segments have undergone some modifications is very common. In Fig. 5-17b, the specialization of the terminal segments is shown using the example of their appendages—the first two segments have fused into the head section, where the legs of the first segment are transformed into antennae, and the second segment into mandibles; the last segment forming the terminal body section does not possess legs at all (this is an abstract example, of course, and does not exactly represent any real group of animals). However, the remaining segments located in the medium part of the body remain (almost) identical. This type of segmentation is called *homonomous segmentation*. It is present in many groups, for instance, annelids (especially polychaetes), trilobites, centipedes and onychophorans (velvet worms).

In many groups of animals, however, further specialization of segments in different body parts occurred (Fig. 5-17c). For instance, in higher insects and crustaceans, the segments lying behind the head fused into one whole (thorax) equipped with well-developed legs, while the legs on the segmented rear part of the body (abdomen) underwent reduction or complete decline. A similar differentiation and specialization also concern internal organs in particular segments; in some segments, they develop strongly, while in others, they completely decline. This type of segmentation, consisting of a high level of specialization of particular assemblies of segments and the division of the body into clear sections, is called *heteronomous segmentation*. In addition to the aforementioned insects and crustaceans, it also appears in arachnids, where an additional fusion of the head and thorax into a cephalothorax occurs.

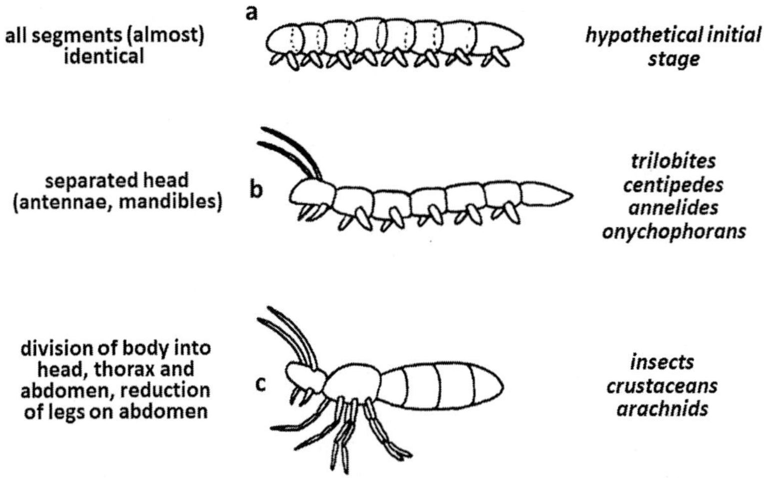

Fig. 5-17. Subsequent stages of the evolution of segmentation: a, hypothetical initial phase with all segments practically identical; b, phase of homonomous segmentation—developed head and terminal body section, medium segments practically identical; c, phase of heteronomous segmentation—differentiation of particular groups of segments fulfilling different functions, division of the body into clear sections: head, thorax and abdomen (alternatively cephalothorax and abdomen in arachnids), within which segmentation frequently declines.

Segmentation illustrates the previously discussed evolutionary rule: "first multiply, then differentiate." After the multiplication of a particular body section to produce a certain number of identical copies, differentiation of particular copies (segments) occurred, resulting in their specialization to fulfilling various functions. In this way, an important increase in the complexity of the whole organism took place.

Chordates (including vertebrates) probably never exhibited such complete segmentation as some invertebrates, e.g., polychaetes. Segmentation in chordates in the first order concerns muscles, gill slits (and arches), peripheral nerves (please recall the architecture of lancelet), and in vertebrates, the vertebrae. In higher vertebrates, segmentation has mostly declined.

Apart from changes in the general construction plan, animals were forced to produce various (systems of) organs. The size of multicellular animals frequently does not enable the efficient transport of various substances through the body through simple diffusion. Therefore, it was necessary to develop a blood circulation system for transporting nutritional substances and oxygen, an excretory system to remove the unwanted by-products of metabolism, and a respiratory system responsible for the intake of oxygen (and the removal of carbon dioxide). To receive stimuli from the environment and react adequately, sense organs, a nervous system and a motor system (primarily comprised of muscles) developed. Each of these systems had its own construction plan, frequently characteristic of particular groups of animals. However, this issue will not be further analyzed in detail here.

Let us return to the "explosion of life" in the Cambrian. The above discussion should facilitate understanding the simultaneous emergence of many different body architecture types (those described above and many others) and the appearance of multicellular tissue animals (the last event occurring in the Precambrian period). In later periods there were very few (if any) principally new construction plans (types of animals)—further developments consisted in the modification and complication of types that already existed. Therefore, almost the entire diversity of the animal world and basic constructive solutions had already been reached in the Cambrian period.

At the end of this subchapter, let us briefly investigate the evolution of vertebrates, which is most interesting for us as we belong to this group. The first chordates, to which vertebrates belong, namely cephalochordates, appeared during the Cambrian period, during the famous Cambrian explosion of life (lancelet is a present-day living example of them). Also, at the end of the Cambrian period, the first "true" vertebrates evolved: agnathans ("jawless creatures"). This was a systematic paraphyletic unit comprising armored (strong external skeleton) "jawless fish" (ostracoderms), the greatest abundance of which took place in the Devonian period, and "bare" conodonts equipped with characteristic "teeth" appearing from late Cambrian to the end of Triassic period (present-day examples of agnathans are lampreys and hagfish). Next, *Gnathostomata* (jawed vertebrates) evolved, of which the earliest forms were armored fish (placoderms) living in the Silurian and Devonian periods (as the very name suggests, their heads and the anterior part of their thoraxes were covered by a thick carapace); "spiny sharks" (acanthodians), whose fossils appear from the Silurian to the Permian periods (they had spines on the front edges of their fins and a row of evenly sized fins between the pectoral and pelvic fin). The latter was possibly the ancestor of cartilaginous fish (sharks, mantles) (having cartilaginous skeletons; they also appeared in the Silurian period) and bony fish (most of the fish living in the present day that appeared in the Silurian period). The latter underwent very early division into ray-finned fish (fin skeletons made of fish bones) and lobe-finned fish (muscled fins with bony skeletons; they respired using both gills and an air sack, through which they took up air from the atmosphere). Lobe-finned fish, in turn, comprise lungfish (long weak fins; of which four to six species remain living today) and coelacanths (solid bony skeleton of fins, abundant in the Devonian period; only the most famous "living fossil"—*Latimeria*—has survived to the present day). The latter "migrated onto land" in the Devonian period and, through a sequence of intermediate forms (*Tiktaalik*, *Ichtyostega*, *Acanthostega*), underwent a transformation into the first land vertebrates (tetrapods)—stem amphibians: labyrinthodonts.

The existence and reproduction of amphibians are strictly associated with water: they have wet skin, which participates in respiration, and they lay eggs in water, where their larvae (tadpoles) develop. In this respect, they resemble spore (cryptogamic) plants—lycopods, horsetails and ferns—for which a wet, watery environment is essential for living and reproduction. In the Carboniferous era, the first (basal) reptiles—cotylosaurs—originated (*Seymuria* constituted a transitional form). Reptiles were already principally independent of water—they are covered with a water-retaining skin, frequently equipped with scales or bony plates, they breathe exclusively using lungs, and they lay eggs surrounded by a thin envelope or hard shell, and they can survive in dry environments (reptiles are amniotes possessing fetal membranes, including an amniotic sac, that maintain the "water environment" inside the egg). Therefore, it can be said that they are, in a sense, the animal equivalent of gymnosperm plants.

Reptiles were divided into several groups, differing *inter alia* in the skull structure, particularly by the number and dislocation of temple openings called *fenestrae* ("windows") in the braincase. The main groups (although not necessarily monophyletic) are anapsids, diapsids and synapsids. A lack of temporal openings characterizes anapsids; cotylosaurs and some other extinct clades (e.g., pareiasaurs) belonged to them, as do some present-day animals—turtles and tortoises (some scientists think that they are, in fact, diapsids that secondarily lost their skull fenestrae). *Diapsids* possess two temporal openings, one above the other. They are divided into two main groups: scaled reptiles (lizards, snakes) and archosaurs ("supreme reptiles"). The latter is characterized *inter alia* by a four-chamber heart, an antorbital fenestra (an additional opening in the skull in front of the eye orbit) and teeth nested in jaws in teeth sockets. Belonging to this group are extinct thecodonts (paraphyletic group, generally resembling crocodiles on more or less erect legs), pterosaurs (flying reptiles) and non-avian dinosaurs and present-day crocodiles and birds (a sub-group of predatory dinosaurs—theropods).

Dinosaurs differentiated into several groups. Theropods (mostly carnivorous, some secondarily herbivorous dinosaurs, including birds) and sauropods (herbivorous giants with long necks and tails, the greatest land animals of all time) belong to saurischian (reptile-pelvis) dinosaurs. Ornithischian (bird-pelvis) dinosaurs (all herbivorous) are comprised of: ornithopods (principally bipedal, but later more massive forms could become partly quadrupedal), ceratopsians (quadrupedal, possessing bony horns and frills on the skull protecting the neck), stegosaurs (quadrupedal, with bony plates on the back and spines on the tail) and ankylosaurs (quadrupedal, covered with a heavy armor of bony plates, some with bony spikes on the back or a bony club at the end of the tail). All of these, apart from birds, became extinct at the break of the Cretaceous and Tertiary periods.

Synapsids are equipped with one temporal opening and the zygomatic arch, beneath which the muscles closing the lower jaw pass, and teeth differentiated into several types (in mammals: incisors, canines, premolars and molars). They are comprised of pelycosaurs, "mammal-like reptiles" (therapsids) and mammals (according to the cladistic taxonomy, mammals are also therapsids; see Subchapter 5.9). Mammals are covered with fur, have three ear ossicles, a four-chambered heart, and a large brain in relation to the body size, they feed their young with milk, they are viviparous (apart from oviparous monotremes), and newborn individuals can be either poorly developed (marsupials) or well developed (placentals).

Mammals came into being in the Triassic period and underwent a huge differentiation at the beginning of the Tertiary era (Paleogene) after the extinction of (non-avian) dinosaurs, which left many ecological niches empty. Many families of placental mammals originated, which belong to the following four main present-day groups that evolved more or less independently on different continents: *Xenarthra*, which are edentates (sloths, armadillos, anteaters) and are of South American origin; *Afrotheria*, which evolved in Africa and to which proboscideans (elephants, mammoths), sea cows, aardvarks, hyraxes and golden moles belong; *Euarchontoglires* evolved in Eurasia and are comprised of rodents (rats, guinea pigs, capibaras), lagomorphs (hares, rabbits), colugos, tree shrews and primates (prosimians, monkeys, apes, humans); *Laurasiatheria* originated in Asia and are comprised of insectivores (shrews, moles, hedgehogs), bats, pangolins, carnivores (cats, dogs, bears), odd-toed ungulates (horses, tapirs, rhinoceroses) and even-toed ungulates (antelopes, pigs, cows), including whales (!) (hippopotami are more closely related to whales, for instant orcas, than to other even-toed ungulates). One of the groups of placental mammals that underwent differentiation in the Paleocene period (first period of the Tertiary) were primates (prosimians, new world monkeys, old world monkeys, apes), which probably appeared already in the Cretaceous period, to which humans belong.

Human's closest living relatives are chimpanzees and dwarf chimpanzees (bonobos). The evolutionary lines of humans and chimpanzees diverged some 6-7 million years ago.[26] From then, the first known hominid appeared: *Sahelanthropus*, presumably resembling the common ancestor of humans and chimpanzees. The capacity of its braincase (~350 cm^3) was similar to that of the chimpanzee, and it was perhaps already partially bipedal. Next, evolution passed through forms similar to *Orrorin* (*Praeanthropus*) (about six million years ago) and *Ardipithecus* (about 4.4-5.8 million years ago) to become *Australopithecus*. This genus appeared over four million years ago and comprised several species, some representing slender forms and some robust forms (the latter is sometimes counted among the genus *Paranthropus*). Australopithecines were decidedly bipedal and slightly over one meter tall; nevertheless, their brains resembled the ape brain by size (400 cm^3) and structure. Therefore, the emergence of typically human features: bipedal walking and having a large brain, did not occur simultaneously.

The first species of our genus *Homo* was *Homo habilis*, with a brain larger (~600 cm^3) than those of chimpanzees and *Australopithecus* and having already a typically human structure, and height little less than modern humans. *Homo habilis* appeared about 2.5 million years ago and probably already produced tools. A further increase in the brain volume (to 800-1000 cm^3) took place some two million years ago in a subsequent form—*Homo ergaster*—in developing modern humans. Undoubtedly, they produced tools. Until then, the whole evolution of hominids occurred in

---

[26] When the last common ancestor of two considered species lived, for instance the human and chimpanzee, can be assessed based on the so-called *molecular clock* (Futuyma 2013, 39-40, 48-49). If we know the pace of accumulation of mutations and the number of mutations by which two considered species (or higher evolutionary units) differ, we can estimate when their evolutionary lines diverged. In the case of man and chimpanzee, the molecular predictions have been confirmed by fossil data. Similarly, the sequencing of mitochondrial DNA (mtDNA) (mitochondria are inherited only by the mother lineage) in various ethnic groups enabled the conclusion that the "mitochondrial Eva," that is the donor of all mitochondria (which does not mean all nuclear genes that undergo shuffling in the result of recombination in sexual reproduction) for all contemporary people lived about 150-200 thousand years ago in Africa. This prediction also seems to have been confirmed by fossil data.

Africa. The first, *Homo erectus*, an Asian variety of *Homo ergaster*, left the "black land" and populated South-East Asia about two million years ago.

Meanwhile, the evolution of hominids in Africa proceeded onwards. About 600,000 years ago, *Homo heidelbergensis* appeared; it was presumably a common ancestor of Neanderthals (*Homo neanderthalensis*) and modern humans (*Homo sapiens*). Around 500,000 years ago, it populated Europe (and the Near East), whereas about 400,000 years ago, it gave rise to the Neanderthal.

Homo neanderthalensis had a stocky body structure, prominent supraorbital banks and a brain that was, on average, slightly larger (1400 cm^3) than the *Homo sapiens*' (1350 cm^3). However, while in modern humans, the prefrontal cortex, responsible for "higher" brain functions—thinking, planning and making decisions—is outstandingly developed, in Neanderthals, the occipital (and parietal) cortex underwent the greatest enlargement (hence the elongated skull of Neanderthals). This structure was responsible for integrating visual stimuli; this was important as Neanderthals lived at high latitudes, where it was relatively dark for a long part of the year (Neanderthals also had relatively large eyes). *Homo sapiens*, characterized by a slender body, delicate skull bones, a high front and a protruding mentum in the mandible, originated in Africa 200-300 thousand years ago. However, about 100,000 years ago, humans began migrating to other continents. Around 90,000-80,000 years ago, they populated the Near East, 70,000-60,000 years ago—South-East Asia, and 50,000 years ago—Australia. Around 45,000-40,000 years ago, humans reached Europe, pushing out Neanderthals, who vanished about 25,000 years ago due to competition or physical extermination. The sporadic crossing of these species occurred—about 2-4% of genes of native Europeans are of Neanderthal origin. Around 40,000 years ago, *Homo sapiens* populated Siberia; 25,000-15,000 years ago, they passed through the Bering Strait to North America and about 14,000 years ago, they reached South America through the Panama corridor.

The ancestors of humans, apes, possessed appropriate pre-adaptations that enabled the origin of our species. These adaptions include prehensile limbs, especially the upper limb, stereoscopic vision (both eyes oriented forwards) and a large brain. These constitute adaptations to arboreal and social styles of life, which are characteristic of most primates. The prehensile limbs and the opposed thumb and big toe enable holding on to tree branches and efficiently moving among them—a fall from a great height for an animal as large as an ape risks disability and constitutes a threat to life. Stereoscopic vision, associated with overlapping vision fields of both eyes to a large extent, enables the precise estimation of the distance and efficient locomotion in the maze of branches, including jumping from one branch to another. In order to efficiently integrate visual stimuli, a well-developed visual and associative (occipital and parietal) cortex, and therefore a larger brain, is needed. Efficient locomotion in the tree crowns also requires excellent coordination of movements (precise spatiotemporal patterns in the contractions of particular muscles). This is ensured by a well-developed motor and premotor cortex (localized in the frontal cortex) and cerebellum. The finding of dispersed clusters of fruits ripening in a given season requires a well-developed episodic memory, situated primarily in the temporal and parietal cortex. Additionally, complex social relations also require a bigger brain, in this case, the prefrontal cortex (planning of strategies of social behaviors) and the temporal cortex (remembering the results of previous contacts, which is useful in, for example, reciprocal altruism; incidentally, the prehensile hand was also very useful in the grooming of other individuals) (Dunbar 1998, 2005, 2016). Generally, anthropogenesis likely involves a network of reciprocal stimulations of various properties/phenomena, leading to the self-driving of the whole process and the overall acceleration of the evolution of humans (Korzeniewski 2020).

After "coming down" from trees due to the cooling of the climate and the drying of the landscape in eastern Africa several million years ago, the above pre-adaptations appeared useful for other purposes. After transitioning to bipedal locomotion and freeing the upper limb from locomotor functions, it could be used, because of its prehensility, for manipulating various objects, for instance, collecting food, the production and use of tools, hunting and care of progeny. The well-developed brain allowed good orientation in the territory, which is useful, for instance, in the search for food, hunting or finding shelter, the production of tools, the planning of future actions (including hunting strategies) and developing complex social relations. The evolutionary development of the brain and the prehensile hand was a self-driving circle (positive feedback loop). The large brain was capable of efficiently integrating and processing sensory data from sense organs, accumulating large amounts of information (memory, learning) and controlling movements, especially hand movements. The prehensile upper limb that enabled the performance of complex movements and manipulation of various objects was able to execute complicated commands of the brain; together, they constituted a self-stimulating to development system promoting the further evolution of each other (Korzeniewski 2020). The developed brain was also useful in achieving success in social relations, which was associated with the efficient propagation of own genes (own identity).

Additionally, spatial (stereoscopic) vision was very useful for the production and manipulation of tools and hunting (predatory mammals, such as canines or felines, also have eyes oriented forwards and see stereoscopically). Undoubtedly, the evolution of humans, particularly social and cultural, underwent dramatic acceleration at the moment of the "invention" of a symbolic language. A very useful pre-adaptation to this phenomenon was social communication in a herd (for instance, warning against predators) by means of emitted sounds. Generally, the process of anthropogenesis comprised a network of mutually-driving mechanisms and phenomena (positive feedbacks) that eventually also led to the emergence of self-consciousness (Korzeniewski 2020).

Therefore, in order for man to originate, his ancestors had to first "climb onto the trees" and develop social relations, which conditioned developing the prehensile upper limb, stereoscopic vision and large brain, and then "go down from the trees" onto open areas. Bearing this in mind, the evolution of man must pass through the *ape* (and earlier—monkey) phase. For many people, this is a cause of psychological discomfort; they prefer to say that man had common ancestors with apes/monkeys. This is obvious hypocrisy because if these ancestors lived today, they would certainly be counted as apes/monkeys.

Was the origin of humans probable or even in some way predestined and inevitable? Of course not! A long sequence of extremely improbable and immeasurably rare events led to the emergence of our species during evolution. From among thousands and millions of such circumstances, I will enumerate four. If, in the process of the origin of life on earth, there had come into being another, perhaps simpler, architecture of the genetic apparatus of the first organisms, probably completely different, maybe much less complex forms of life would have evolved. If, again, due to unpredictable random events (e.g., mutations), a completely different set of construction plans emerged during the explosion of life in Cambrian, chordates and vertebrates, consequently also, humans would not have come into being. If the collision with an asteroid 65 million years ago had not occurred, most probably dinosaurs (non-avian), which fortunately for us became extinct due to this impact, would still be ruling earth today. If not for the cooling of the climate a few million years ago and the steppe formation in the place of forests in east Africa, our ape ancestors would not have come down from trees to the ground and given birth to the genus *Homo*. Generally, if evolution on our planet was run again from the beginning, completely different life forms would most probably come into being than those currently known and whose fossils from previous geological epochs have been excavated.

Therefore, the origin of the species *Homo sapiens* was extremely, even astronomically, improbable. “Well then”, somebody could ask, “why has our species come into being at all?”. The question is wrongly posed. SOMETHING had to originate due to evolution; fortunately, this something just happened to be humans. The origin of any other concrete form of life (species) was equally improbable. It makes no sense to ask why our species evolved on the earth. If we randomly picked one number from the range of one to one billion (1-1,000,000,000), for instance, 750,234,598, then afterward, *post factum*, one could wonder why just this number was drawn, not another, and deliberate about how improbable this event was. However, after all, SOME number had to be drawn, and it just happened that it was this one and not another. On other planets, the process of the origin and evolution of life most probably proceeded completely differently than on earth.

It should also be emphasized that while the origin of just the species *Homo sapiens* was extremely improbable, then the origin of some beings endowed with intelligence, subjective psyche and self-consciousness was somewhat more likely. Nevertheless, in my opinion, while the spontaneous emergence of life from inanimate matter is very likely (perhaps even unavoidable) when there appear appropriate conditions, reason and consciousness—especially of a kind that allows developing civilization, fine arts and science—come into existence rather seldom in the process of evolution (for instance on other planets in the universe). Apart from primates, there seem to be no other candidates on earth for ancestors of creatures endowed with intelligence and self-consciousness. A completely improbable coincidence of circumstances led to the coming into being of reason on our planet. For instance, dolphins, regarded as the most intelligent animals apart from primates, do not possess (and, as water animals, cannot develop) the prehensile hand necessary for manipulating tools, without which a technical civilization cannot develop.

First multiply, then differentiate

An immensely important rule related to increasing the degree of the complication of the structure during biological evolution that has often been stressed previously is a very efficient and frequently applied principle (its name comes from me): “first multiply, then differentiate.” It consists of the fact that initially, we have a single element of structure (or associated with the genesis of structure); next, we multiply it to obtain a certain number of identical copies, and in turn, each of these copies evolves in a different direction, and as a result, we have many copies that now have different structures that can fulfill different functions. This simple mechanism can quickly lead to the increased complexity of the structure (and function) and the specialization of its particular elements. Indeed, the principle "first multiply, then differentiate" results, at least on the level of macroscopic morphological properties, from the previously described cascading and hierarchical mechanism of the generation of structure. Some elements of structure can be copied due to multiple triggering of the same part of the morphogenetic cascade in different places within the organism, and next, the fragments of the cascade leading to the production of particular copies of this element of structure can undergo modification, leading to the differentiation of these copies. The discussed principle ubiquitously appears in the process of biological evolution on different levels of the hierarchy of the organization of structure and frequently leads to an increased amount of these levels.

To avoid being groundless, I will give several examples of the discussed rule, to some extent already discussed above. Many proteins, especially enzymes, are composed of so-called domains fulfilling various functions, for instance, catalysis, binding substrates, binding metal ions, anchoring in cellular membranes, binding to DNA, connecting other domains, *etc*. Of course, domains in the protein chain composed of amino acids correspond to concrete fragments of the gene (sequences of nucleotides in DNA) encoding this protein. Strong evidence exists that many proteins originated through the "shuffling" of particular domains (more strictly—gene fragments corresponding to them—sometimes these are exons) (Futuyma 2013, 555-556)—see the righthand scheme in Fig. 4-3. Particular domains have various properties and fulfill the role of building blocks—various configurations of the arrangement of domains in the polypeptide chain forming proteins are predestined to fulfill a broad range of functions. Many cues suggest that at least some domains come from a "common ancestor"—a primeval domain that underwent multiplication in the genome to create many copies that evolved in divergent directions. The shuffling of domains enormously increases the number of proteins occurring in the organism.

Additionally, alternative splicing can produce many proteins on the matrix of a single gene (and appropriate mRNA) (Futuyma 2013, 190-191). In this process, particular exons of a gene may be included within or excluded from the final mRNA produced from that gene (from pre-mRNA synthesized on the matrix of the gene, still containing introns). Therefore, the proteins translated from alternatively spliced mRNAs will contain different combinations of amino acid chain fragments and, consequently, will perform different biological functions. This phenomenon additionally increases the number of proteins produced and the degree of complexity of the whole organism.

A similar situation took place in the case of whole proteins. The genes encoding them underwent duplication or multiplication, due to which two or more identical copies originated. One such copy, as before, coded for a protein fulfilling a certain function in the organism. On the other hand, the remaining copy underwent numerous mutations and thus accumulated genetic changes (Futuyma 2013, 73-74, 556-562)—see the lefthand scheme in Fig. 4-3. This did not threaten the loss of functionality of the encoded protein as it was still produced based on the first, unchanged copy of the gene. However, it could turn out that the protein encoded by the second modified copy of the gene is useful for performing other functions in the organisms; for instance, it could catalyze a different reaction than the original protein. This way, the new copy can become a new gene encoding a new protein. Such a process of duplication and differentiation of genes can occur many times. Proteins that separated from the common protein-ancestor relatively recently (in the evolutionary scale of time), and therefore exhibit a large similarity in the sequence of amino acids, form the so-called families of proteins (and the genes encoding them—families of genes) (Futuyma 2013, 73-74, 556-562). An example of such a family is myoglobin (being a monomer protein carrying oxygen within muscle cells) and subunit α and subunit β of hemoglobin (protein being a tetramer, composed of four subunits: two α and two β, carrying oxygen in the blood) (in fact, more genes encoding subunits of hemoglobin, active in the fetus or not active at all, belong to this gene/protein family). Of course, the origination of the discussed families leads to an increased degree of complexity of the organism and the diversity of its functions.

Similarly, complexes of homeotic genes HOM/Hox containing up to about 13 homeotic genes originated, determining the differentiation of subsequent body sections during embryogenesis—particular genes in these complexes most probably derived from a common gene-ancestor (Twynam 2001, 235-243; Futuyma 2013, 566-571). In vertebrates, the principle “first multiply, then differentiate” was applied twice to Hox genes on two levels. The entire Hox complex was multiplied to create four copies (they participate in developing the entire body and limbs) (Futuyma 2013, 569). Therefore, the multiplication and differentiation of a single homeotic gene led to the origin of the Hox complex, which in turn was multiplied and differentiated once again in vertebrates producing four copies of the entire complex fulfilling separate (although to some extent overlapping) functions.

One of the most evident and spectacular examples of the rule “first multiply, then differentiate” are different types of cells and tissues built from them in more complex multicellular organisms. During the origination of multicellular organisms, the copying and differentiation occurred of perhaps the most fundamental element of the structure of living systems—the living cell. This was presumably initiated by a mutation causing a lack of the separation of cells of unicellular organisms after the cell division. The first multicellular organisms were composed of identical cells—in the present day, a pellet of several/several tens of identical cells represents a free-living phase (*Eudorina*, *Pandorina*). However, the cells underwent differentiation relatively soon. First, vegetative (somatic) cells (responsible for feeding and other life functions) and generative cells (responsible for reproduction) separated (Maynard Smith and Szathmáry 1995, 210-215). This phase is represented by volvox and *Mesozoa* (primitive parasitic multicellular animals).

In the next stage of biological evolution, various types of vegetative cells (and tissues), nervous, muscle, secretory, epidermal *etc*., separated; these cells became very different from each other and were specialized in fulfilling specific, strictly determined functions in the organism. Some hundreds of different cell types can be distinguished in the highest organized organisms, including humans. As a rule, different cell types possess identical genetic information, and their dissimilarity results from the fact that different assemblies of genes are expressed in them (see Fig. 5-8). The direction in which a given cell develops during embryogenesis is determined by the positional information specified by an appropriate combination of morphogen concentrations (Fig. 5-6). Usually, the cell cannot turn back from its developmental path once it has been determined—during cell division (mitosis), the "fate" of a given cell is passed on to its progeny (for instance, due to chemical modification of appropriate DNA sections). This can occur very early in embryonic development. For instance, a closed pipe called the neural tube is formed in an early human embryo (phase of neurula) due to herniation of the embryo wall into the embryo body above the notochord along the front-back axis. From this, the central nervous system (brain + spinal cord) develops, while in order for peripheral nerves to be formed, the protoplasts of the nerve cells must migrate from the neural tube to the appropriate parts of the embryo (Twyman 2001, 269-322). Therefore, this is the neural tube where the determination of the fate of the cells entering into its composition takes place. Similarly, determining cells developing into epidermal and muscle tissue also occurs early in embryogenesis (Twyman 2001, 8-12, 187-204).

Generally, it is the case that evolution and an increase in the complexity of the structure of living organisms occur through modifying the existing structure and, more strictly—through modifying the hitherto-acting developmental cascade. For this reason, embryos frequently possess organs of ancestors of a given organism that vanish in the adult stage—for instance, at a certain stage, the human embryo possesses a tail and gill slits and arches (these are, of course, properties of our fish ancestors) (Futuyma 2013, 54). This can be regarded as an additional rule: "instead of creating anew, modify that which already exists" (see Subchapter 5.7).

As discussed above, with slight simplification, it can be said that the entire general structure of the overground organs of vascular plants—stems, branches, leaves and flowers—together with their complex architecture, originated through the multiplication and differentiation of one very simple element: the telome. Initially, the scheme of the structure of the telome was copied in such a way that some telomes grew from the endings of other telomes. The first land vascular plants, which were the most primitive *Psilophytales* (as I mentioned, this is an umbrella taxon in which various distantly related primitive [pre-]vascular plants are grouped), represented just this construction plan—their overground parts were comprised of several branched shoots of identical thicknesses (most of the "free" endings of the shoots were terminated by sporangia). Later, during evolution, the modification and differentiation of particular telomes led to forming of various overground organs of more advanced vascular plants. The stem was formed through the loss of equivalence in bifurcations for a sequence of telomes, one ramification of which was thicker (superior) and entered into the composition of the main axis of the stem, while other, thinner offshoots became side branches. The leaf, in turn, was formed due to the arrangement of the bifurcations of particular telomes in one plane. Next, the pith underwent flattening and filled the space between the vascular bundles forming the blade of the leaf; vascular bundles became the "veins" of the leaf.

However, this is not the end of the whole story. Here, the principle "first multiply, then differentiate" again acts on various levels—it applies twice to the same structural elements as leaves also underwent multiplication, modification and differentiation. Many of them still fulfilled photosynthetic functions, while others were transformed into various parts of the flower: sepals, petals, filaments and styles (styles, as a rule, originated due to the fusion of a few leaves) (the leaves being "ancestors" of filaments and styles had sporangia on their ends or surface). It is difficult to believe today, watching a magnolia, oak or tulip, that they are, in fact, only modified assemblies of telomes.

While the most general element of the structure in plants can be identified with the complex of equivalent ramifications—telomes, then the most fundamental scheme of the structure of animals is the pipe—an elongated body penetrated by the gastrointestinal tract, beginning with the mouth and terminating with the anus. Various internal and external organs have developed in more complex animals, e.g., brains and nerves, heart and blood vessels, skeletal muscles, excretory organs, legs, *etc*. Here the principle “first multiply, then differentiate” can also significantly increase the degree of complexity and specialization of structure. As mentioned earlier, various organs and structures of the body (the pipe) can be multiplied in two different ways. First, repeating copies of a particular structure element can be arranged radially around the digestive tract; the animal then gains radial symmetry. This appears, for example, in echinoderms and is associated with sedentary organisms or those of limited mobility. Second, particular organs and body segments can be multiplied one after another along the pipe (the mouth-anus axis). The construction plan of the animal then becomes segmented (Fig. 5-16). Initially, particular segments (apart from the first and last) are almost identical, which is manifested, for example, by their very similar size and shape and the possession of the same internal organs and legs. Next, particular segments can undergo differentiation and specialization during evolution; their size and shape change, some fuse, some internal organs are enlarged, and others become smaller or are eliminated. Some segments may also develop wings. Generally, the body undergoes differentiation into the head, thorax (which can additionally fuse into a cephalothorax) and abdomen; each of these sections fulfill specific, specialized functions (Fig. 5-17). Due to the above process of multiplication and the differentiation of segments and organs “contained” in them, increasing the complexity of the body structures and efficiency of their functions takes place.

In chordates, the segmentation is relatively limited and concerns, for example, gill slits, muscles, nerves, blood vessels and later, the vertebral column. Primeval vertebrates (agnathans) possessed more or less identical gill arches—elements of the skeleton sustaining the septa between particular gill slits. Gill arches have gone through extensive modification during the evolution of vertebrates, particular arches considerably changed their structure and function, and in mammals, became objects as different as maxilla (upper jaw) and mandibula (lower jaw) bones (palatine and dentary bones), hyoid bone, numerous cartilages of the larynx and trachea and three ear ossicles: the malleus, incus and stapes (Kardong 2002, 236-237; Reece et al. 2011, 705, 721; DeLaurier 2018). This is illustrated in Fig. 5-18.

The teeth of amphibians and reptiles are identical or almost identical. However, teeth of mammals that evolved from stem reptiles are differentiated into incisors, canines, premolars and molars, fulfilling different functions. This is also an example of the diverse evolution of particular items of some multiplied, initially identical structure.

The vertebrae in fish are almost identical (they differ to some extent by size). However, the vertebrae in most mammals, including humans, are differentiated in structure and form different segments of the vertebral column: cervical spine, thoracic spine, lumbar spine, sacrum and coccyx. In whales, vertebrae and teeth have undergone secondary uniformity; of course, in *Mysticeti* (which are toothless whales), the teeth disappeared and have been replaced by baleens (Futuyma 2013, 63-63).

I think that the above-quoted examples are completely sufficient to support the thesis that the principle “first multiply, then differentiate” constitutes a ubiquitous and enormously efficient mechanism for increasing the degree of complexity of the structure and the functioning of living organisms. It also introduces some order to the huge diversity of this structure, demonstrating that biological systems are not as hopelessly complicated as they appear to be.

Finally, the principle “first multiply, then differentiate” also refers to the “super-organism” constituted by the colony of social insects, e.g., ants or termites (Wilson 1971; Maynard Smith and Szathmáry 1995, 255-278) (as discussed in Chapter 6, such a colony constitutes a living individual in the cybernetic sense). Here we encounter various casts differentiated by the structure of particular insects. These casts include: sexual casts—the queen and males fulfilling reproductive functions, and “somatic” casts—workers of different sizes that get food (forage), take care of progeny and sexual casts, build the nest, *etc.,* and soldiers that defend the colony. In many species, workers and soldiers are differentiated into several sub-casts differing by their size and functions (workers and soldiers are sterile; in termites, they can be both sterile males and females, while in hymenopterans—wasps, bees and ants—they are exclusively sterile females) (Wilson 1971). The differentiation of particular insects within the colony corresponds quite strictly to the differentiation of cells in a multicellular organism (into reproductive cells and various somatic cells fulfilling various functions) (see Chapter 6).

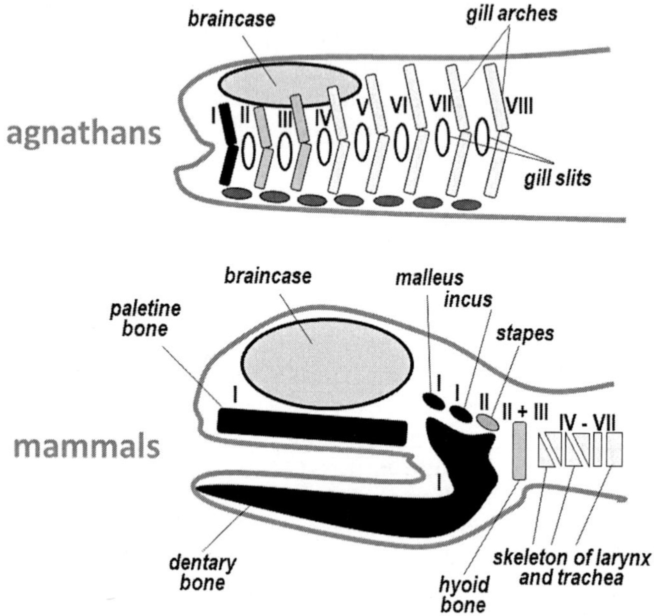

Fig. 5-18. Differentiation of initially almost identical gill arches (skeleton of gill slits) in agnathans into various structures, such as bones of jaws, hyoid bone, the larynx and trachea cartilages, and three auditory ossicles (malleus, incus and stapes) in mammals. The Roman numerals from I to VIII indicate consecutive gill arches in agnathans and the structures in mammals evolved from them.

Symbiosis—the exchange of services

A subsequent evolutionary strategy can be formulated as follows: "if somebody has something you do not have, and you have something they do not have, establish cooperation with them." At the same time, the balance of gains and losses counts—the benefits drawn from the cooperation must surpass its costs. The result of this principle is the broadly-understood phenomenon of *symbiosis*. In fact, a more appropriate term for the discussed phenomenon is *mutualism* (reciprocation) (Futuyma 2013, 529-530), rather than symbiosis (common living), but as the word "symbiosis" is so deeply rooted in common language, I will simply use that. As already discussed, the origin of life was, at least in my opinion, the result of establishing "cooperation" between the autocatalytic cycles of proteins and nucleic acids, which led to forming hypercycles. The endosymbiosis of prokaryotic (bacterial) ancestors of mitochondria with proto-eukaryotes (archaea?) leading to the origin of eukaryotes proper has already been discussed. Bacteria brought a very efficient bioenergetic system to this cooperation (membrane phosphorylation—ATP synthesis driven by the proton gradient), while proto-eukaryotes contributed an efficient and complex genetic apparatus and a complex cell structure. Finally, lichens provide a flagship example of symbiosis, as they are an assembly of two components: fungi and algae. The algae cells synthesize organic substances in photosynthesis, while the fungi efficiently take mineral substances from a very poor bed (rock, tree bark) and protect the algae against harmful environmental influences. Symbiosis quickly and efficiently leads to the increased complexity of living systems.

Some more rules…

Some additional rules of the course of biological evolution can be quoted. Among these, the rule "slow and steady wins the race" means that even huge evolutionary changes are mostly the sum of tiny changes occurring step by step, for example, in proportions of various body parts and organs, while large "jumps" in an accidental direction resulting from macroscopic mutations most frequently terminate with breaking of legs. The reasons underlying this rule have been discussed above. Another previously mentioned rule is: "instead of creating anew, modify what already exists." This concerns, *inter alia*, the origin of bat wings from the front legs of their ancestors, formation in the brain of the gland secreting hormones regulating, for example, the cycle of circadian activity—the pineal gland—from the third parietal eye or the above-mentioned adjustment of particular gill arches to fulfilling very different functions.

Figure 5.19 presents modifications, in various groups of mammals, to the primeval construction plan of the front limb of land vertebrates, present first in tetrapods—labyrinthodonts (descendants of crossopterygians), and also in many present-day organisms, e.g., salamanders, lizards and crocodiles. The primitive, stocky, five-finger limb underwent various changes. In horses, all fingers apart from the middle finger (number 3) retracted or underwent far-going reduction, while the third finger was strengthened, the bones of the forearm and the third finger fused, and a well-developed hoof appeared at its end. This is, of course, an adaptation to persistent locomotion, e.g., to run on open grass areas. In bats, the fingers underwent extreme elongation—the flying membrane is stretched between them. In whales, the bones of the arm and forearm became very sturdy, while the fingers underwent elongation, and their phalanx bones, through multiplication, formed the skeleton of the pectoral fin. Finally, in humans, a slender hand developed (the upper limb) with an opposable thumb (inherited from our arboreal ape ancestors), useful in manipulating and producing tools. Despite their apparently very different appearance, all of these "variations on a theme" of the front limb of labyrinthodonts have the same relative arrangement of the arm, forearm, carpus, metacarpus and fingers bones, a similar course of nerves and blood vessels, and are characterized by comparable development during embryogenesis. Therefore, they are *homologous organs*, coming from one organ-ancestor and adapted through various modifications to fulfill different functions.

Structure makes structure

Finally, it is worth mentioning one more very important rule of biological evolution. Before this, let us recall some facts. Life originated on the earth at least 3.8 billion years ago. For at least half its history, it only had the form of prokaryotic organisms—bacteria and archaea. During that time, probably nothing important happened in the evolution of the general structure of living organisms. Only about 1.5 billion years ago (perhaps slightly earlier) did the first "nuclear" organisms appear—eukaryotes. From this moment, a clear acceleration of the evolutionary increase can be observed in the complexity of structure in some developmental lines. Multicellular organisms (animals) originated about 600 million years ago in the late Proterozoic era (Precambrian) (so-called Ediacaran fauna); it was only about 540 million years ago that their clear differentiation took place, resulting in an increased degree of complexity and the appearance of many different construction plans (the Cambrian "explosion of life").

Around 400 million years ago, more or less, life "came ashore" and populated the land, where the multiplicity of environments promoted the fast differentiation of organisms. Subsequently, increasingly complex amphibians, reptiles, supreme and mammal-like reptiles, and finally, birds and mammals evolved. In the end, our own species (*Homo sapiens*), endowed with a disproportionately large brain, intelligence and psyche, and the ability to create civilization, culture, science and fine arts, appeared on the earth "at the last moment", some 250,000 (200,000-300,000) years ago. Generally, the above facts (and also many others) constitute an example of the overall regularity that the higher structurally organized living organisms are, the faster they become (or at least have the potential to become) yet more complex. In other words, we are dealing with the principle (also named by me) "structure makes structure", analogous to the famous economical law: "money makes money." This results from the cumulativeness of biological evolution. New properties originate based on old properties—therefore, the more properties (aspects of structure and function) already exist, the more new properties (aspects of structure and function) can develop. If the increase in the degree of complexity of the construction plan occurs through modification in the course of embryonic development (the hierarchically-organized morphogenetic cascade), the possibility of modification is proportional to the current degree of complexity of this development.

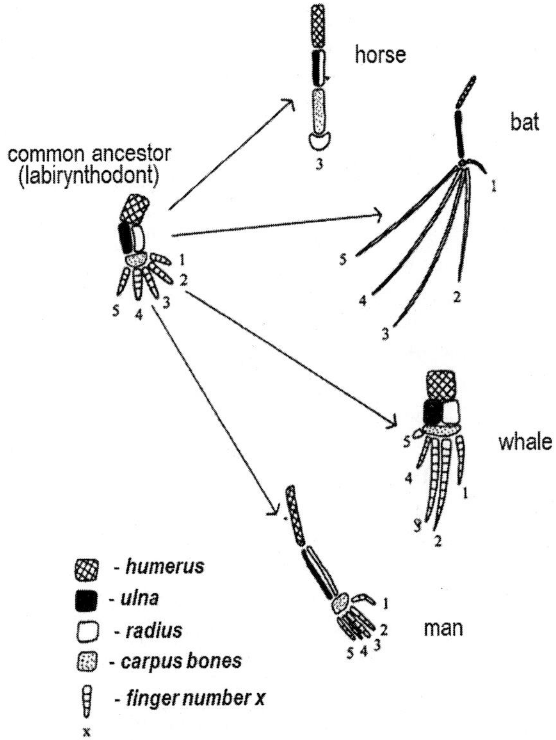

Fig. 5-19. Skeleton of the front limb in land vertebrates—the primeval construction plan in the common ancestor (left) and examples of modifications of this plan in different mammals (right). Despite changes in the size and proportions of particular parts, their mutual spatial arrangement remains the same.

It is also difficult to resist the impression that the recent explosive development of science, technology and civilization constitutes a simple continuation of the very quick evolution of the human brain and psyche over the last millions of years.

On cladistic taxonomy

It has been mentioned several times within the pages of this book that birds are dinosaurs (a sub-group of predatory dinosaurs—theropods), while all mammals are reptiles, reptiles belong to amphibians, *etc*. Such statements seem to contradict intuition, common sense and traditional systematics (taxonomy). Nevertheless, they have a deep meaning. They constitute a derivative of *cladistic taxonomy* (the discipline dealing with the methods of classification of living organisms) that is commonly used nowadays, and that best reflects the course of the evolution of the living world and mutual kinships between living organisms (Futuyma 2013, 23-30).

Cladistic taxonomy is based on the following postulates:

- every group of organisms possesses only one *sister group*, which is closest related to it;
- together, they form a higher systematic unit, which in turn possesses its own sister group; and
- a systematic group must fulfill two conditions:
    - ▶ all organisms belonging to a given group must come from one common ancestor (the group is monophyletic rather than polyphyletic—see below);
    - ▶ a given group must contain all organisms coming from the common ancestor (the group is monophyletic rather than paraphyletic—see below).

The cladistic system is organized hierarchically: (sister) groups of a lower order enter into the composition of systematic units of a higher order, which in turn have their sister groups, with which they form systematic units of yet higher order, *etc*. Many levels of hierarchy are present here. Sometimes, for the sake of simplicity, the strict rigorousness is not followed, and many units (e.g., species) are bracketed into one group (e.g., genus) without indagation regarding which of them are more and which are less closely related to each other. This concerns in particular fossil forms, for which it is impossible to carry out tests of similarity of the sequences of nucleotides in DNA and RNA or amino acids in proteins.

An advantage of cladistic taxonomy is its consistency and evolutionary rigorousness. However, its disadvantage is that it is rather awkward in practical application, creating a huge number of systematic units (sister groups) and levels of hierarchy, and uses the names of these units in ways that are different from the traditional meanings.

Thus, reptiles in the common understanding (e.g., lizards, snakes, tortoises and crocodiles from among present-day forms and non-avian dinosaurs, pterosaurs, ichthyosaurs and plesiosaurs from among the extinct forms) is an incorrectly created group from the point of view cladistic taxonomy, as it does not comprise all groups deriving from the common ancestor—basal reptiles (cotylosaurs). Such a group is called a *paraphyletic group*. A correct group should also contain mammals (and their ancestors) and birds. Birds are a sub-group of theropods (predatory bipedal dinosaurs), and more broadly, reptile-pelvis dinosaurs (saurischians)[27] and dinosaurs in general. All traditionally understood amphibians and reptiles, together with birds and mammals, belong to amphibians in the cladistic meaning as all of these groups derive from primeval amphibians—labyrinthodonts.

On the other hand, the protozoans and algae are incorrect cladistic systematic units, comprising either non-photosynthesizing or photosynthesizing unicellular (or forming thallus) eukaryotes, which are not related to each other at all and differ principally by cell structure. Such groups are called *polyphyletic groups*. For instance, unicellular collar flagellates (*Choanoflagellata*) constitute the sister group of animals (poriferans and higher animals). This entire group is much more closely related to fungi and land plants (and chlorophytes, especially charophytes, from which the latter come) than with all other eukaryotes, in particular, most "protozoans" and "algae" (once photosynthetic bacteria—cyanobacteria—were counted as algae, and consequently as plants).

According to cladistic taxonomy, nothing like apes or great apes exists (unless we count humans). A sister group is formed here by man and two species of chimpanzees—the gorilla is slightly less related to this group, forming together with it a greater group. The orangutan, in turn, constitutes its sister group; a few species of gibbon form the subsequent sister group, and so forth. For most of its history, humanity evolved in Africa, where its greatest diversity was found (including genetic diversity). The descendants of only a small sub-group of native Africans populated all continents except for Antarctica.

---

[27] It can sound paradoxical that birds belong to reptile-pelvis (saurischians), and not bird-pelvis dinosaurs (ornithischians), but the apparent similarity of the pelvis of the latter to the pelvis of birds results from convergent evolution, and not from close evolutionary kinship.

CHAPTER 6

WHAT ARE LIFE AND LIVING INDIVIDUALS?

Introduction

The previous chapters primarily covered the popularization of achievements of science, although from my own personal perspective. This chapter is mainly devoted to the cybernetic definition of life (a living individual) as proposed by myself. The subsequent chapter deals with the cybernetic model of self-consciousness (a psychical individual) , also of my authorship. Although, to a large extent, this definition/model is aimed at popularizing contemporary knowledge in biology, neurophysiology and cybernetics. It goes beyond strictly popular-science frames and thus demands a greater intellectual effort from the reader. I do not expect that all, or even most of the receivers of my ideas, will fully, or at least partly, accept them. Nevertheless, I hope that regardless of whether this happens, the reading of these chapters will supply a certain dose of intellectual satisfaction.

As mentioned before, life can be characterized from the point of view of its various aspects. These are:

- the genetic aspect
- the structural aspect
- the thermodynamic-energetic aspect
- the biochemical aspect
- the metabolic aspect
- the redox aspect
- the enzymatic aspect
- the signal/information transduction aspect
- the regulatory aspect and
- the developmental (morphogenetic) aspect

All of these contribute important information concerning the phenomenon of life. All are discussed, at least superficially, within the pages of this book. Nevertheless, some of these aspects are "excessive" in the sense that they are not necessarily important for insight into the heart of the essence of life, although they can highlight conditions that must be fulfilled by biological organisms in order to function in the real, physical world (Korzeniewski 2005). Of the enumerated aspects, the closest to me is the genetic aspect, as it refers directly to reproduction and the inheritance by descendant individuals of the identity of parental individuals and (the possibility of) its evolution from generation to generation. It also seems that this aspect could be the first, together with the structural aspect, which emerged during the spontaneous origin of life, although very quickly, other aspects also had to appear (on the other hand, some scientists think that the metabolic aspect appeared before the genetic aspect, which also cannot be excluded). However, the genetic aspect lacks one very important element, namely the purpose of living individuals and the organs and cells they are composed of: directing them toward the execution of a certain determined task, their instrumental intentionality expressed as a tendency toward ensuring the survival and reproduction of living individuals, and the value and purposefulness of the information related to their structure and function.

A great achievement of physics (thermodynamics) was the explanation of the genesis of dynamic structures organized in space and time—the dissipative structures—originating and existing in the presence of sufficiently high thermodynamic gradients, sustained and driven by the dissipation of energy and the production of entropy. This enabled the understanding that living organisms, whatever else they would be, are also dissipative structures (and more strictly—hierarchically organized, immensely complicated complexes of many dissipative structures). However, life is not only an assembly of dissipative structures. After all, biological phenomena differ by something very important from convective currents and whirls in liquids or gases.

As discussed in Chapter 2, thermodynamics univocally defines the amount of information and the complexity of structure that characterizes a given system. The problem lies in the fact that this is completely different information and structure than that which is of interest to us in the case of living organisms. According to the laws of thermodynamics, matter, just after the Big Bang (in which our universe originated), which was distributed homogeneously in the gravitational field (mostly due to the phenomenon of inflation), was characterized by significantly lower entropy and by a much greater amount of information and complexity of structure in a thermodynamic sense than the living organisms formed from this matter many billions of years later. In order not to reach so far back in time, it is enough to consider developing the human organism from the stage of the zygote to the stage of the adult individual. The matter, energy and thermodynamic information (negentropy) for this development are delivered in the form of organic substances and oxygen taken up from the surroundings. These substances transform either into the matter of the body in the process of biosynthesis of various, more or less complex organic compounds or into carbon dioxide and water in the process of gaining energy through the oxidation of organic compounds contained in food, or into the unnecessary by-products of metabolism and the remains of undigested food. Additionally, the adult organism needs nutrients and oxygen to sustain its functioning through, *inter alia*, continuously gaining energy in the form of ATP. It also needs nutrients and oxygen to replace the components of the living cells that have undergone decomposition (proteins, RNA, lipids, sugars, *etc.*).

In every spontaneous process, entropy must increase. Therefore, the food (sugars, fats, proteins, *etc.*) and oxygen that serve the building and sustaining of the structure and function of our bodies possess lower entropy and thus more thermodynamic information, structure and complexity than these bodies themselves. In the face of this, the biological complexity, structure and information characteristic of living organisms, determining their essence, completely differs from the complexity and information in the thermodynamic sense. Indeed, life in the universe cannot exist either at minimal entropy (the previously mentioned homogeneous distribution of matter in the gravitational field) or at maximal entropy (systems in thermodynamic equilibrium) (ultimately: matter/mass concentrated in black holes) but only at intermediate values of entropy, in systems where significant thermodynamic gradients exist, and where intensive dissipation of energy and production of entropy occurs (the current situation on the surface of the terrestrial globe). Of course, low entropy, a large amount of information, and high complexity in the thermodynamic sense are necessary for the emergence of life, but they do not reach the very heart of its essence. Apart from this, as could be seen in the case of the genesis of thermodynamic structures, what is important is not low entropy itself but the high entropy production rate. Therefore, to understand the very heart of the essence of the phenomenon of life, it seems necessary to define (the amount of) the biologically-relevant information and complexity of structure specific to living organisms.

How should a definition of life look?

In order to speak reasonably about some sets of phenomena and separate them from other phenomena, it is necessary to define and delineate the phenomena of interest in as simple, clear and univocal ways as possible. Of course, this also concerns the complex phenomenon called life. A satisfactory definition of life, and in fact of the living individual, which in my opinion is a decidedly distinguished level of organization in the hierarchy of biological systems, should fulfill five fundamental criteria (Korzeniewski 2005):

(1) A living individual should be defined using the language of formal sciences, such as mathematics, cybernetics or information theory. This would allow us to demonstrate the essence of life that cannot be reduced to (and expressed in terms of) the physical level of reality.
(2) A definition of a living individual should be based on possible primary and fundamental properties and concepts that cannot be derived from other, more primary properties (concepts). A definition based on properties that can be derived from other properties is not a true definition but a consequence of a certain sequence of logical operations. A true definition should be based on axioms that cannot be reduced to anything else.
(3) A definition of a living individual should clearly and univocally distinguish a living individual from systems that are not living individuals, including:
    (a) inanimate systems, such as stones, planets or convective currents,
    (b) psychological and sociological systems, for instance, individual human minds and human societies,
    (c) biological systems that are not living individuals but parts or assemblies of them, such as the muscle cell, liver or ecosystem; the concept of a living individual is central to the phenomenon of life; for instance, it is possible to "fertilize" a certain distant planet with a single individual of bacterium and in this way give birth to a whole biosphere, but one cannot have life without at least one living individual.
(4) A definition of a living individual should determine the unit of Darwinian natural selection—the subject of biological evolution. Evolution constitutes an inseparable feature of life; nevertheless, life as a whole certainly is not a subject of evolution; therefore, a good definition of life should determine what this subject is.
(5) A proper definition of a living individual should allow for determining the amount of purposeful and valuable biological information strictly related to a living individual. It should explain in general terms the internal purposefulness of the structure and function of a living organism (individual).

Generally, the point is to specify what distinguishes life from inanimate phenomena. A decisive criterion should be quoted, which without a shadow of a doubt, would allow the separation of *all* living organisms (individuals) and include *only* living organisms (individuals). Otherwise, physicists and other representatives of strict sciences would still regard life as only some kind of configuration and movement of atoms, distinguished somehow only due to the subjective illusion of the human mind, or at best, as one more dissipative structure, and biologists will not have a good argument at hand to oppose such an opinion. In other words, a biologist must have something more to say than: "I know intuitively what life is as long as nobody asks me about this." A good definition of life should explain the origin of the internal purposefulness of living individuals as directed-to-something systems, their functionality directed to the realization of determined tasks, and the specific quality or value of information associated with living systems. Such a definition would have to apply not only to the forms of life currently existing on our planet but also to life-like phenomena presumably existing on other planets in the universe and to the process of the spontaneous origin of life on earth almost four billion years ago. Additionally, it should refer to artificial life and virtual life functioning only in the operational memory of a computer. Generally, an adequate definition of life should reach the very heart of the essence of life.

Many definitions of life have been proposed, which usually overlap to a large extent and concern the above-mentioned aspects of life (see, e.g., Muller 1935; Muller 1966; Gánti 1975; Prigogine 1980; Schrödinger 1992; Maynard Smith and Szathmáry 1995; Bedau 1998; Koshland 2002). These definitions mostly concentrate on how living organisms reproduce, inherit genetic information and evolve. Living organisms are frequently also characterized as systems built of organic compounds, as dissipative structures (or at least dynamic systems of low entropy, significantly displaced from thermodynamic equilibrium), as hierarchically-organized open systems, as complexes of redox reactions, as nonlinear and or computing systems, and as systems capable of supple adaptation (Tsokolov 2009). Of course, all such definitions refer, to a greater or lesser extent, to the phenomenon of life currently known on the earth and or quote the necessary physical conditions that must be fulfilled for life to originate and exist in the real world. However, in my opinion, the mentioned definitions do not reach the very core of the essence of life and do not fulfill all five of the conditions enumerated above. They are most frequently, in a sense, excessive as they quote the characteristic properties of life that can be derived from other, more fundamental properties. Additionally, many features of known life—for instance, being a dissipative structure, containing organic compounds or carrying out redox reactions—are most probably inseparably associated with the phenomenon of life in the reality of our universe, although they also appear in inanimate systems and are accidental with regard to the very core of the essence of the phenomenon of life.

The issue of the problems that are met by most definitions of life can be illustrated with one example. The definition of life quoted by NASA will be used as a frame of reference[1]. According to this definition, life is: "A self-sustaining chemical system capable of Darwinian evolution." Let us consider whether this definition is "the truth, the whole truth and nothing but the truth" in the context of the above-formulated conditions that an appropriate definition of life should fulfill. Undoubtedly, all contemporary living systems on planet earth known to us (and I suppose all comparable systems on other planets in the universe) fulfill this definition, which is, therefore, "the truth." However, (1) this definition is rather more general and intuitive ("a self-sustaining chemical system") than precise and formal; (2) it can be based on more fundamental terms (e.g., cybernetic terms), explaining what "self-sustaining" means; (3) it does not define the living individual—the cell or liver in the antelope body or the whole savannah ecosystem in which it lives are in a sense "self-sustaining," and depending on understanding, many inanimate systems can be regarded as "self-sustaining", e.g., a cyclone or the sun; the human mind and societies are in a sense "self-sustaining"; (4) it does not determine what the subject of biological evolution is—maybe it is the "naked" gene, the liver cell, the whole liver, the whole antelope or the whole savannah; (5) it says nothing about the purposeful, biologically-relevant information essentially. Therefore, we can conclude that the NASA definition certainly says neither "the whole truth" nor "nothing but the truth".

To a large extent, the above-mentioned criticisms of the NASA definition of life also applies to all other definitions of life known to me. Generally, they do not fulfill all (or even most) of the five criteria formulated above. For instance, as is discussed below, the capability of evolution and subtle adaptation to current environmental conditions commonly used in various definitions of life can be derived from other, more fundamental properties. Nevertheless, in my opinion, life, and especially the living individual, can be defined in an adequate, clear and univocal manner and maximally parsimonious, although perhaps (at first glance) it may appear anti-intuitive, extremely minimalistic and abstract. Such an example is offered by the cybernetic definition of the living

---

[1] https://astrobiology.nasa.gov/research/lifedefinition/about/

individual (Korzeniewski 2001, 2005, 2011, 2013), briefly characterized, developed and justified below.

Features of terrestrial life vs. features of inanimate systems

Let us start with a general characterization of biological systems on the terrestrial globe. The living organisms that we currently know exhibit many characteristic properties. As a rule, the set of features quoted in this context effectively enables the clear separation of life from inanimate phenomena; there are two main reasons for this. First, despite the enormous diversity of terrestrial living forms, their most fundamental biochemical, genetic and cellular principles of structure and function are astonishingly similar (in fact, they are identical) for all organisms currently existing on the earth. Second, even the smallest and simplest free-living biological forms, to which some bacteria and archaea belong, still exhibit a very high degree of complexity (some even simpler forms, including viruses and viroids, are not able, being parasites, to live independently, and therefore they need free-living, much more complex organisms as their hosts). For this reason, complexity (of a certain determined sort) seems to be a good determinant of life. However, the simple enumeration of the properties characteristic of living organisms on the earth can lead us astray. On the one hand, there is a risk that we will consider the accidental features of life—characteristic for terrestrial life, but not for what we would like to regard as life in general—as intrinsic attributes of life itself. On the other hand, properties that characterize complex biological systems can be taken as immanent properties of every life. Therefore, the point is to define the very essence of life, its deepest characteristic properties. Below, I will try to extract the unique attributes of life and next reduce the definition of life (the living individual) to the minimal form possible, which will contain only the necessary elements sufficient to classify a given system as living or inanimate.

As discussed above, many properties characterize the current life on earth. Living organisms are comprised of various *organic compounds* (proteins, nucleic acids, carbohydrates, lipids *etc.*); they possess specific complex, hierarchically-organized *structures*; they have complex and hierarchical sets of *functions*; the transformation of matter and energy—*metabolism*—occurs in them, driven by and directed by *enzymes*; *redox reactions* associated with the transfer of electrons/hydrogens are common in them and are essential for their existence; they are hierarchically-organized (complexes of) *dissipative structures* sustained by significant thermodynamic gradients; they are characterized by their ability for *growth* and *autonomous movement*; they can *perceive stimuli* from the external world and adequately *react* to them; they reproduce, that is, they produce descendant systems (individuals); they *inherit* genetic information, and thus their characteristic properties (forming their *identity*) are transferred to progeny; and finally, they undergo *evolution*—the transformation of (the identity of) some forms into (the identity of) other forms in subsequent generations.

Nevertheless, many of these properties also characterize inanimate systems. For instance, organic compounds can be synthesized outside living organisms; they (e.g., glycine, the simplest amino acid) are also formed in cosmic space (found in meteorites, and a few hundred were discovered in clouds of interstellar gas and comets). Additionally, there are no reasons known to us for which life on other planets could not be built of slightly different organic compounds than those from which terrestrial life forms are composed.

There are inanimate systems of great complexity, natural (e.g., our planet) and artificial, created with human intelligence (robots, computers, the Internet). They can be more complicated than the simplest free-living organisms (e.g., bacteria), already leaving aside parasitic viroids or viruses. Therefore, the point here is more the special type of complexity than the degree of complexity.

Similarly, inanimate objects can have a hierarchically-organized structure (again, the earth and the computer are effective examples) and be characterized by complicated internal processes ("functions"). As a matter of fact, some regularity of repeatable structures (e.g., cells) seems to be a specific property of living organisms. However, the same property refers to electronic systems (e.g., microprocessors in computers), crystals or Bénard cells in silico oil. One of the very typical properties of living organisms consists of the internal purposefulness of their functions and is realized, *inter alia*, by various regulatory mechanisms, chiefly negative feedbacks, which aim to maintain homeostasis (constancy of internal conditions), counter-balance various disturbances from the surroundings, maintain proper functioning in order to ensure survival and reproduction, supervise correct embryonic and individual development and control cognitive functions and behavior. However, negative feedbacks are also inbuilt into many devices constructed by man, including thermostats in refrigerators, rockets and robots.

The transformation of matter and energy occurs in many natural inanimate systems. The carbonate-silicate cycle in the crust and mantle of the earth (see Subchapter 3.1) can serve as an example here. Inanimate objects can also be capable of growth (crystals in a saturated salt solution, snowflakes, "flowers" "painted" by frost on windows) and autonomous movement, where the cause of movement lies within the considered system (the mantle and crust of the earth, cars, robots). Additionally, some devices (e.g., robots, computers) have some equivalent of perception of stimuli from the external environment and can react appropriately to them (e.g., a computer, in response to the stimuli from the keyboard, generates text, saves it on the hard disc or prints via the printer, not to mention the complicated calculations and model simulations carried out by it). Summing up, none of the properties mentioned above seem unique in themselves for the phenomenon of life, is the reason for its uniqueness, and therefore reaches the very core of the essence of life.

Life on earth (and, in my opinion, life in general) seems to possess four properties strongly linked to each other and constitute different aspects of the same thing and which are absent in inanimate systems. Namely, life is (1) composed of particular, well-separated *individuals*, which (2) *reproduce*, which is associated with (3) *inheritance of identity*, and (4) *evolve* (their identity can change from generation to generation). Therefore, this set of properties can constitute my *genetic-individual definition of life*. Although I decidedly like it, it lacks one key element—the satisfactory specification of the living individual, especially with regard to its *purposefulness* (aiming for something).

As already discussed, the reproduction of individuals associated with heredity consists in the production of descendant individuals of the same (or similar) identity possessing the same specific complex of features. In the case of present-day terrestrial organisms, this identity is recorded genetically (in the sequence of nucleotides in DNA) and expressed (first of all) as the synthesis of appropriate proteins (possessing specific amino acid sequences) in the appropriate place, time and quantity. Their progeny inherits the identity of parental individuals. In this respect, living organisms differ from, for example, simple (monomeric) autocatalytic chemical molecules, as the latter possess no identity (they are as they must be—their structure results univocally from physicochemical laws, primarily quantum mechanics), and therefore they cannot inherit it. On the other hand, the structure of linear-polymeric compounds, which are composed of different organic monomers and upon which terrestrial life is based (namely nucleic acids and proteins), is by no means determined by physical laws—there can be an enormous variety of molecules of these compounds differing from each other by their sequence of nucleotides and amino acids, respectively.

In order to describe the sequence of amino acids in a protein molecule, a huge amount of information must be delivered. On the other hand, in a sense, no information is needed to describe the structure of a simple monomeric chemical compound—it results from the "intrinsic" (based on physical laws, described in Chapter 1) predisposition of matter to self-organization and forming structures of different kinds. In my opinion, the ability of individuals endowed with a clearly-expressed identity, associated with transferring this identity to progeny (reproduction), is a very characteristic property of living organisms, although it is not unique as it is also possessed by simple *replicators*. Of course, the essence of this identity, based on just nucleic acids and proteins, can only constitute a specific attribute of terrestrial life but not of life in general. We do not know whether slightly (e.g., proteins and nucleic acids containing somewhat different sets of amino acids and nucleotides)[2] or very different organic linear heteropolymers could be "carriers" of life on the molecular level. On the other hand, the question arises of whether the geochemical processes in the earth's mantle, hydrothermal vents and hydrosphere, did not impose a specific chemistry of life (Smith and Morowitz 2016). We know that the genetic apparatus of very early life was based on RNA and not DNA as the genetic information record and involved fewer than 20 amino acids.

The identity of particular individuals can change slightly in a random direction between parental individuals and progeny due to mutations (and recombinations) in the genetic record (in sexual reproduction, both parents make a comparable contribution to the identity of progeny). As already mentioned, mutations are unavoidable because of the physical nature of this record: perfect preservation in the original form and copying of the DNA strand is not possible for purely thermodynamic reasons (quantum mechanics can also play an important role in the generation of mutations; Al-Khalili and McFadden 2014, 286-304). Therefore, different individuals possessing different identities will compete for the available resources from the environment (at least potentially, more progeny is produced than can

[2] The record of information in DNA is "quaternary" in this sense that four nucleotides: A, C, G, T participate in it. There is a theoretical possibility that in some organisms that may have spontaneously originated on other planets, the code is binary, like in computers, which means that within the linear heteropolymers (being the carrier of the genetic record, the equivalent of DNA) in these organisms, there are only two types of monomers present (the equivalent of nucleotides); let us call them X and Y. They would correspond to 0 and 1 in computer code. The "codons" in these organisms would have to contain at least four or five monomers X/Y, in order to encode several-some tens of "amino acids": a sequence of four monomers of two kinds gives $2^4 = 16$ combinations, while the sequence of five monomers—$2^5 = 32$ combinations. The record of genetic information could also be ternary, based on three "nucleotides" X, Y, Z, where X would form a pair with Y, while Z would form a pair with itself.

survive). Statistically, those individuals best fitted to the current environmental conditions have the greatest chances for survival and the transfer of their identity to progeny. Therefore, the identity of living individuals will evolve over time, from generation to generation. For this reason, the occurrence of biological evolution is not an assumption or "axiom" in the definition of life, as it can be derived from other, more "primeval" properties of living individuals.

The fourth (in fact, the first) aspect of life distinguished by me that cannot be separated from the remaining three aspects is the fact that life does not form some continuous mass, an undisrupted pudding-like layer covering the surface of the earth but is composed of particular discrete individuals. This is just the individual that constitutes the carrier of identity that reproduces, inherits identity and evolves (by an individual, I understand the whole life cycle of a given organism, from conception and birth or division of the parental individual to the adult phase and death or division into two descendant individuals). The individual and reproduction associated with heredity and evolution are four strictly connected "faces" of the phenomenon of life.

In my opinion, a properly defined living individual is equivalent to the subject of evolution, the unit of Darwinian natural selection. Therefore, to define the latter, it is absolutely necessary to define the living individual—in some important sense, which I will explain in detail below, the concept of life is secondary in relation to the concept of the living individual. Life can be specified as a complex of living individuals. The classical paradigm of a living individual treats it as a structurally and functionally integrated whole. However, this paradigm encounters serious difficulties. It refers equally well to a living organism and a robot. It can be applied to a single cell in the human body, to a body organ (e.g., the liver or heart), or on the other hand, to some ecosystem or even the whole biosphere. It is commonly used in relation to particular insects in a eusocial insect colony. Additionally, it is not able to univocally decide whether objects/phenomena of such dubious status as colonial cnidarians, slime molds, insect societies, viruses and viroids, cancer cells, prions, plasmids, parasitic DNA and symbionts, including mitochondria, chloroplasts and hydrogenosomes should be classified as living individuals or not.

Cybernetic definition of the living individual

In my opinion, a definition of a living individual can be formulated in cybernetic terms while fulfilling the criteria presented above, namely a strict, rigorous formulation, being based on the most fundamental concepts, univocity, decidability, specification of the subject of evolution (unit of selection), explanation of the internal purposefulness of living organisms as aimful systems, lack of redundancy, determination of the amount of biologically-specific information and reaching to the very core of the essence of life. As people are likely not to remember what the term "cybernetics" means, I will present some fundamental elements of this discipline below, particularly concerning living organisms.

Cybernetics[3] (Wiener 1948; Ashby 1962), in the strictest sense, is the science of regulation appropriate for steering a certain system to achieve some prespecified desirable effect, despite various disturbing external influences. In a more general sense, it is also concerned with transforming and relaying related information (communication). Systems in which the realization of certain goals takes place, with the aid of appropriate regulatory mechanisms, are called *purposeful (aimful) systems*. Such systems can be devices built by humans, such as robots, a simple fridge thermostat, or living organisms. A basic concept in cybernetics is *negative feedback*. This regulatory mechanism aims to maintain the value of a given parameter on an assigned level, profitable from the point of view of the functioning of the whole system. Such a parameter can be, for example, temperature, the rate of chemical reaction, the concentration of a given compound, the color of a certain object in a given place, a particular movement trajectory, humidity, exposure to sunlight, the size of a given structure of an entire body in a given moment of embryonic development, the acuity of a picture formed on the retina, adequate behavior in response to concrete signals from the surroundings, and many, many others.

Negative feedback counter-balances (counteracts) any deviations of the regulated parameter value from the assigned value caused by various disturbances from the surroundings, elicits an action leading to a change in the value of the parameter in question in the opposite direction (hence the name) to that of the deviation. At the same time, the negative feedback should be clearly distinguished from the dynamic equilibrium with disturbances. The former is equipped with special mechanisms entering into the composition of the steering loop, while the latter is a simple physical phenomenon, for instance, protuberances on the sun. They are arches of plasma ejected from the surface due to the radiation pressure, magnetic field orientation and turbulent convective currents in its interior. The ejected plasma is drawn back to the surface of the sun through the force of gravity. Examples of negative feedback are the maintenance of a constant temperature in a fridge, the correction of the deviations of a rocket trajectory

[3] See: https://en.wikipedia.org/wiki/Cybernetics

from the assigned course, the maintenance of a constant sugar level in the blood or the adjustments of the eye to see either distant or near objects. These examples are discussed in more detail below.

Positive feedback has the opposite action—the further the parameter's current value is already displaced from a specific determined value (most frequently, this is simply zero), the faster the recession of the value of a given parameter from this assigned value. In brief, the greater the parameter's value, the faster it increases. Therefore, this mechanism enhances something already to some degree present (hence the name "positive"). Thus, while negative feedback is a stabilizing, homeostatic mechanism, maintaining the system near to some determined state, positive feedback leads to an immanent instability, to explosive and or chaotic behavior. Examples of positive feedback are the nuclear chain reaction, bacteria reproducing in a medium or the cascade of blood coagulation. These are also discussed below.

A system without any feedback, a system with negative feedback and a system with positive feedback are compared in Fig. 6-1. In Fig. 6-1a, we have a system devoid of any regulation. The value of a given parameter, say temperature, results from a passive reaction of the considered system to surrounding influences. If we leave a fridge in a warm room, the temperature inside it will quickly equalize with the temperature in the room. If the unit's walls are thick and built with a good heat isolator, then the process of temperature equalization will be slowed down, but the final effect will be the same. If some food is put in our kitchen unit, the effect of the prevailing temperature will be the rapid decomposition of this food. This effect is harmful from the point of view of our goal, namely maintaining the freshness of food, possibly for a long time. Therefore, in a system without regulation (non-cybernetic), the value of a given parameter passively reacts to interactions with the environment and, due to the disadvantageous influence of the surroundings, can elicit an undesirable effect with regard to a prespecified goal.

The regulation of a parameter value (temperature in this case) consists of maintaining near the optimal assigned value eliciting the best effect regarding a preassigned purpose. In our case, such an assigned temperature is low enough to prevent food from rotting for a long time and, at the same time, sufficiently high so as not to cause excessive energy costs for temperature maintenance (and not to freeze food if not required). The mechanism allowing the parameter (temperature) to be maintained at the assigned (optimal) level is just the negative feedback. Just this feedback constitutes the basis of the functioning of the thermostat in a refrigerator, which, as we know, is a device providing food storage through the ability to maintain a constant (with tiny oscillations) low temperature.

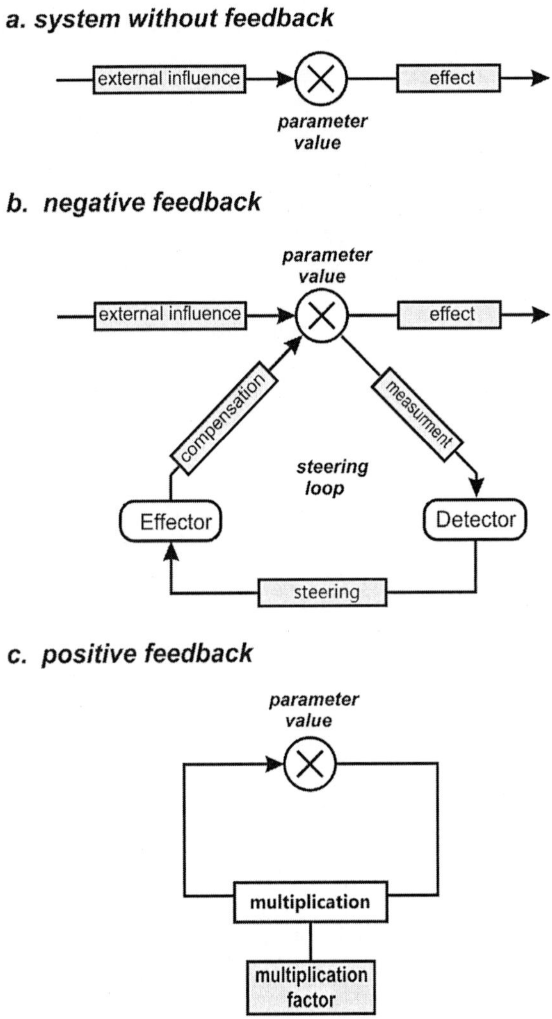

Fig. 6-1. General scheme of negative and positive feedback. (a) system without any feedback, passively reacting to influences from the surroundings; (b) system with negative feedback, regulating (maintaining on a constant level) the value of a certain parameter and compensating disturbances from the surroundings—the steering loop is shown with a detector measuring the parameter value and sending a signal to the effector in the case of detection of deviation of the parameter value from the assigned value, which returns the parameter value to the assigned value; (c) system with positive feedback, anti-regulatory, multiplying the value of a given parameter by a determined multiplication factor during a determined time interval.

The general scheme of negative feedback action is presented in Fig. 6-1b. An appropriate detector (in the case of a refrigerator: a bimetallic strip—two glued strips of two metals of different thermal expansion) performs continuous monitoring (detection) of the value of the relevant parameter (temperature) by bending the bimetallic strip. Suppose the detector encounters an increase in temperature above the assigned value. In that case, it sends a steering signal (by closing the electrical circuit) and boots an appropriate effector (in our case: cooling device), which counter-acts the deviation of the parameter value from the assigned value caused by influences of the surroundings and therefore compensates the disturbances caused by these influences (the temperature returns to the optimal assigned value). Once the detector measures the return of the parameter value to the assigned value, the effector (cooling device) is turned off. The whole causal sequence of events is called the *steering loop*: increase the parameter value above the assigned value—measurement, detector, steering, effector, compensation—the return of the parameter value to the assigned value.

By contrast, positive feedback, shown in Fig. 6-1c, acts, as the very name indicates, opposite to negative feedback. While negative feedback constitutes a mechanism stabilizing the system, drawing the value of the regulated parameter to the assigned value, positive feedback is strongly destabilizing; it "repulses" the parameter value from the assigned value—the stronger the deviation is, the greater the further deviation from the assigned value (usually this is zero). As a result, the parameter value very quickly (in an exponential way) tends to infinity, and the resultant effect can be called anti-regulatory or explosive.

Classic examples of positive feedback are nuclear chain reactions and the reproduction of bacteria in a medium. An unstable nucleus of radioactive uranium isotope ^{235}U emits three neutrons during decay, while a bacterium divides into two descendant bacteria. Each released neutron fragmentizes the next unstable nucleus, while each descendant bacteria divides again (after duplication of its mass due to feeding and growth) into two bacteria. Therefore, after a certain period, the number of neutrons and the disrupted nuclei undergo triplication, while the number of bacteria—duplication (this is the multiplication factor). This process can be repeated, at least potentially, an infinite number of times (in reality, it is limited by the amount of unstable uranium isotope or the medium for bacteria). Other examples of positive feedback are certain chemical compounds catalyzing their synthesis from other compounds (so-called autocatalysis discussed previously) and rock avalanches, in which a rock rolling down a slope sets in motion more and more rocks.

Let us return, however, to negative feedback. It is the basic regulatory mechanism in such purposeful (aimful, targeted at realizing a certain effect) systems as various devices constructed by humans and living organisms. In the case of manufactured devices, negative feedback can express itself in the maintenance of constant pressure in the steam boiler of a steam engine, the correction of deviations from an assigned rocket trajectory, the optimization of operations performed by some robots and in the maintenance of constant temperature in a fridge as discussed above.

The pressure in the steam boiler of the steam engine cannot be so low that the steam cannot work or so high that the boiler may explode. When the steam pressure rises above the assigned pressure, two rotating arms on the safety device ending with balls, rotate faster, and they rise due to centrifugal force and expose openings through which the steam escapes from the boiler. When the steam pressure falls, the rotating arms fall back down and close the openings, which prevents too great a decrease of pressure in the boiler.

Regarding crewless space crafts (e.g., in missions to Mars), the flight trajectory is precisely fixed. Appropriate devices detect any deviations of the actual rocket course from this trajectory. This results in the activation of correction engines, which bring the rocket back to the proper course—otherwise, it would never reach the scheduled place on the Martian surface or may even fail to reach Mars at all. More advanced robots use negative feedbacks for the precise control of their movement.

Countless regulatory and steering mechanisms occur in living organisms, first and foremost, negative feedbacks. With regard to functional control, biological systems are, in fact, "built of" hierarchically-organized networks of mutually inter-connected and co-operating negative feedbacks (as in the thermodynamic respect, they constitute a complex of hierarchically-organized dissipative structures). Particular negative feedbacks provide homeostasis (the constancy of internal conditions) in the whole organism and its particular organs and cells with regard to the compensation of various disturbances coming from the environment, the coordination of the correct functioning of various parts of the organism, the perception of the current and anticipation of the future variability of the environment in space and time and adequate reaction to it; control of the whole hierarchical morphogenetic cascade leading to forming an adult individual from a fertilized egg cell—this last example in particular being of immense importance. Generally, this entire network of reciprocally co-conditioning negative feedbacks shaped in the course of biological evolution constitutes a dynamic (regulatory) representation of a given individual's external and internal environment (together with its entire diversity and variability). This network is directed at the survival and reproduction of the individual in its ecological niche (its "place" in the ecosystem and physical surroundings).

A major (and relatively simple) example of negative feedback in living organisms on a biochemical level is the regulation of the synthesis rate of a given amino acid based on the current concentration of this same amino acid (Umbarger 1978). As is illustrated in Fig. 6-2, such an amino acid (marked as product P) inhibits the first enzymatic reaction in the metabolic pathway leading exclusively to its synthesis. Namely, the amino acid lowers the activity of an appropriate enzyme by binding to an appropriate regulatory place—an allosteric center. This causes the maintenance at a relatively constant level of the concentration of both this amino acid and other metabolites entering into the composition of this pathway, regardless of the variations in the current consumption of the given amino acid (e.g., by the process of protein synthesis) and its delivery through the circulatory system from the food that is taken up and digested in the gastrointestinal tract. Without such a regulation, the reserve of the amino acid could be very quickly depleted or, conversely, the amino acid could accumulate to excess and thus disturb the functioning of the cell. The very fact of the existence of this feedback and the fact that just the first enzyme (catalyzing the $A \rightarrow B$ reaction) in the metabolic pathway leading exclusively to the synthesis of a given amino acid is inhibited constitutes a subsequent example of the cleverness and purposefulness of biological evolution. If another enzyme (for instance, catalyzing the $B \rightarrow C$ reaction) is inhibited at lower amino acid P consumption, and thus with a high concentration of this amino acid, metabolite B, which is not needed by any other process than P production, would accumulate in the cell. If the enzyme catalyzing the production of metabolite A, being a substrate for not only P production but also for other metabolic pathways in the cell, was inhibited at high P concentration and intense A usage by other processes, the concentration of A would essentially fall to zero, which would disable the functioning of these processes. Therefore, the actual solution found by evolution seems very logical from the engineering and logical point of view.

Another example on the metabolic-energetic level is the production of the universal carrier of energy in the cell—ATP—by oxidative phosphorylation in mitochondria. High ATP concentration, occurring in the rest state, inhibits this compound's metabolic block of production. A sudden increase in ATP consumption due to muscle contraction (mechanical work requires energy expenditure) causes hydrolysis of a certain amount of ATP to ADP and P_i. The resultant increase in ADP and P_i concentration activates ATP production (especially by oxidative phosphorylation; Chance and

Williams 1956) and partly compensates for the fall in ATP level, in the result of which, ATP is not quickly depleted, which would prohibit further work of the muscle cell. However, it should be emphasized that in this case, the described negative feedback is accompanied by another purposeful regulatory mechanism—parallel direct activation during muscle contraction of both ATP hydrolysis by mechanical work (actomyosin-ATPase) and the ATP production block (all oxidative phosphorylation complexes and the NADH supply block, including glycolysis and TCA cycle—compare with Fig. 2-10). Both mechanisms cooperate in order to ensure the possibly effective homeostasis of ATP and other metabolite concentrations (Korzeniewski 2017b).

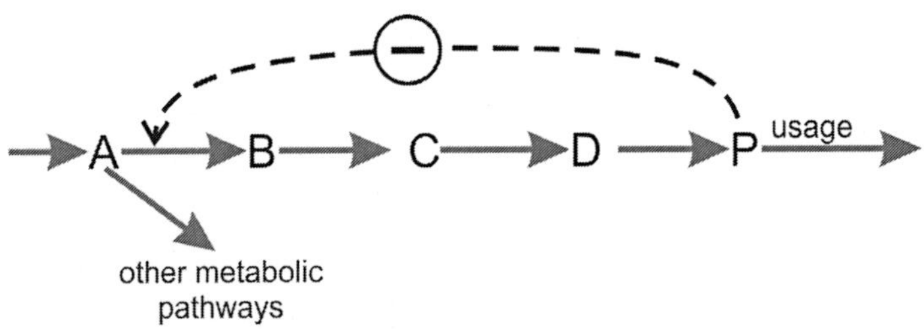

Fig. 6-2. Negative feedback in the metabolic pathway leading to synthesis of amino acid *P*. High *P* concentration inhibits the first enzymatic reaction leading exclusively to the production of this amino acid, thus preventing the complete depletion or excessive accumulation of *P* and intermediate metabolites *B*, *C* and *D*.

On a genetic level, the synthesis rate of particular proteins, including enzymes, is adjusted to the rate of their decomposition (because they are damaged, for example, by free radicals) so that their concentration remains constant. This also concerns other components of the cell. Metabolism must "know" how fast particular compounds should be produced in order to replenish their deficits. Appropriate negative feedback ensures this (Toyama and Hetzer 2013).

The optimal protein concentration frequently depends on circumstances. An example is the regulation of lactose operon (Jacob and Monod 1961; Maynard Smith and Szathmary 1995, 219-220), the first to be discovered of many kinds of operons occurring in prokaryotes. This operon, being a segment of DNA in the circular molecule of bacterial DNA, is composed of the *promoter* that can bind the enzyme RNA polymerase responsible for the transcription of DNA to mRNA, the *operator* that can bind the *repressor* protein inhibiting the functioning of the whole operon, and three genes encoding three enzymes (strictly speaking, two enzymes and one carrier) participating in the catabolism (decomposition) of lactose. This is shown in Fig. 6-3. In the absence of lactose, the repressor, synthesized non-stop on the matrix of the regulator gene, binds to the operator and blocks the action of RNA polymerase, and thus the expression of the three mentioned genes, their transcription to mRNA and the synthesis of the proteins encoded by them (the upper scheme in Fig. 6-3). When lactose appears in the surroundings (especially in the absence of other, better-assimilated sugars, primarily glucose), it binds to the repressor and deactivates it. That being so, the repressor no longer inhibits the activity of RNA polymerase, and the three genes of the operon undergo expression—first, their nucleotide sequence in DNA undergoes transcription to the sequence of nucleotides in mRNA (in prokaryotes, all genes of one operon are transcribed to one common mRNA), and next translation to the sequence of amino acids in the proteins that metabolize lactose (the lower scheme in Fig. 6-3). Therefore, the assigned value of a given parameter (in this case, the concentration of enzymatic proteins), which functions as an element of negative feedback on a lower level of hierarchy (maintaining a constant delivery of building material and energy in the easily-accessible form—ATP– the source of both of which is lactose, and through this, constant metabolite concentrations) can be changed through some other negative feedback mechanism on a higher level of the hierarchy, limiting the energetic and material expenses for the production of proteins that are not currently needed.

Sigma (σ) factors (Alberts et al. 1998, 216-217) are bacterial proteins that bind to RNA polymerase and, in turn, cause its binding (increases its affinity) to specific operons. In *Escherichia coli* commonly occurring in the digestion track in humans, factor σ70 is present under normal conditions, and factor σ32 appears in conditions of heat shock and other types of stress, exerting protective action on the cell components through favoring, for example, the preferential production of heat-shock proteins (HSP), while factor σ28 is responsible for the production of flagella and chemotaxis, that is moving in the direction of growing or falling concentration of a given substance. On the other hand, in *Bacillus subtilis*, factor σ45 occurs in normal conditions, but when conditions become disadvantageous, different σ factors lead to sporulation (the transformation of bacteria cells into spores that are practically devoid of water and metabolism and are very resistant to a lack of food and water and to high and low temperatures). Of course, all these actions represent regulatory mechanisms (including negative feedbacks) that maintain internal homeostasis, compensate for disturbances from the surroundings, ensure food delivery, and avoid harmful substances.

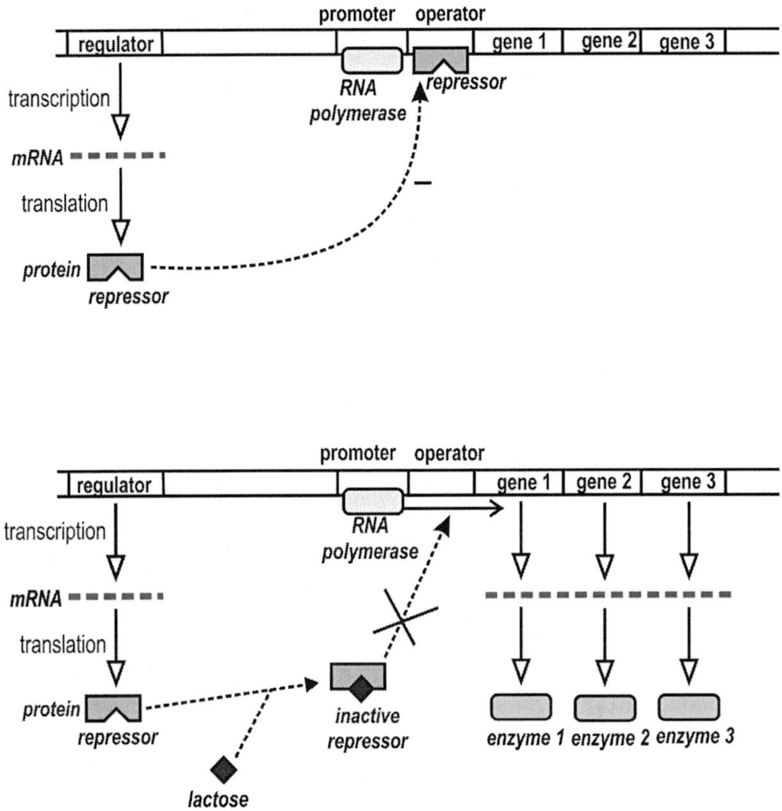

Fig. 6-3. The principle of the action of the lactose operon. In the absence of lactose, the repressor binds to the operator, because of which, the RNA polymerase bound to the promoter cannot carry out transcription of the three genes of the operon to common mRNA and its subsequent translation to proteins participating in lactose metabolism (the upper panel). When lactose appears in the surroundings, it "turns off" the repressor, through which the transcription carried out by RNA polymerase, and consequently protein synthesis is unblocked.

Alarmones (Rinke et al. 2013) are small-molecule signaling chemical substances causing a change (increase or decrease) of the affinity of RNA polymerase to promoters of appropriate operons that are produced in bacteria under conditions of carbon deficit (lack of carbon-rich compounds, particularly carbohydrates) or nitrogen deficit (lack of nitrogen compounds, e.g., amino acids) and lead to the synthesis of these compounds and or the limitation of their consumption. The alarmone of amino acid deficit (ppGpp) causes activation of amino acid synthesis/uptake genes and inhibits tRNA, rRNA and ribosome production (participating in protein synthesis). The alarmone of carbon deficit (cAMP) activates all catabolic operons encoding proteins responsible for the uptake of available organic compounds that can serve as carbon sources (e.g., lactose operon) from the environment and their decomposition.

The shortage of nutrients also frequently stimulates the production of flagella and the active search for food in bacteria[4]. This is, of course, also an example of negative feedback. Bacteria can move rectilinearly or chaotically, unceasingly randomly changing their direction of movement. The rectilinear (straight) movement takes place when, due to this movement, a bacterium flows in the direction of a growing concentration of beneficial substances (e.g., nutrients: amino acids and sugars) or the direction of a falling concentration of harmful substances (e.g., poisons or acids). By appropriate receptors in the cellular membrane, bacteria can detect changes in various substance concentrations over time during movement. The flagella (sing. flagellum) of bacteria are built of the protein flagellin, which has the form of a laeotropic (left-handed) helix (stretched "spring"). When the flagella rotate in a counter-clockwise direction, they intertwine into one "super-flagellum", which results in the rectilinear movement. When rotating clockwise, flagella work independently, leading to chaotic movement. When the concentration of beneficial substances increases over time due to the movement of a bacterium, the receptors of these substances stimulate the synthesis of a "rectilinear movement factor", which switches chaotic movement to rectilinear movement. This factor is spontaneously decomposed, so its concentration tends to zero if it is not continuously synthesized anew. This decomposition can be accelerated by the receptors of harmful substances when they detect an increased concentration of such substances over time due to movement; this is illustrated in Fig. 6-4. In the absence of the "rectilinear movement factor", the movement again becomes chaotic until the bacterium finds, through trial and error, the direction in which the concentration of beneficial substances grows and or the concentration of harmful substances declines, which results in the renewed intense synthesis of the "rectilinear movement factor" and the bacterium switch to rectilinear movement.

---

[4] see https://en.wikipedia.org/wiki/Bacterial_motility.

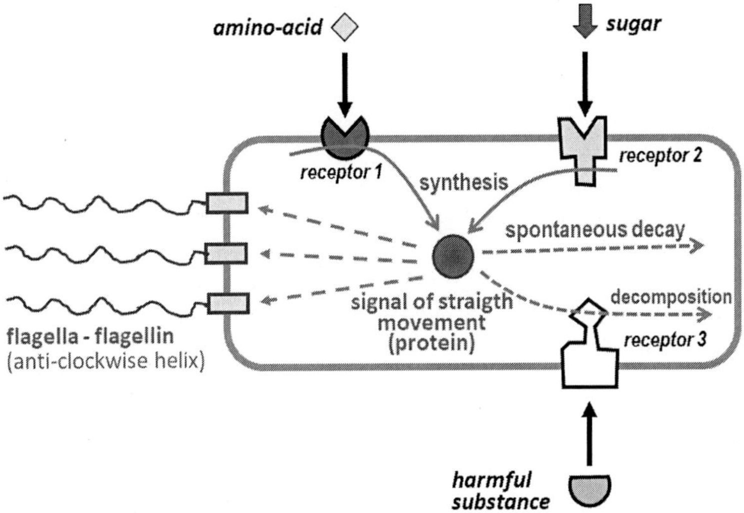

Fig. 6-4. Steering of bacteria movement. A bacterium can move either rectilinearly (straight movement, moving along the gradient of increasing or decreasing concentration of some substance), or chaotically (movement in random direction, the detection of potential concentration gradients). It detects, by means of appropriate receptors, changes in concentrations of beneficial substances (e.g. nutrients) or harmful substances (e.g. poisons) over time (during movement). Bacterial flagella are built of the protein flagellin, which has the form of a left-handed (counter-clockwise) helix. When they rotate in a counter-clockwise direction, they interweave forming one thick "super-flagellum", which results in rectilinear movement. In the case of rotation in a clockwise direction, flagella work independently, which leads to chaotic movement. The "switch" between these two types of movement is the "rectilinear movement factor", which as its very name indicates, switches chaotic movement to rectilinear movement, while its absence has the opposite effect. Its synthesis is stimulated by receptors of, for example, amino-acids or sugars when their concentration increases over time as a result of moving in a given direction, while its decomposition can occur spontaneously or be accelerated by a receptor of a harmful substance if its concentration grows over time as a result of movement in a given direction. Generally, if the bacterium moves "upwards" the concentration gradient of profitable substance or "downwards" the concentration gradient of harmful substance, the rectilinear (straight) movement is maintained, while in every other case, it passes to chaotic movement until the moment when it finds the direction in which the concentration of beneficial substances increases and or the concentration of harmful substances decreases. It then starts to move rectilinearly again.

In eukaryotic organisms, in particular in multicellular and tissue organisms, a very complex and hierarchical regulatory system, based to a large extent on a network of negative feedback, regulates gene expression to a high degree of precision (Maynard Smith and Szathmáry 1995, 219-223; Alberts et al. 1998, 257-273; Futuyma 2013), in other words, it determines which concrete proteins will be produced in a given cell compartment, at a given time and to which amount. As discussed above and below, this has fundamental significance for cell differentiation and functioning of the hierarchically-organized morphogenetic cascade, and through this, for embryonic development, and thus for the ultimate structure and function of a given organism (see Chapter 5). In eukaryotes, in contrast to prokaryotes, an activation/repression of a given gene constitutes the resultant action of many activators (binding to enhancers in DNA and stimulating gene expression) and repressors (binding to silencers in DNA and inhibiting gene expression)[5]; frequently, the same activators/repressors participate in the regulation of the expression of many genes, and the intensity of the expression of particular genes constitutes a derivative of the action of numerous activators and repressors. Eukaryotes lack operons—only one protein is synthesized on the matrix of one mRNA.

On a physiological level, two hormones—insulin and glucagon—constitute antagonistic signals in the steering loop of the negative feedback mechanism maintaining a constant concentration of glucose in the blood (Cryer 1991)—this is illustrated in Fig. 6-5. When hyperglycemia occurs after a substantial meal, that is, the rising of blood glucose above the assigned level (optimal with regard to the needs and internal regulation of an organism), B (beta) cells in the pancreas are activated, while A (alpha) cells are inhibited. B cells produce the hormone *insulin,* a protein that lowers glucose concentration in the blood by stimulating the liver to uptake this sugar and polymerize it to glycogen, a polysaccharide that is insoluble in water and deposited in the cells of this organ. Thus, glucose is restored to the assigned level. Excessive glucose in the blood can lead to diseases, primarily diabetes. When, in turn, due to prolonged starvation, the glucose level in blood falls below the assigned optimal level, there is a glucose deficit known as hypoglycemia. This can be pernicious for the organism, for example, because glucose is the only respiratory substrate for the brain (other organs can use fatty acids, lactate, amino acids or glycogen deposited in them), and a low level of this sugar can disturb its functioning. At low glucose levels, the activation of B cells in the pancreas is absent, while A

[5] Enhancers and silencers are frequently situated in places in DNA that are distant from the location of the regulated gene, or, on the other hand, within introns of the gene whose expression they regulate.

cells are unblocked and start to produce the hormone *glucagon* (also a protein), which is antagonistic to insulin. This hormone stimulates the decomposition of glycogen in the liver into glucose molecules, which are then released into the bloodstream. Again, the assigned level of blood glucose is restored. This constitutes an evident example of negative feedback (or, if preferred, a complex of two cooperating negative feedbacks). Because of the key significance of the precise regulation of the level of glucose in the bloodstream for the physiology of the organism, it uses two opposing signals—insulin and glucagon—to maintain the value of this parameter to near the exact assigned level.

Another hormone, adrenaline, regulates, *inter alia*, the rate of blood pumped by the heart (Reece et al. 2011, 1302-1304). It is released under conditions of sudden stress, primarily as a signal preparing for fight or flight, or contrariwise, to the pursuit of prey, and also during physical exercise, for instance, running. Any physical activity is associated with ATP hydrolysis driving muscular contraction. ATP must be quickly resynthesized, in most cases primarily by oxidative phosphorylation, which requires oxygen and respiratory substrates (sugars, fatty acids). Intensive consumption of these compounds requires their fast delivery by the circulatory system, especially for oxygen from the lungs. Adrenaline, together with neural signals, accelerates the heart rate and increases the rate of blood pumping by it and, at the same time, elevates the general brain arousal level, which is useful during hunting, fight or flight. This constitutes a subsequent negative feedback loop, acting on many levels and ensuring a relatively constant concentration of respiratory substrates and oxygen in the blood, and as a consequence, for example, avoiding a threat (flight), defending territory or winning a sexual partner (fight, mating display), or satisfying hunger (hunting).

The muscles of the iris regulate the amount of light entering the eye by changing the pupil diameter (Reece et al., 2011, 1096-1097). With intensive illumination, the pupils narrow so that light of too high an intensity does not damage photosensitive cells in the retina. Additionally, the smaller the pupil opening, the sharper the vision, as the photons emitted/reflected by a small fragment of the observed object are focused on a relatively small area of the retina. By contrast, the pupil diameter is enlarged at a low light intensity so that as many photons as possible can reach the retina. As a matter of fact, this lowers vision acuity, but it is better to see blurrily at twilight than not to see at all. Here, the negative feedback regulating the size of the pupil opening ensures possibly acute and "bright" vision under given conditions.

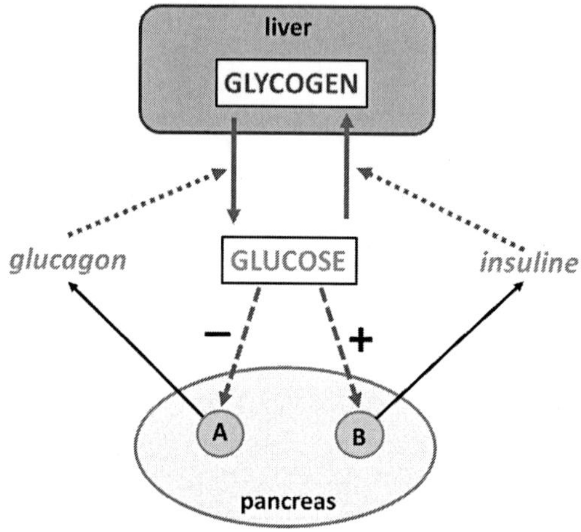

Fig. 6-5. Regulation of the level of glucose in blood. A high glucose level stimulates the production of the hormone insulin by B (beta) cells in the pancreas while inhibiting the synthesis of hormone glucagon by A (alpha) cells. A low level of glucose exerts the opposite action—it stops activating B cells and stimulates A cells. Insulin lowers the concentration of glucose in the blood through activation of the uptake of this sugar by the liver and its polymerization to glycogen deposited in liver cells, while glucagon elevates the concentration of glucose in blood through the stimulation of glycogen decomposition into glucose molecules and their release to blood. In this way, the antagonistic action of two hormones leads to the maintenance of the level of glucose in blood near the assigned level that is optimal for the organism's physiology.

The eye accommodation, consisting of changes in the lens's curvature, ensures the accurate perception of both close and distant objects (Reece et al. 2011, 1101). The contraction of muscles supporting the lens causes a decrease in its curvature (it becomes flatter), which enables the accurate visual perception of distant objects, while relaxation of these muscles leads to an increase of the lens curvature (it becomes more rounded), which enables the accurate visual perception of objects from a short distance (e.g., reading). Therefore, an appropriate negative feedback mechanism ensures accurate visual perception depending on whether we focus our sight on distant or nearby objects.

With regard to the behavior of the whole organism, the exhaustion of nutrient reserves (fats, sugars) in the body of an animal following a long period of starvation or after intensive exercise stimulates the animal to undertake appropriate actions directed at the acquisition of food (an antelope goes to pasture on the savannah, while a leopard begins its hunt for antelopes). Numerous negative feedback loops comprising the action of the nervous and motor systems (strict control of the strength and speed of the contraction of particular muscles ensuring the smoothness of movements) are involved in realizing this task. On the other hand, when the animal is satiated and the level of nutrients (glucose, fatty acids) in the bloodstream is sufficiently high, it lies down in an appropriate place to rest and digest food. The level of nutrients in the blood is sensed by the hypothalamus (a structure within the diencephalon of the brain), which controls the activity of an animal directed toward the acquisition of food (and also reproduction, fight and flight) (Banich and Compton 2018, 11-12, 369-371). Another signal relating to food acquisition is its physical pressure on the stomach walls. This steering loop ensures sufficient building material, respiratory substrates, and, thus, energy in the organism.

Negative feedback is a key element responsible for the coordination of muscular contractions, their strength and speed, and through this, for making smooth and precise movements and frequently complex sequences of movements (Banich and Compton 2018, 28-29, 100-124). In humans, they allow, for example, efficient bipedal locomotion (the muscles of the lower limbs are mostly responsible for this), complicated manual operation executed with the aid of the arm, especially the hand (muscles of the upper limb), speech generation (muscles of the lips, tongue, mouth and larynx) and facial expressions (face muscles). It is commonly known how much it takes for a child to learn to walk, precisely manipulate various objects and speak (of course, in order to speak, it is necessary not only to possess efficient apparatus for the purely mechanic generation of sounds but also to learn a language and to "have something to say"; in a child, all these abilities develop gradually and in parallel).

Thus far, no robots have been constructed that can efficiently and smoothly locomote, especially on two legs, on uneven terrain or possess "manual" abilities for the precise and diverse manipulation of various objects, at least comparable with human abilities. This testifies to the complexity of the whole negative feedback system responsible for coordinating muscle contractions and supervising the smoothness and precision of movements. They occur in such a way that receptors of internal sensing (proprioceptors, kinesthetic receptors) located in muscles and tendons measure the strength and speed of contractions and send appropriate feedback information to the brain, where the premotor and motor cortex situated in the frontal lobe initiate and steer movements, while mostly the cerebellum is responsible for the coordination of movements (Banich and Compton 2018, 100-124). The latter checks whether the currently realized spatiotemporal "pattern" of particular muscular contractions corresponds well to the proper, "assigned" pattern recorded in procedural memory. Its seat is mostly just the cerebellum and the pre-motor and motor cortex. Suppose the multi-dimensional "trajectory" of movement determined by appropriate muscle contractions deviates from the "assigned trajectory." In that case, the cerebellum sends feedback to muscles by correcting signals (as elements of a network of negative feedback loops) that modify the strength and speed of contraction to realize an appropriate pattern of movements. Such patterns are mostly learned in humans and acquired during individual development (walking, riding a bicycle, driving a car, operating cutlery, playing tennis or articulating speech). Only very primeval motor patterns are innate, such as alternating limbs inherited from the first land quadrupedal vertebrates. We can observe that during marching or running, humans alternately push forward the left and right leg and, in the opposite phase, the left and right hand, which is steered by the medial motor pathway (Banich and Compton 2018, 112). The latter serves nothing—it constitutes our evolutionary heritage after our quadrupedal ancestors, which moved forward alternating the front and hind limbs on the principle: left hind limb—right front limb forward and inversely—of course, all quadrupedal animals still move in this way today.

A broadly understood perceptive, cognitive and decisive functions of living organisms constitute elements of an immeasurably rich network of negative feedbacks engaged in the transformation of stimuli coming from the environment, through a greater or lesser amount of intermediate steps, into the behavior of the animal, which ultimately leads to satisfying its various biological needs. In the case of animals with a more developed nervous system, especially in humans, an important role in this process is played by memory (Banich and Compton 2018). This is because the enormous variability of the environment in space and time requires an appropriate and supple repertoire of reactions by a living individual in order to maintain as constant as possible the parameter values related to its internal homeostasis (e.g., appropriate temperature or concentrations of sugars or mineral salts) ensuring survival in good condition and realization of reproductive functions (satisfaction of the sexual drive). Obviously, to preserve the stability of internal conditions and reproduce, it is necessary to feed sufficiently, avoid predators, parasites and wounds, find appropriate shelter or nesting place, *etc*.

The negative feedback involved in broadly understood perception, cognition and decision-making can comprise, on the one hand, simple phenomena such as (already mentioned) lactose operon and eye accommodation, and on the other hand, complex processes like the brain operation directed to coordinating current and future (planned) behavior of a living individual with past, current and anticipated properties of the environment and events occurring in this environment. This activity comprises, *inter alia*, the perception of stimuli from the environment, the integration of sensory data into complex properties and neural objects, the association of neural representations of various features, objects and aspects of the surroundings, the detection of regularities and relations in the complexes of sensory data, the formation and usage of memory records, planning and decision making and similar processes (any readers who are interested in the general functioning of the brain in the cognitive aspect and some of its philosophical consequences can refer to my book *From neurons to self-consciousness. How the brain generates the mind* (Korzeniewski 2010), while a much more detailed description of the brain's cognitive functions is presented by Banich and Compton (2018)).

The reaction to stimuli from the surroundings and properties/aspects of the environment can be simple, automatic and instantaneous ("unthinking"), as it takes place in the case of the *unconditioned reflex*, when very quickly and without the participation of our will, we withdraw a hand after casually touching a hot or sharp object. Slightly more complex is *classical* (Pavlovian) *conditioning*, in which there takes place the association of a neutral stimulus (sound, flash of light) occurring immediately before exposure to a profitable stimulus (such as the appearance of food) or a harmful stimulus (e.g., irritation by electric current). After forming such conditioning, an animal reacts to the neutral stimulus in a manner appropriate for the profitable/harmful stimulus associated with it, for instance, by producing saliva or stress hormones and the desire to escape. In *instrumental (operant) conditioning*, the animal tests different options by a method of trial and error, for instance, by eating different leaves and insects in order to identify which are edible and encoding the results of such trials to memory and from then on, only eating those leaves/insects that are good as food, while avoiding those that are poisonous[6] (Korzeniewski 2010). It can also concern the correct/incorrect identification of, for example, a predator hidden in a bush (Korzeniewski 2010). Many behaviors steered by the neural system, especially by the brain, are even more complex, such as the planning of hunting strategies. The ultimate purpose of all of these is the preservation of internal homeostasis and survival in a possibly good condition up to the reproductive period (or subsequent reproductive periods, depending on the strategy applied by a given species).

---

[6] see https://www.britannica.com/science/animal-learning/Classical-and-instrumental-conditioning.

Therefore, these behaviors are involved in a very complex network of negative feedbacks that supervise the animal's behavior.

Finally, as has already been indicated several times, a rich network of negative feedback controls the course of embryonic development based on morphogen gradients and the activation of strictly determined genes in strictly determined places and times, which leads to forming a particular "target" adult living individual (Hall 1999; Slack 2001; Twyman 2001). Appropriate regulatory mechanisms check the course of morphogenesis, and if its multi-dimensional "trajectory" deviates from the "assigned", correct, and optimal course regarding the fitness of the adult individual, it undergoes correction so that the correct trajectory of developing the structure and the function of an individual is restored. When these mechanisms fail, deformities develop. Fortunately, this happens relatively rarely.

All of the discussed negative feedbacks are hierarchically organized, mutually connected (directly or indirectly), and mutually co-condition each other. For instance, the hormones insulin and glucagon, which supervise the maintenance of an appropriate, possibly constant level of glucose in the blood, are proteins. Proteins are synthesized through the polymerization of amino acids. Therefore, their undisturbed production is possible thanks to negative feedbacks adjusting the rate of amino acid synthesis to the rate of their consumption and delivery from the blood, which allows the maintenance of a relative concentration constant (homeostasis). The regulation of gene expression in eukaryotes consists of synthesizing the proteins encoded by them, including insulin and glucagon, in an appropriate place, time and quantity. A whole network of negative feedbacks using various activators and repressors of transcription as signaling factors, themselves proteins containing homeodomain, is involved. As we recall, such proteins on different levels can be organized into a whole, hierarchical cascade of the regulation of gene expression. Additionally, the production of proteins can be controlled (inhibited) on the translation level by miRNA.

In turn, glucose in the blood can serve, when its level is appropriate, as "fuel" for oxidative phosphorylation enabling ATP production and as a substrate for amino acid synthesis, driven by ATP hydrolysis. In this way, the described causal sequence of regulatory events forms a circle: all negative feedbacks in this circle enable the functioning, directly or indirectly, of all other feedbacks. Additionally, this generally concerns the entire assembly of negative feedbacks constituting the cybernetic identity of a given individual. It forms a hierarchically organized network, essentially comprising all functional aspects of the living individual.

A similar relationship exists between the cybernetic system coordinating muscle contractions and thus movements, the system adjusting the rate of energy (ATP) supply to the rate of its consumption, the system of stimuli perception, and the planning, decision making and steering of behavior. At the basis of the latter lies the whole nervous system, primarily the brain, and the complex of sense organs and muscle receptors. For example, when a leopard's hypothalamus detects a lowered glucose level in the bloodstream, it stimulates the brain for action. The animal decides to initiate the hunt for herbivorous prey. It elaborates its strategy (e.g., lurking in bushes, prowling, pursuit) based on previous experiences (episodic memory), the shape of the terrain, the vegetation covering it and the prey encountered. It moves smoothly and coordinately (thanks to the precisely-controlled spatiotemporal pattern of muscle contractions), incessantly observing its surroundings. It approaches the prey and finally starts to run in its direction, trying to kill it (predators hunting in packs, for example, wolves, additionally cooperate during the hunt). In all of this, sense organs participate (those concerning sight, hearing, smell, internal sensing) as does the brain—its occipital, parietal and temporal lobes integrate stimuli of different modalities coming from receptors, the temporal lobe stores episodic memory, the prefrontal and frontal lobes deal with planning, decision-making and their execution through the stimulation of particular muscles, and the cerebellum is the seat of procedural memory and the center of movement coordination. ATP for muscle contraction is supplied at an appropriate speed, so it is not depleted. After a successful hunt, the leopard can eat at will, which elevates the level of nutrients in the blood, which in turn causes the hypothalamus to "feel satiated" and inform the rest of the brain that the animal can lie down and quietly digest its intake of food. Each stage of this entire process (the purposeful sequence of events) is based on numerous negative feedbacks, which are all necessary for the whole—if at least one of them were lacking, the animal would simply die of starvation. Again, we are dealing with the mutual co-conditioning of regulatory mechanisms and forming a closed loop, or rather a network of relationships by them. Ultimately, all negative feedbacks condition the action of all other negative feedbacks. This constitutes a special case of the *causal closure*—that is, the mutual support of all aspects of structure and function of living individuals, first and foremost, the regulatory mechanisms that are specifically directed at the survival and reproduction of the individual. This property of "internal closure" characterized the description of a mysterious object sent by extraterrestrials in the excellent novel by Stanisław Lem *His Master's Voice* (Lem 2020).

It is easy to find many similar examples of the mutually functional interconnection of various negative feedbacks. Generally, all negative feedbacks in a living individual form a hierarchically-organized network, within which all feedbacks are reciprocally dependent on and mutually co-condition each other. Of course, it is not only the case that low-level negative feedbacks condition negative feedbacks on high levels of the hierarchy; the opposite also occurs. For example, food uptake on the behavioral level ensures the delivery of respiratory substrates for the biochemical (bioenergetic) level. Therefore, in a functional sense, the purpose of every particular negative feedback mechanism is determined by the context of the whole complex of negative feedbacks constituting the cybernetic identity of a given individual.

On a side note in the discussion of directed-on-something negative feedbacks, it is important to remember that while living individuals are certainly characterized by an internal purposefulness directed at survival and reproduction, from the external point of view, represented by both realities on a physical level and the psyche, life is a process that is as equally senseless and purposeless as the falling of stones in the gravitational field of the earth, the growing of crystals in a saturated salt solution and the painting of ice flowers on windows by frost.

I would like to emphasize one more thing clearly. Although negative feedbacks are undoubtedly the purposeful internal regulatory mechanism in living organisms, they certainly do not represent the entire purposefulness of the structure/function of biological systems. Generally, there are two types of the internal purposefulness of living individuals directed ultimately on self-survival and reproduction in many copies similar to the original. *Passive purposefulness* is various passive (non-regulatory) structural properties and functions on different hierarchy levels. Examples of passive purposefulness include the physicochemical properties of the catalytic niche of enzymatic proteins responsible for the efficient catalysis of determined biochemical reactions, the impermeability to many substances of the lipid-protein membranes surrounding the living cell and its various compartments and their selective permeability to other substances; complex metabolic pathways serving the production/decomposition of many substances; the heart pumping the blood that provides all cells in the body with oxygen and nutrients; the skin protecting the whole organism; the teeth devoted to the obtaining and fragmentation of food; the capability of contracting muscles executing commands from the brain; the entire previously discussed morphogenetic cascade leading to forming an adult individual during individual development.

In contrast to passive purposefulness, *active (regulatory) purposefulness* comprises maintaining homeostasis and the compensation of various disturbances coming from the surroundings (and from the organism's interior). In particular, it comprises the perceptive-cognitive functions discussed earlier in this book. However, this is not all because active purposefulness leads to the production, maintenance and supervision of passive purposefulness. Such processes are responsible for the regulation of the reading of genetic information, which results in the synthesis of various proteins at the appropriate place, time and amount, including enzymes that are in turn responsible for the organization of the chemical reactions they catalyze into metabolic pathways; the product of a given metabolic pathway frequently inhibits its own synthesis. Regulatory processes ensure an appropriate composition and area of bilayer lipid-protein membranes, and they adjust the rate of blood pumped by the heart to the current needs of the organisms, for example, by means of hormones such as adrenaline. Finally, which is of immense importance, they supervise the production of various structures of the body, such as the heart, skin, spots on a shell, eyes, brain and teeth during the action of the developmental (morphogenetic) cascade during embryogenesis and ontogenesis, compensating any deviations from the correct (assigned by the genetic record) course of morphogenetic procedures. In other words, active purposefulness controls the formation and maintenance of passive purposefulness (although the former is also underlain structurally and functionally by the latter). Therefore, in some important sense, active purposefulness is superior to passive purposefulness, determining the target structure and function of the organism. Such systems as, for example, a computer, possess only passive purposefulness imposed on them by humans (therefore, they tend to go down and sooner or later are spoiled). Most of the active (regulatory) purposefulness is in the form of negative feedbacks. Therefore, without the risk of excessive simplification, one can say that from the functional-regulatory point of view, living organisms (individuals) are built of negative feedbacks that mutually co-condition each other and are linked with each other into a coherent, hierarchically organized system (network), whose only superior aim is to survive and multiply its own (cybernetic) identity.

Of course, the cybernetic aspect is just one of many aspects of the phenomenon of life. The living individual (organism) can be treated as a hierarchical structure composed of atoms and chemical molecules, particularly organic carbon compounds. Equally well, or maybe even better, the organism can be treated as a dynamic, hierarchically-organized complex of metabolic processes of transforming matter and energy. It can be considered as an assembly of various dissipative structures on different levels of the hierarchy of complexity. The phenomenon of life can be illuminated from the point of view of, *inter alia*, its genetic, structural, energetic, enzymatic and redox aspects. It can also be presented as a system of regulatory mechanisms (chiefly negative feedbacks) and pathways of the transfer and transformation of biologically-valuable information, purposefully directed toward its own survival and copying its identity to a potentially vast number of copies.

In summary:

- ❖ Ultimately, negative feedbacks in a living organism form a network in which *all negative feedbacks are functionally inter-connected* and co-condition each other, either directly or indirectly.
- ❖ This implies that negative feedbacks mean by *connotation* (relation to each other): their sense is determined by other negative feedbacks and, ultimately—by the whole network of negative feedbacks of a given individual.
- ❖ The negative feedback network constitutes a *functional (operational) representation* of various aspects and properties of the world (external and internal) and its disturbances and variability.
- ❖ The negative feedbacks of the living individual are (internally) *intentional* (in the instrumental sense), purposefully directed toward the survival and reproduction of the whole cybernetic identity of the living individual constituted by particular networks of negative feedbacks.

Self-copying, which is the reproduction of living individuals, is, as previously discussed with regard to bacteria in a medium, an example of positive feedback. In the case of living individuals, this is the *superior* positive feedback[7] to which the whole purposefulness of the individual's structure and function is subordinated. It is aimed at the unrestricted, exponential increase of living individuals to infinity. Of course, this increase is normally limited by the finite capacity of the environment. In stable environments, the number of individuals in a given species is relatively constant. Therefore, in such a case, no expansion of living individuals (increase in their numerical force), so characteristic of the positive feedback, takes place. Nevertheless, even here, positive feedback resulting from the network of negative feedbacks underlying it sustains the potential of expansion (imposed by reproduction) that can express itself as the

[7] An inferior positive feedback mechanism, subordinated to the functioning of the whole individual and inbuilt as an element of the network of negative feedbacks is, for example, the blood coagulation cascade.

competition of a given individual with individuals endowed with different identities (of the same or other species) for limited resources of the environment or the population of new areas (or environments/habitats). Therefore, even in a population remaining in stasis, the realization of superior positive feedback remains the main internal purpose of all regulatory mechanisms of a given individual (and, of course, of the living individual itself or its whole cybernetic identity).

The ultimate goal of the identity of every living individual (comprising its structure and complex of functions) is therefore not surviving for a possibly long period, but the reproduction and multiplication of its identity into possibly a large number of copies that are similar to the original. Every living individual will eventually die[8] due to old age (a rather rare case in nature), being eaten by other individuals (of other or even the same species), being killed by parasites, environmental changes (e.g., the water reservoir drying out), and random events *etc.* Of course, to reproduce efficiently, such an individual must aim to survive in good condition until reproduction. Some organisms (annual plants, salmons, and one-day mayflies) only once allocate all of their energy reserves to reproduction, and just after progeny production, they die. Other organisms distribute their reproductive effort over many occasions (e.g., years or broods—this is what most mammals do). In such a case, they must rationally allocate energetic resources each time between the current reproduction event and survival to the next reproduction event. Nevertheless, the ultimate purpose is always maximizing the production of a possibly large amount of progeny well suited to survival and reproduction.

Sometimes it "pays" for the living individual to sacrifice its life during (or just after) the act of reproduction, even if there is a chance for subsequent reproductive acts—that is, to maximize the probability of the transfer of its own identity to progeny. From the biological-evolutionary point of view, this pays infinitely more than safe and effortless, quiet existence to death without undertaking any reproduction trials. Let us consider the male mantis (we could also consider similar situations with many spiders). The female mantis, in many cases, eats the male just after copulation and sometimes begins his consumption starting from the head during the sexual act. If the male cared only, or at least primarily, about his own survival (and living in good condition for a long time), he could "think" (or his genes could "think"): "what do I need progeny for, I want to live?" He would give his possible mate a wide berth, enjoying his *dolce vita* to old age. However, the

[8] Leaving aside the simplest organisms, for example, bacteria that reproduce through division and are potentially immortal and eternally young.

male mantis, directed toward self-propagating the identity of the living individual, is crying out for multiplication to a potentially large number of copies. Therefore, the imperative of spreading his own identity "forces" the male to serve as a snack for the lady of his heart (and thus become a chevalier of her stomach). Nevertheless, such behavior remains in his well-understood evolutionary interest, although this may seem decidedly obnoxious to us. The nutritious substances from his body will better equip a greater amount of his progeny. Statistically, such a strategy may appear to be more efficient evolutionarily (giving a greater expected amount of progeny with higher levels of fitness) than searching, after successful copulation, for another partner which he could fertilize, especially if the chances for this are small.

Finally, the time has come to quote the previously announced cybernetic definition of the living individual. First, a brief definition: *The living individual is a network of negative feedbacks subordinated to superior positive feedback, reproduction.* The entire complex of negative feedbacks (more broadly—purposeful regulatory mechanisms) working on various levels of hierarchy and representing the cybernetic aspect of the functioning of the living individual has the purpose of sustaining the identity of this individual, at least until the time of reproduction. In turn, the only goal of this identity is self-multiplication to a potentially large number of copies. From the cybernetic point of view, the identity of a given individual is a particular unique complex of negative feedbacks. Of course, the structure, function, and passive and active aspects of the purposefulness of living individuals have their material basis in the form of atoms and molecules of chemical compounds and their complexes.[9] Therefore, referring to the subtitle of this book, the living individual can be presented as a specific complicated dynamic configuration of atoms intentionally directed (in an instrumental sense) on itself, its survival and its reproduction.

Considering the discussion in the present chapter, the above brief definition can be somewhat extended. According to the extended definition: The living individual is a network of inferior, meaning by *connotation* negative feedbacks, being a *functional representation* of various aspects of the world, *intentionally* (in the operational sense) directed on the realization of the superior positive feedback: *copying of itself*, that is reproduction. Such a recursive directing of a system or procedure on itself is called *recurrence*.

---

[9] Their construction during embryonic and individual development is guided by the genetic record, in any manner in a very complicated way that involves a huge number of elements.

A conceptual scheme of the cybernetic living individual is presented in Fig. 6-6.

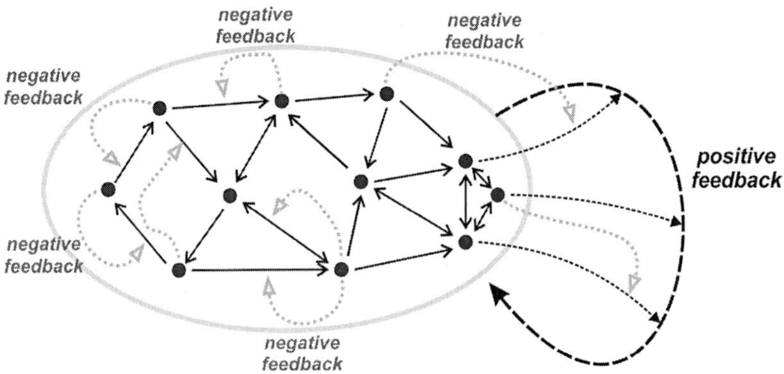

Fig. 6-6. A conceptual scheme of the cybernetic living individual. A network of negative feedbacks supervising the functioning of the living individual, maintaining the constancy of internal conditions (homeostasis) and compensating various disturbances coming from both outside and inside the organism, steering the morphogenetic cascade during embryogenesis and ontogenesis and controlling the behavior of the living individual is ultimately directed on the realization of superior positive feedback: the multiplication of the cybernetic identity of the living individual, which consists of a specific combination of negative feedbacks, in a potentially large amount of copies that are similar to the original. Black solid arrows indicate various "passive" processes in the living individual; grey dotted arrows are negative feedbacks; black dashed arrows indicate positive feedback, i.e., self-copying of (the identity of) the individual; black dotted arrows indicate processes directly supporting self-copying, i.e., reproduction of the individual.

In summary, it is important to emphasize that:

- The cybernetic living individual refers to the whole *life cycle*, not only the adult stage.
- The cybernetic living individual constitutes the *subject of evolution, a unit of Darwinian selection.*
- In the case of saturation of the environmental capacity, the superior positive feedback is expressed as the *potential of expansion* in the context of competition with other individuals of a different identity.
- The living individual constitutes a certain *causally-closed entity* in which all elements co-condition each other (causal closure) and are sufficient for the individual's independent survival and reproduction in its environment. Dawkins' "selfish genes," which in the majority

are not selfish at all but cooperate with other genes for the good of the whole individual,[10] only constitute their undoubtedly very important elements.

- The cybernetic living individual contains *all* and *only* negative feedbacks that are directed toward the realization of its superior positive feedback.

Therefore, the cybernetic living individual is a network of *all* negative feedback purposefully directed *only* and *exclusively* on the realization of the superior positive feedback that is copying its own identity and is *self-dependent* in carrying out this task. This self-dependency means that in a given physical and biological environment (comprising properties of the physical world and other individuals of the same or other species), the living individual does not need any purposeful external mechanisms (particularly negative feedbacks) which would intentionally support it (in the instrumental sense) in the realization of the superior task of reproduction. This requirement excludes, for example, cells in multicellular organisms or their organs and particular insects in eusocial insect colonies from the circle of cybernetic individuals (this is discussed in detail below). The above definition means that the cybernetic identity of an individual comprises all those and only the negative feedbacks that participate in sustaining the potential of the expansion of this identity, and also that particular negative feedbacks of the living individual cooperate and do not compete or fall into antagonisms. For this reason, the population, the ecosystem of the whole biosphere, "saturated" with competition and hostility between particular individuals of the same or different species, cannot be regarded as living individuals.

In different developmental stages, and in a shorter term in different seasons or times of the day and slightly different environmental conditions, a given individual has at its disposal (at least slightly) different complexes of negative feedbacks. Therefore, the network of negative feedbacks constituting the cybernetic identity of the living individual is a dynamic system relativized to the environmental conditions and developmental

[10] In fact, they "have no other choice," as their replication, expression and its regulation is carried out by proteins. Apart from this, any trial of such a gene to be "selfish" would be associated with harm for the whole organism and therefore for the gene itself. Only parasitic DNA, such as transposons or chromosomes B, which are discussed below, can allow themselves to be "selfish." "Regular" genes simply constitute an element of developmental instructions, construction procedures that lead to the creation of the structure (morphogenesis) and supervise current functioning of the living individual.

stage. Negative feedbacks also supervise the transitions between different developmental stages, including embryogenesis and maturing.

Briefly speaking, the cybernetic definition characterizes the living individual as a *dynamic homeostat.* It is constituted by a complex network of negative feedbacks and mechanisms of regulation and signal/information transduction in general that undergoes modification over time, both during individual development and in varying external conditions. Its superior task is to be an *effective replicator* (realizing the "target" positive feedback—self-copying). In a yet shorter (although significantly simplified) manner, the cybernetic living individual can be characterized as a *homeostatic replicator.*

As mentioned, some positive feedbacks, for example involved in the blood coagulation cascade (Reece 2011, 911-912)[11] , can be inbuilt into a given individual's negative feedbacks network. However, they are positive feedbacks of a low level and are subordinated, like negative feedbacks, to the realization of the superior positive feedback, that is replication.

Treating the matter even more widely, various mechanisms associated with signal transduction and information transformation in a purposeful way underlie and regulate the functioning of a living individual on different levels (metabolic, genetic, cellular, physiological). The very process of DNA replication, its transcription to mRNA and translation to proteins constitute examples of the copying, transfer and transformation of information. Of course, the functioning of the whole nervous system is associated with the transduction, transformation, processing, integration and association of signals/information. Additionally, hormones secreted by glands to blood and distributed by the circulatory system to all cells of the body participate in information transfer; however, only some of the cells, which have appropriate receptors, can bind a given hormone and adequately react to it. The molecule of a hormone activates a protein receptor (either in the cell membrane or within the nucleus), which triggers a whole cascade of signal-transduction processes in the cell, frequently in the form of inferior positive feedback (associated with signal amplification, acting within the network on negative feedbacks) leading to desirable effects—the activation/inhibition of various enzymes or participation in the regulation of gene expression (Martin 1987).

In many cases, the intermediary between the hormone receptor in the cell membrane and the intracellular cascade of the regulatory processes is the so-called protein G and such small-molecule compounds as cAMP (cyclic AMP, adenosine monophosphate), IP_3 (inositol triphosphate) or Ca^{2+}

---

[11] https://www.osmosis.org/answers/coagulation-cascade.

(calcium ions), which can, for example, activate protein kinases causing the phosphorylation of various proteins and through this, their activation or inhibition (Martin 1987). Apart from blood glands and hormones distributed by the circulatory system over the whole body that constitute the *endocrine system*, there is also the *paracrine system*, which involves the transfer of chemical signals over short distances between neighboring cells. For example, NO (nitric oxide) causes smooth muscle relaxation and, as such, increases the diameter of blood vessels (Reece 2011, 1350). On the level of entire organisms, some volatile chemical substances, namely pheromones, can constitute communication between individuals of the same or opposite sexes. They can also serve the internal integration of the behavior of particular insects within whole eusocial insect colonies (see below) (Wilson 1971; Wilson 1975; Maynard Smith and Szathmáry 1995, 255-278). In honey bees, the famous "waggle dance" performed by workers in the hive after the discovery of a food source informs other workers about the direction (the inclination angle of the straight part of the dance in relation to the vertical, corresponds to the angle of the direction of the food source in relation to the current position of the sun) and the abundance and proximity of the source (intensity of abdomen vibration) (Wilson 1975). All of these pathways/mechanisms of signal transduction/information transfer constitute the supplement of negative feedbacks, which constitute broadly-understood purposeful cybernetic mechanisms, and as such, they, of course, enter into the composition of the (extended) cybernetic definition of the living individual.

Cybernetic living individual and biologically purposeful information

My cybernetic definition of life is primarily a cybernetic definition of the living individual. What is the relationship between these phenomena? In my opinion, a system can be regarded to be alive (in the conventional meaning of this term) if (a) it is composed of living individuals (e.g., a population, an ecosystem or the whole biosphere) or (b) it constitutes a part of a living individual (e.g., a cell, body organ), which possesses at least some integrated complex of negative feedback (this excludes, for example, atoms or simple molecules of chemical compounds, including macromolecules of organic compounds from the domain of living systems; certainly "naked" genes and the whole double helix of DNA and proteins considered in isolation are not alive). Therefore, the phenomenon of life also comprises phenomena on both higher and lower levels of the organizational hierarchy than the living individual. On the other hand, the living individual constitutes a clearly distinguished level in this hierarchy, as it represents a self-sustainable (in determined environmental conditions) regulatory entity (unit) purposefully directed at its own survival and reproduction, constituting the subject of evolution and the unit of selection. In the face of this, the living individual represents the most important element and level in the hierarchy of life—no life can exist without living individuals, and life can be defined as a complex of living individuals.

The living individual (in fact, its identity) characterized above becomes, in a sense, automatically, the subject of evolution. Only the identities of living individuals compete with each other for limited resources of the environment and undergo evolution from generation to generation. The subject of evolution through natural selection S is something that: (1) can differ from another subject of evolution Z; (2) can independently, without the purposeful support of other systems, transfer these differences to descendants Ss; (3) competes with different Zs for limited environmental resources; (4) can undergo an accidental change into S', which is slightly different from S. Identities of living organisms differ from identities of other living individuals; they compete with other identities for the limited capacity of the environment and can change from generation to generation. On the other hand, genes, for example, usually cooperate with other genes within a given individual, and they (their complexes) do not compete directly (but only through produced phenotypes) with genes (complexes of genes) of other individuals, by themselves (without, for example, the whole genetic apparatus and metabolism) they mean nothing and can do nothing, they constitute only some elements of self-copying systems (living individuals). This is why only the cybernetic individual (its identity) is the subject of Darwinian evolution, while the "naked" gene is not. As is discussed below, in many cases, the cybernetic individual is not equivalent to the classical biological individual, which is usually defined according to structural criteria as a certain internally (structurally and functionally) integrated whole that is clearly separated from the surroundings.

One of the main advantages of the cybernetic definition of life is the possibility of the sensible defining of (the amount of) biological information that is purposeful, meaningful, directed on something, and valuable from the biological point of view. Being dissipative structures, living organisms constitute systems displaced from thermodynamic equilibrium. As previously mentioned, they are characterized by low entropy, and, therefore, by a large amount of information. However, the thermodynamic information is not strictly related to the essence of the phenomenon of life as it does not concern the internal purposefulness of living organisms (although a great amount of thermodynamic information is necessary for biological systems coming into existence). Therefore, it does not allow life to be distinguished from other dissipative structures, for instance, convective currents, or more generally, from among physical systems displaced from thermodynamic equilibrium, which are also characterized by a large amount of thermodynamic information. Therefore, a definition of biologically relevant, purposeful, "intentional" (in the operational sense) information is needed. Such information would measure not the (very) general level of order, which the thermodynamic definition does, but the degree of order specific to life. As previously mentioned, the lowest entropy and, therefore, the highest thermodynamic ordering occurred just after the Big Bang in the homogeneous distribution of matter in the gravitational field. The system of nutrients plus oxygen is endowed with lower entropy and, therefore, higher thermodynamic information than the organism of humans, based on which it is formed and its existence is sustained. A large amount of thermodynamic information is necessary for the origin and "occurrence" of the phenomenon of life as a complex of hierarchically-organized dissipative structures. However, this is not the kind of information that matters when we want to speak about the essence of life, particularly the essence of living individuals.

The internal purposefulness of living individuals is strictly associated with the network of regulatory mechanisms (especially negative feedbacks) serving their main purpose: survival and reproduction. The information related to a given system is usually defined as a binary logarithm of the number of states of this system. Therefore, the amount of purposeful, biologically-relevant information has been defined (Korzeniewski 2001) as follows:

$$I_B = \log_2(n \cdot 2) \tag{1}$$

where I_B is the amount of biologically-relevant, purposeful information (in bits), and n corresponds to the amount of negative feedback present in a given biological individual. The number 2 in parentheses corresponds to the minimal number of states a given negative feedback can recognize (and adequately respond to); the value of the regulated parameter either differs (state 1) or does not differ (state 0) from the assigned value.

These two possible states correspond well to, for example, the action of the thermostat in a refrigerator, where the temperature is either higher than the assigned temperature and the cooling device is turned on, or the temperature is lower than or equal to the critical temperature, and the cooling device is turned off. The situation in living organisms is more complex—most of the negative feedback appearing in them detects the continuous range of the value of the regulated parameters and has at its disposal a continuous range of strength of response (steering) of the appropriate negative feedback loop (the strength of which is proportional to the current deviation of the parameter value from the assigned value). This takes place, for example, in the case of the regulation of amino acid concentration in the cell and the regulation of the glucose level in the bloodstream. Therefore, there can theoretically exist an infinite number of states distinguished by a given negative feedback mechanism, which would make the above equation completely nonsensical (the infinite number of possible states would, in each case, give an infinite amount of information). However, this problem can be overcome quite easily. It is enough to introduce a parameter corresponding to the (in)accuracy of the regulation (or the tolerance of the system to regulation inaccuracy) r_i determining the maximal variation (fluctuation) of the parameter *i* that has no noticeable effect on the functioning of the system (organism) (or, more generally, on its fitness) and or is not detected by the appropriate negative feedback loop (e.g., because of intrinsic noise in the system, such as thermal noise). The expression for the amount of biologically-relevant information can then be re-formulated (Korzeniewski 2005) as follows:

$$I_B = \log_2 \sum_{i=1}^{n} R_i / r_i \qquad (2)$$

where R_i corresponds to the range of the variability of the parameter *i*. In this formulation, R_i/r_i corresponds to the number of states recognized by the negative feedback regulating the value of parameter *i*. Although this expression is certainly more realistic than Eq. 1, it would be rather difficult to apply it in practice (to calculate the amount of the purposeful information of a given organism) because of the huge complexity of living organisms and taking into account the limitations of our knowledge. Therefore, the point is to present a general conception rather than obtaining strictly quantitative results.

While the cybernetic definition of the living individual enables us to decide whether we are dealing with life or not in a given case, the amount of biologically relevant (purposeful) information tells us "how much" life we are dealing with (how functionally complex a given individual is). The amount of biological information also seems useful to measure the degree of integration of two symbionts or polyps in the coelenterate colony (see below).

Which systems are living individuals?

I aimed to propose a concise and adequate definition of life (living individual), which would univocally separate biological systems from inanimate phenomena, and also living individuals from biological systems that are not living individuals (composed of the latter or being their parts). It seems that it would be difficult to elaborate a more "minimal" definition of life than the cybernetic definition. However, the question remains whether this definition constitutes a sharp criterion that enables the unequivocal classification of some phenomenon, frequently of dubious status, as a living being or an inanimate object, and strictly speaking, as a system that either is or is not a living individual. This also concerns biological systems that are undoubtedly "alive," but it is not certain whether they are living individuals or not.

As has already been discussed, examples of many negative and positive feedback loops can be found in inanimate systems. However, they never appear in the above-specified configuration corresponding to the cybernetic living individual. Let us consider positive feedback in the case of a rock avalanche. Here, the parameter for which the value increases exponentially over time (at least at the beginning) is the number of rolling stones. However, rolling stones do not possess any purposeful regulatory mechanisms (negative feedbacks) directed at sustaining their movement, and ultimately, the whole avalanche freezes at the bottom of the slope. Additionally, the identity of the rolling stones adopts only a rudimentary form—this is just their movement. Therefore, of course, the stones in a rock avalanche are by no means alive according to the cybernetic definition of life. However, as discussed below, from the formal point of view, the case of the rock avalanche is strictly analogous to prions (infective protein factors).

Negative feedbacks seems absent in inanimate nature, but they are present in many devices constructed by humans. Some robots can even possess a hierarchically-organized network of negative feedbacks, allowing them to, for example, move efficiently and precisely manipulate various objects and "cleverly" navigate their surroundings. Nevertheless, the artificial products of humans lack the superior purpose—the "target" positive feedback in the form of reproduction. Instead, they serve purposes imposed on them by humans and not their own interests. While they possess some identity (at least robots, although many are produced serially), they cannot propagate this identity by themselves (and cannot evolve spontaneously). As such, they are not autonomous in their purposeful action; they constitute a part of some bigger directed-on-something system, namely human, technical civilization.

Thus far, the cybernetic definition of the living individual seems to separate living individuals univocally from inanimate objects. If so, let us see what it can say about various phenomena of dubious status currently known on earth.

First, one should ask what the relationship is between the cybernetic individual to the "classic" living individual, distinguished (frequently intuitively) mainly based on structural criteria, as some complex systems clearly separated from their surroundings. In most cases, the classic individual is largely (but not entirely) identical to the cybernetic individual. Thus, to a large extent, the bacterium cell, cactus and ladybird are all cybernetic individuals. However, if we look more closely at many biological systems/phenomena, the matter appears more complex.

Genetically identical colonial coelenterates (e.g., corals, sea anemones or siphonophores) that originate from one individual or larva (planula) differ little from clones of bacteria or parthenogenetically reproducing plants. However, the clone of bacteria or asexually-reproducing plants is certainly not a living individual (particular bacteria or plants do not cooperate and do not support each other in reproductive success, and their networks of negative feedbacks are not integrated for the common benefit, but rather compete for space, minerals, nutrients, light, *etc.*). The situation is not so clear in the case of the, to some extent, physically integrated colony of "individuals" such as polyps of cnidarians. While being genetic clones, the sedentary polyps of corals in a colony or forming colonies sea anemones look, in principle, identical, then siphonophores are morphologically and functionally differentiated, the division of work occurs in them (Mackie et al., 1988). We have feeding polyps (gastrozoids), excretory polyps (palpons) and reproductive polyps (gonozoids), while all of them are connected to a stem that is underslung in a floating bell (pneumatophore) which lifts the whole colony in the water. Solitary (non-colonial) cnidarians exist, whose polyps are largely separate cybernetic individuals; such cases are hydra or solitary sea anemones. However, particular polyps in the colony of corals, or especially siphonophores, exhibit highly advanced cooperation. In corals, every polyp that has just hunted something shares nutritional substances with its less lucky colleagues (therefore, negative feedbacks act between particular polyps and bind them into a functional network). All organisms without a separate generative (germ) cell line, for instance, any tree, can be regarded as a colony of groups of somatic cells (e.g., branches), which produce their own sexual cells (and organs). There is no problem as long as the whole colony is genetically identical. However, every somatic mutation within it can disturb this state of affairs (Futuyma 2013, 193). Therefore, the question arises what is the living individual in corals and siphonophores: a single polyp or the whole colony?

To be or not to be a biological individual does not necessarily assume the principle of the excluded middle—it is possible to be an individual only partly, to some degree. In the case of a colony, for example, the above-mentioned colony of corals, it is possible to quantitatively define the degree of "individuality" of polyps, D_i, which can be expressed as the ratio of the biologically-purposeful information (defined above in Eq. 2) associated only with internal negative feedbacks of a polyp I_{Bi} to the total biologically-purposeful information (being the sum of information associated with internal negative feedbacks of a polyp I_{Bi} and information associated with external negative feedbacks that integrate the polyp with the rest of the colony I_{Be}):

$$D_i = I_{B_i} / \left(I_{B_i} + I_{B_e}\right) \tag{3}$$

The same measure of the degree of "individuality" can also be applied in the case of symbiosis (see below). In this way, the cybernetic definition of the living individual not only supplies a criterion allowing decisions of whether, in a given concrete case, we deal with a biological individual but also allows determination of the degree of integration or separation of various quasi-individuals.

The situation in siphonophores seems much simpler than in corals. As in this case, particular polyps are specialized structurally and functionally, and none of their types is independent with regard to survival and reproduction. Only gonozoids can produce gametes but can still not feed or defend themselves. Moreover, all polyps in the colony are lifted by a common pneumatophore. Therefore, only the whole colony of polyps of siphonophores can be regarded as a cybernetic individual.

Slime molds (mycetozoans) are distantly related to fungi organisms that have the form of unicellular amoeboids (amoeba-like cells) for most of their life cycle, while with a deficit of nutrients, they join into plasmodium in order to form a sporangium and produce spores. The sporangium is situated at the top of a stalk, and its chamber contains spores that are disseminated when the walls of the sporangium break (Maynard Smith and Szathmary 1995, 212-214). In this case, it is doubtful whether single amoeboids or whole plasmodium is the cybernetic individual (subject of evolution). This depends on the genetic identicalness of amoeboids in the plasmodium. It is quite likely that the whole local population of amoeboids comes from one spore (it is also possible that genetically identical amoeboids recognize each other, and only they cooperate in sporangium formation). The problem then vanishes: the whole population of cloned amoeboids simply constitutes an "individual in dispersion", which undergoes consolidation into plasmodium to produce and disseminate spores. However, if a few genetically different amoeboid groups enter plasmodium composition, we are dealing with an intra-species reproductive symbiosis. Nevertheless, an interesting possibility of "reproductive parasitism" arises here; one of the genetically different groups of amoeboids could acquire a mutation that would cause all amoeboids from this group to transform themselves selfishly into spores instead of honestly co-creating the stalk and sporangium walls with other groups. Of course, only amoeboids transforming into spores will transfer their identity (genes) to the progeny; therefore, such a situation seems evolutionarily unstable.

Viruses are infective factors in which the carrier of genetic information is a single- or double-stranded nucleic acid: RNA or DNA (Reece 2011, 381-395; Harris and Hill 2021). They can parasitize on bacteria (so-called bacteriophages), plants (e.g., tobacco mosaic virus) and animals (e.g., influenza virus or HIV, or the currently infamous coronavirus). Outside the host cell, they appear as *virions*, strands of nucleic acids enclosed within a protein envelope—*capsid*. Virions are usually presented in books and journals as brilliantly colorful (those are, of course, artificial colors), and we identify just virions, more or less consciously, with viruses as such, with their whole life cycle. Are viruses alive? Are they cybernetic individuals? Certainly, virions outside of the host cell do not exhibit any living activity—they do not possess any metabolism or exhibit any movement. In this respect, they very much resemble spores of bacteria (or the slime molds mentioned above) produced by them in unfavorable conditions. Does this, however, mean that spores are dead (inanimate)? After all, they constitute a certain stage of the life cycle of otherwise living organisms.

The same applies to viruses. A virus "becomes alive" only when its RNA or DNA enters the host cell. This is where it manifests its dexterity, purposefulness and directing on its reproduction. The nucleic acid of the virus (and its protein products) enters the host cell's metabolism into its network of biochemical-genetic processes and switches it to a course appropriate for the virus. The infected cell, instead of taking care of its own interests and homeostasis, becomes re-directed to the production of a potentially large amount of virus particles (the above-mentioned virions), which can infect other cells and dramatically disturb their functioning by depriving them of building substances and energy and regulatory mechanisms, leading to their death. The strand of the nucleic acid of a virus encodes some enzymes responsible for its replication and other functions related to self-copying and the proteins of the capsid, whose synthesis is strictly controlled. Therefore, viruses certainly have germs of their own network of negative feedbacks directed toward the superior purpose—the reproduction of the virus. The network of negative feedbacks from the host manages the rest. Simply put, viruses can enter the system of negative feedbacks of the host cell and modify it appropriately for them, changing the optimized parameter values. The purpose of the whole system now becomes the production of a potentially large amount of the copies of virus particles and not the regulation of such parameters as ATP or amino acid concentration. The entire sequence of nucleotides in the RNA or DNA of the virus evolved for millions of years to elevate their infective efficiency. Viruses, despite the lack (in principle) of their own network of negative feedback, can exploit (after an appropriate modification) the network on negative feedbacks of their hosts. Such usage of the system of negative feedbacks from the host constitutes the essence of parasitism. This resignation of own negative feedbacks in favor of those of the host adopted an extreme form in viruses—virtually all negative feedbacks used by viruses for their reproduction come from the cells of the host (one should, however, remember that virions must penetrate, despite various defending mechanisms, the host cell and the nucleic acid must appropriately switch its metabolism, i.e., change the parameters that are optimized).

From this point of view, one can conclude that viruses outside the host cell (virions) are not alive. On the other hand, the infected host cell (and therefore the system: virus + host cell) represents a system of negative feedback subordinated to the superior positive feedback, although this is no longer the host's reproduction but the violent multiplying of the particles of the virus. Therefore, the virus acquires properties of the living individual at the moment when it is situated in the context of the network of negative feedbacks of the host that is modified by it. It becomes "alive" when it is "dipped" in the host cell's metabolism. For this reason, viruses should, despite all reservations, be acknowledged as cybernetic individuals.

Even simpler than viruses are *viroids*; these are the parasitic naked strands of RNA devoid of the protein hood (capsid) (Reece 2011, 393). They parasitize on plants and, like viruses, can force the host cell to produce a potentially large amount of viroid particles instead of taking care of its own interests (as a part of the whole plant individual) and homeostasis. Their RNA can, of course, evolve and therefore undergo selection and modification. Generally speaking, practically, the above reasoning concerning viruses can also be applied to viroids. Therefore, they can be regarded as cybernetic individuals, undoubtedly "alive" in the "context" of the host's metabolism. They constitute the simplest known parasites able to infect living cells (and are therefore not exclusively parasites of the genome—compare the discussion of transposons presented below).

It is also worth mentioning that viruses parasitize other viruses. For example, the "sputnik" virophage exploits particles (virions) of one of the largest viruses, *Mimivirus*, to transfer itself and reproduce in the cells of the host of the latter. Linear or circular naked strands of nucleic acids, RNA and DNA, which are somewhat analogous to viroids, also parasitize on viruses.

Prions (Prusiner 1998) are pathogenic misfolded protein molecules that cause neurodegenerative diseases of the brain in animals and humans (e.g., "mad cow disease" [BSE] in cattle or "laughing sickness"—kuru in cannibals in New Guinea).[12] The prion proteins occur in two forms: the normal form, with the structure of α-helix, and the pathogenic form, adopting the harmonica-like β conformation. Normal molecules are produced in the brains of animals (and humans), fulfilling some useful (although still unknown) functions there. Infective pathogenic molecules can transform normal molecules into other infective molecules, and in turn, they stimulate the transformation of yet more "healthy" molecules and so on, thus creating a typical positive feedback loop. Of course, prions are not cybernetic individuals—they are strictly analogous to the rock avalanche. The "identity" of prions simply reduces to their pathogenic conformation, such as the only "identity" of the stones in the rock avalanche being their movement. Prions have at their disposal no network of negative feedbacks that would increase the efficiency of their infection as their structure (sequence of amino acids) is recorded in the DNA of the animal that produces them, over which they have no control (after all, this is not the prion protein that reproduces in the brain but only the infective/pathogenic conformation at the cost of the normal conformation). The activity of prions does not bear any traits of purposefulness which would favor them in achieving any evolutionary benefits, particularly their survival and reproduction. Therefore, prions certainly do not fulfill the criteria of the cybernetic living individual.

Cancers (Reese 2011, 243-245) constitute another interesting case. Traditionally, cancer cells are not treated as biological individuals. But why is this? Perhaps primarily because they come from the "normal" cells of their "host," which mutated and escaped the control of the organism and therefore are almost genetically identical to their "carrier." Aside from this,

[12] Cattle and sheep are infected with prions due to the addition of minced body remains, including brains, of already infected individuals to the feed of healthy individuals, while cannibals in New Guinea—through eating the brains of their dead ancestors.

their lives terminate together with the organism of their host. However, from the cybernetic point of view, they are undoubtedly living individuals. The tumorous transformation of a normal cell can be treated as breaking out from the superior purpose of the whole organism and establishing its own superior purpose—its own survival, reproduction and spreading in the "environment" of the host organism. Cancer cells are subject to normal Darwinian selection and evolution. Only those cells survive and produce "progeny" that create (due to mutations) some purposeful mechanisms (including negative feedbacks) causing, for example, resistance to the immunological system and other defensive mechanisms of the organism of the host, the adaptation of metabolism to lowered oxygen levels, the stimulation of angiogenesis (the formation of blood vessels) in the neoplastic tumor and the maximization of the rate of division and migration to other tissues and organs (which results in metastases of cancer). Therefore, cancer cells possess a rich negative feedbacks network directed toward their own reproductive-evolutionary success. Within the organism, they undergo normal natural selection and only the most efficient quickly grow, divide and populate other "habitats" (by forming metastases). Therefore, in the "environment" of the host organism, they behave as regular living individuals. Of course, cancer cells "do not know" that their evolution will terminate at the moment of the death of the host organism (however, non-avian dinosaurs also did not know that they would become extinct due to an asteroid impact at the break of Cretaceous and Tertiary periods). The impossibility of the infection of other "hosts" constitutes the main difference (apart from the origin) between cancer cells and normal parasites.

Nevertheless, there are known examples of cancers that have become regular parasites. In the case of venereal sarcoma in dogs, cancer cells can infect new host organisms, transferring from dog to dog during sexual contact (Barski and Corneferd-Jensen 1966). Another very interesting case is tumor cells being mutated lymphocytes in the Tasmanian devil (a marsupial living in Tasmania) (Woods et al. 2015). Thus, these cells induce a growth on the snout and pass from one animal to another due to individuals biting during fights for territory or mates (sexual partners). The rabies virus spreads fairly similarly, although it induces unspecific aggression in its "carriers" to all animals (and humans) in the surroundings. On the other hand, it is not known whether the "parasitic lymphocytes" cause aggression of Tasmanian devils toward representatives of other species (this would not make much sense, as these cells coming from normal Tasmanian devil cells would most probably not survive in individuals of other species, being instantly detected and killed by their immunological systems). This kind of cancer cell (snout cancer) most probably originated relatively recently and perhaps would still kill the whole population of Tasmanian devils today if a small healthy sub-group were not isolated in a corner of Tasmania by humans.

For all the above reasons, cancer cells should certainly be regarded as cybernetic individuals, although the evolution of most (but not all) of them terminates blindly at the moment of the death of the host organism.

Another doubtful case is *parasitic* DNA (Doolittle and Sapienza 1980; Orgel and Crick 1980). Its two main classes are some repeatable mobile DNA sequences occurring in normal chromosomes, *transposons*, and supernumerary *chromosomes B*. Transposons can move from one place in the genome to another place (localized in either the same or another chromosome), and as far as it is known, they do not fulfill any useful function in the cell. The replication of transposons (during the doubling of chromosomal DNA or copying of transposons themselves) needs significant energy investment and building material; too great a size of chromosomes containing transposons can handicap the process of nucleus division (karyokinesis), while the insertion of transposons into sequences within genes encoding proteins (exons) can disable or significantly disturb the functioning of these genes. Therefore, the problem arises of why organisms do not simply get rid of their transposons. The only rational explanation seems to be the supposition that transposons possess some purposeful mechanisms allowing them to avoid their removal from the genome.

Transposons (Maynard Smith and Szathmáry 1995, 132, 171; Futuyma 2013, 192-195) are not only copied together with chromosomal DNA replication; some can also independently and actively copy themselves within the genome (cells) and encode enzymes responsible for this. These are primarily *retrotransposons* (Futuyma 2013, 196-197). When inbuilt into the host genome, they have the form of a normal double strand of DNA. They also undergo normal transcription (catalyzed by regular RNA polymerase), synthesizing appropriate mRNA on the matrix of one of the DNA strands. Retrotransposons possess one distinct, purposeful and evolutionarily "clever" property. Namely, they encode the enzyme *reverse transcriptase* (RT) (along with a few other enzymes supporting their proliferation), which catalyzes the synthesis of a complementary DNA strand on the matrix of an RNA strand. An analogous enzyme is encoded by some viruses (so-called retroviruses,[13] to which HIV belongs) and was

[13] In fact, the RNA of retrotransposons is analogous to the RNA of retroviruses (it is composed of the same elements and encodes the same proteins) with one exception being that the former does not contain the gene encoding the capsid protein.

also possessed by LUCA (compare with the discussion in Chapter 3). In the case of retrotransposons, on the matrix of the retrotransposon RNA strand, one of the two DNA strands of the double helix of transposon DNA is synthesized by RT, to which the complementary second strand is added in the process of normal DNA replication (with the participation of DNA polymerase). The double-stranded DNA segment formed in this way is next integrated into chromosomal DNA (this process is also catalyzed by one of the enzymes encoded by the transposon), and in this way, a new copy of the transposon originates.

Various transposons can most probably compete for limited space in the genome, and their nucleotide sequences have a long evolution behind them, which optimized their efficiency in this rivalry (in fact, they likely evolved from retroviruses). Additionally, transposons can propagate not only within a given chromosome or genome but also on other chromosomes and genomes (e.g., on the whole haploid genome coming from the other parent), which for some reason, does not yet possess a given kind of transposon, which is possible because of, for example, the "shuffling" of chromosomes and their fragments during sexual reproduction or DNA deletions. They are therefore characterized by a potential for expansion (with positive feedback underlying it) and negative feedbacks underlying this potential. Unlike viroids (which are also strands of nucleic acids), transposons cannot infect other hosts and therefore remain intra-genome parasites that cannot spread to genomes of other cells horizontally. Nevertheless, in the environment of a given genome, they constitute probably the simplest cybernetic individuals (beside viroids). However, as in the case of viroids and viruses, a living individual is constituted not by an isolated transposon itself but by a transposon "dipped" in the context of the network of negative feedbacks of the cell of their "host."

Supernumerary chromosomes B (Jones 1975; Bougourd and Jones 1997) occur in many species of plants and animals. They most probably come from normal chromosomes of a given species. They are not needed for the correct functioning of their "carriers". This can be concluded at least from the fact that individuals devoid of them do equally well, if not better, than individuals containing them. Chromosomes B are copied with other chromosomes during the cell cycle and are transferred to progeny, which is associated with substantial energetic expenses. It seems that organisms try to eliminate chromosomes B, and that chromosomes B possess a special mechanism enabling them to counteract this. Therefore, chromosomes B seem to be, like transposons, cybernetic individuals and internal parasites of the genome (the main difference is structural: a chromosome is a much more complex object, containing, for example, histones and much more DNA than a simple segment of DNA).

It seems that chromosome B is usually relatively harmless for its hosts. Nevertheless, chromosome B of wasps, *Nasonia vitripennis* (Nur et al. 1988), is characterized by an exceptional malignancy.[14] It is transferred only through the male line and carried only by spermatozoa. Hymenopterans have the haplodiploid mechanism of sex determination: haploid males develop from unfertilized egg cells, while diploid females develop from eggs fertilized by spermatozoa; these are zygotes. Therefore, it could seem that this is a death sentence for chromosome B, which has got to the zygote genome through a spermatozoon, as the female that develops from this zygote will not produce spermatozoa. Here, however, chromosome B exhibits a truly deep ingratitude concerning the genome it was recently a part of. Namely, it causes the elimination from the zygote of all chromosomes coming from the father delivered to it by the spermatozoon. Due to this, the zygote becomes haploid, develops into a male and chromosome B can further propagate. Therefore, chromosome B is totally lethal here for the genome in the composition of which it entered before a while, for its former fellow chromosomes. It would be difficult to avoid the conclusion that, as a cybernetic individual, it possesses a set of purposeful regulatory mechanisms (negative feedbacks) directed toward the propagation of its identity.

Cancer cells, infective lymphocytes in the Tasmanian devil and parasitic DNA are examples of what can be called "life within life." They constitute a certain partly autonomous system within the whole organism that escaped from the organism's control (although it is likely that retrotransposons come from retroviruses and not from their hosts' DNA fragments). Multicellular organisms evolved from unicellular organisms, and therefore, cancer cells are, in a sense, cells that have re-gained the freedom of their unicellular ancestors. Chromosomes B are most probably mutated, "insurgent," and undergoing evolution normal chromosomes of their hosts.

The above examples describe the situation in which one individual undergoes splitting into two or more individuals. However, there is possible an opposite process: forming a certain whole by two initially totally independent individuals. In the case of *symbiosis* (strictly speaking—*mutualism*), the network of negative feedbacks of two partners of different origins becomes partly mutually interconnected so that both partners benefit from the cooperation (with regard to the balance of gains and costs). The intensity of symbiotic cooperation can range from relatively loose

---

[14] I owe this example to Andrzej Joachimiak.

facultative relationships to inextricable mutual dependence as "a matter of life and death." The degree of "individuality" of each of the partners can be, at least in principle, quantitatively expressed with the aid of the expression presented above (Eq. 3), quantifying the degree of "individuality" of a polyp in a colony of coelenterates.

Cleaner fish (or shrimps) and their hosts (most frequently predatory fish) remain in a relatively loose partnership. The cleaner gains food by eating food remains from between the teeth of its host or parasites from its skin. The cleaner can feed easily, while the host benefits from healthier teeth and skin. Nevertheless, both partners are not absolutely dependent on each other; they can manage alone, although not in symbiosis. The cleaner can gain food in another, although more difficult way, while the host can survive in a worse condition with less healthy skin and teeth.

Significantly stronger integration between partners in symbiosis occurs in lichens; as is commonly known, they are composed of two components: algae and fungi (Reece 2011, 549-550). The algae cells supply organic compounds produced through photosynthesis, while the fungi partner is more effective in gaining water and mineral substances dissolved in it from very poor bedrock, for example, bare rock or tree bark. Therefore, lichens populate these habitats—they grow on just such surfaces. In most cases, both components are indivisibly bound to each other and cannot survive independently, at least under natural conditions (many can be cultivated in a laboratory). However, cases are known where the algae or fungi component, or both, can be encountered separately, although not in such hostile environments. Therefore, two separate cybernetic individuals have fused largely (although in many cases, not completely) into one cybernetic individual.

Mitochondria in eukaryotes exemplify a very tight, inextricable co-existence of two partners in symbiosis. Initially, mitochondria were free-living respiring (performing oxidative phosphorylation) bacteria (most probably α-proteobacteria) that entered into endosymbiosis with ancestors of eukaryotic cells.[15] Proto-eukaryotes contributed a large, well-protected genome "operated" by proteins to this cooperation and a complex cell structure, while the ancestors of mitochondria contributed a very effective mechanism of ATP synthesis in the form of oxidative phosphorylation. Most mitochondrial genes have either been transferred to the cell nucleus or lost (although these organelles still possess a small circular DNA double-

[15] Controversies exist as to whether those were almost eukaryotes with the structure of "true" eukaryotes, only lacking mitochondria, or remained relatively large archaea with a typically prokaryotic cell structure (see above).

strand containing—in mammals—several genes encoding some subunits of the respiratory chain complexes and their own genetic apparatus: ribosomes, tRNAs and so on), while mitochondria divisions are under the strict control of the cell. Mitochondria never originate anew but only through the division of maternal mitochondria.

Nevertheless, the mitochondrial DNA still can mutate, and therefore, particular mitochondria can possess their own identity. This can lead to so-called intra-genomic conflicts. For instance, in angiosperms (in particular in flowering plants), there are selfish mitochondrial genes that cause the sterility of male flowers and the stronger development of female reproductive organs, and thus support their own propagation (mitochondria are transferred to progeny exclusively through the female line) (Futuyma 2013, 446). On the other hand, mammalian mitochondria, for example, have such a small DNA that no selfish genes are found there. Therefore, mitochondria remain cybernetic individuals, although to a rudimentary extent.

Hydrogenosomes, encompassed by a double lipid-protein membrane, are organelles producing energy in the form of ATP in the process of a certain kind of fermentation converting pyruvate into acetate (molecular hydrogen and CO_2 are by-products of this process) in some protists living in anaerobic environments (Yarlett and Hackstein 2005). These have advanced even further than mitochondria. Their ancestors were most probably mitochondria. However, hydrogenosomes do not possess their own DNA, RNA, ribosomes or the rest of the genetic apparatus. All of their genes have either been transferred to the cell nucleus or lost. Therefore, in this case, the complete fusion of two cybernetic individuals has occurred.

Organisms are known that have originated due to a few subsequent events of endosymbiosis through the fusion of four initially independent living individuals and contain four separate genomes; *Guillardia* and *Bigellowiella* belong to them. They are unicellular flagellates that possess "their own" nucleus (containing 20-25 thousand genes) and mitochondria (about 35 genes) that are former α-proteobacteria. Additionally, they contain a eukaryotic symbiont carrying out photosynthesis (a descendant of a free-living "alga"). In turn, this symbiont contains a dwarfed, reduced nucleus, a nucleomorph (280–500 genes), and a plastid (50–150 genes), a descendant of cyanobacterium. Therefore, there are three episodes of endosymbiosis occurred here in the order of an α-proteobacterium with a proto-eukaryote (proto-flagellate), a cyanobacterium with a proto-eukaryote (proto-alga), and an alga with a flagellate. In what was probably the distant past, one more episode of endosymbiosis occurred—an α-proteobacterium with a proto-alga, due to which mitochondria of the proto-alga originated that was then lost when the alga "settled" in the flagellate.

Plasmids (Kado 1998) are small circular DNA molecules occurring in bacteria cells and can transfer from one cell to another. They can contain genes responsible for sex (conjugation) or antibiotic resistance and can be useful for their hosts. Their replication is independent of copying the bacterial nucleoid, but their number seems to some extent controlled by the cell. Plasmids can also transfer genes from one bacterium cell to another and thus constitute a substitute for the sexual recombination of genetic information, serving as an agent in horizontal gene transfer, both within one species and between individuals of different species. On the other hand, their replication is associated with energy costs, and in many conditions, they are not needed by the bacterium, and in the face of this, they are eliminated from the cell (after all, not all bacteria contain a given type of plasmid). Plasmids possess some limited reproductive autonomy and can "infect" other bacteria cells. Therefore, they are undoubtedly cybernetic living organisms, and their status changes, depending on conditions, between symbionts and parasites.

Social[16] insects (some wasps, many bees, all ants and termites) (Wilson 1971) and one rodent species (the hairless mole) form colonies containing sexual forms: the queen (a fertile female) in hymenopterans (the males die after copulation) or a queen and king (a fertile male) in termites and hairless moles, and sterile (unable to reproduce) forms. The latter may be workers (females or males), frequently a few kinds of different sizes and morphology, and soldiers (females or males), which also frequently occur in different forms. Insects of one type form casts, whose numbers can exceed ten in termites. While in hymenopterans (wasps, bees, ants), sterile casts are composed exclusively of females, in termites and hairless moles, they can contain insects/rodents of both sexes. These are so-called *eusocial* organisms (see the footnote). The role of queens and males is, in principle, limited to reproduction. The queens of ants and termites often have a monstrously expanded abdomen filled with eggs; they cannot move independently and simply constitute machines for producing (fertilized) eggs. On the other hand, sterile casts fulfill all other tasks: taking care of queens (and kings), eggs and larvae, feeding them, obtaining food, building the nest, defending the colony and so forth. According to the classical (structural) paradigm of the living individual, both sexual forms and insects/rodents representing sterile casts of

---

[16] In fact, these are *eusocial* societies characterized by the following three properties: division of labor between reproductive casts and non-reproductive casts, co-existence of subsequent generations and contribution to the care of progeny (Wilson 1971).

social insects and hairless moles are of course, regarded as individuals (they constitute a separate entirety, possess a head, abdomen, legs and so on, like solitary insects or rodents, they only lack a functional reproductive system).

However, according to the cybernetic paradigm, they are not cybernetic individuals, as they are not self-sufficient in the realization of the superior aim of such an individual; this superior aim refers to self-copying (propagation of own identity), requiring self-dependent sustaining of own existence (at least to the reproduction period) and reproduction. This conclusion obviously concerns all sterile insects (and rodents). However, the queen and its progeny would also die very quickly without the delivery of food, care and protection performed by workers and soldiers. Therefore, sexual and sterile forms serve the superior purpose of a certain greater whole—the entire colony, its survival and reproduction. Between insects in a colony, there is coordinated cooperation, communication and division of labor; they are interconnected by a dense and rich negative feedbacks network. These include (Wilson 1971):

- exchange of food; a fed worker returns food to another, hungry insect by the method "mouth-mouth" (ants) or "anus-mouth" (termites)—this is called *regurgitation* and ensures an abundance of nutritional substances to all members of the colony;
- cooperation in building and the repair of the nest, requiring communication and coordination of actions;
- regulation of temperature and humidity in the nest; if it is too hot and too wet inside the nest, especially in the chambers with the queen and progeny, workers open the ventilation canals and thus increase the air circulation; in the opposite case, ventilation canals are closed;
- marking odor traces, for example, leading to a newly discovered source of food;
- releasing chemical alarm substances in situations of danger;
- organized defense of the nest or a wandering column of insects;
- common expeditions for food, forming floating islands or hanging bridges from coupled bodies of ants;
- regulation of the numerical force of particular casts and sub-casts through pheromones or appropriate feeding of larvae according to the current needs of the colony.

Therefore, the functioning of all sterile insects in the colony is mutually co-conditioned. Additionally, the queen, although capable of reproduction, depends on the negative feedbacks network in which the sterile casts participate.

We have to come to the inevitable conclusion that only the whole colony of eusocial insects or hairless moles, composed of both sexual and sterile forms, is self-dependent in survival and reproduction. Therefore, just the colony is the cybernetic living individual because it constitutes an autonomous, self-copying entirety, self-sufficient and independent in realizing this task and equipped with a rich network of negative feedbacks.

It is difficult not to notice a certain analogy between the colony of social insects containing various insects and their casts and a multicellular animal composed of different types of cells and organs. The similarities between these systems are presented in Fig. 6-7. The ancestors of multicellular and tissue animals were unicellular eukaryotes, while the colonies of social insects evolved from single solitary insects. In both cases, the protoplast, both unicellular and "uni-insect," was a fully-fledged living individual. In the process of the evolution of multicellular animals, particular cells underwent differentiation into liver, muscle, nerve cells *etc*., which in turn entered into the composition of organs such as the liver, kidney or brain. The separation of the cells of the generative line also occurred, leading to gametes—egg cells and spermatozoa—and somatic cells fulfilling all functions other than reproduction. None of these alone can realize the superior purpose of a living individual: survival and reproduction. In a sense, the organism of, for example, man is only a colony of particular cells, which admittedly are highly integrated and cooperating in a coordinated way while remaining under the control of the whole individual,[17] but it is the whole multicellular organism (and not its somatic or generative cells) that constitutes the cybernetic individual. Various cells and organs in the body of an animal, fulfilling various functions (a far-going division of labor can be seen here), mutually support each other and condition the existence and survival of themselves and the whole individual (animal). Similarly, particular insects and their casts in a colony of social insects serve each other and the whole colony, supporting its existence. In both cases, we are dealing with the causal closure that has already been mentioned several times.

Additionally, the superior purpose of the whole system in both cases is reproduction—the multiplication of its identity in a possibly large number of copies. Animals realize this purpose through the production of gametes (eggs and spermatozoa), while colonies reproduce through the production of reproductive forms: virgin queens and males. The whole system of the cybernetic living individual—the animal or the colony—is purposefully directed toward this process.

---

[17] Cancer cells are an example of that which happens when this control fails.

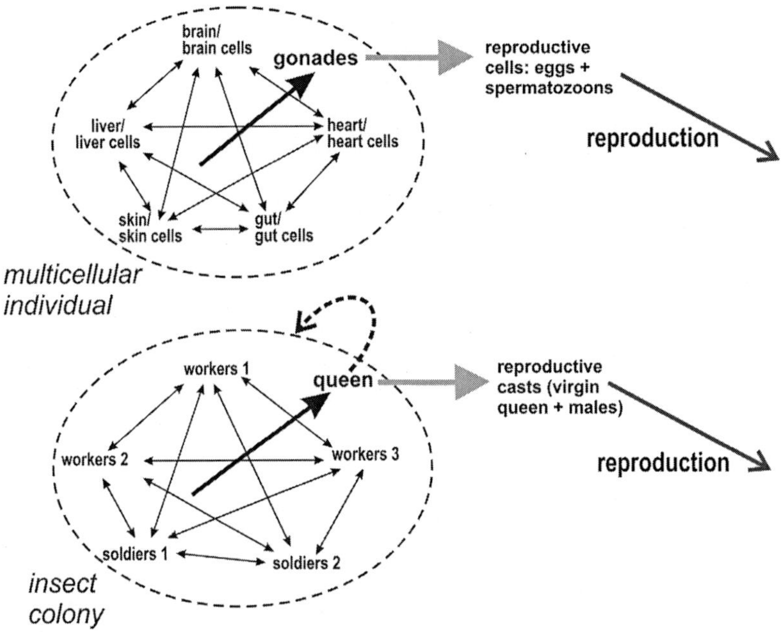

Fig. 6-7. Formal similarity between complex multicellular animals and colonies of social insects. Various types of cells and organs in multicellular animals and particular insects and their casts in insect colonies reciprocally support the functioning of and serve each other. The whole system is ultimately directed toward the production of generative (reproductive) elements; these are eggs and spermatozoa, and virgin queens and males, respectively. Thus, they are directed toward reproduction and the propagation of own identity. Neither particular cells/organs, nor insects/casts, either somatic/sterile or generative/reproductive, are self-sufficient in survival and reproduction. Therefore, only the whole system—the multicellular animal or social insect colony—is the cybernetic living individual.

Of course, in the case of an animal, the cybernetic individual is (almost) identical to the classical (structural) individual (by contrast, consider the above discussion concerning cancer cells, mitochondria and parasitic DNA). The main difference between the animal body and social insect colony is the fact that in the former, cells are strongly integrated both structurally (they occupy a fixed location in relation to each other, and neighboring cells are physically connected) (by contrast, consider blood cells moving passively through the whole body with the bloodstream) and functionally, while the insects in the colony are primarily integrated functionally, and to a much smaller extent, structurally (spatially). Although they often remain within the colony and contact each other, they incessantly change their position in space and do not form any fixed structure. Therefore, the colony of social insects can be called a "dispersed individual". However, this is the functional (regulatory) integration that plays a key role in the cybernetic definition of the living individual. Therefore, from a formal point of view, the colony of social insects is strictly analogous to an animal. The human brain possesses an innate predilection to perceive spatial-structural integration, while it is much more difficult for it to notice and embrace functional integration. This constitutes the main reason why many people oppose treating the whole social insect colonies, and not particular insects, as living individuals.

Therefore, to sum up, it should again be emphasized that the whole colony and not its "members" constitute the cybernetic living individual. At the same time, it is worth stressing that frequently the answer to the question "to be or not to be a living individual" is not at all simple and univocal.

The cybernetic definition of life also helps us to clearly understand why the whole biosphere on earth is not an individual or super-individual that can be treated as a whole that realizes some purposeful tasks, as was proposed by Lovelock in the so-called Gaia hypothesis (Lovelock 1979). First, the biosphere does not reproduce or expand in any other way beyond the earth's surface (although perhaps it will in the future due to human activity). Second, which is much more important, most properties of the biosphere on various levels of its organizational hierarchy do not result from cooperation or purposeful mutual co-conditioning but from competition and antagonism between its different elements—primarily between individuals of the same or different species, and also from the exploitation of one individual by another, for instance, exploitation of prey by a predator, a host by a parasite or a plant by a plant-eater.

Using the cybernetic paradigm, while the negative feedbacks within a living individual cooperate for the sake of the superior common profit (survival and reproduction), the negative feedback networks of different individuals within the biosphere fall into numerous cases of antagonism, compete with each other and exploit each other. Of course, in the case of symbiosis (mutualism), cooperation rather than rivalry takes place; nevertheless, the partners in symbiosis only partly constitute separate individuals as they partly form one individual (see above). Apart from this fact, competition, herbivory, carnivory and parasitism are much more common in nature than symbiosis. The purpose of the particular negative feedbacks present in a given cybernetic individual (and, of course, of their whole integrated assembly of negative feedbacks) is to expand the cybernetic identity of this individual (its network of negative feedbacks) and not the interest of the whole biosphere. Just the proliferation of the cybernetic identity constitutes the guidepost of biological evolution by natural selection. The biosphere is a derivative of the antagonistic actions of different cybernetic individuals and some inanimate physicochemical processes. Therefore, the terrestrial biosphere can be treated as a "super-organism" (whatever this would mean—the concept of an organism seems rather vague and imprecise in contrast to the concept of the cybernetic individual) and is "alive" only in the above-specified sense—it is composed of living individuals—but not as a "super-individual" (in the cybernetic meaning). Similar reasoning can be applied to populations and ecosystems—they are also decidedly not living cybernetic individuals, although they are composed of such individuals.

What about people who do not reproduce for social, cultural, religious or biological (because of inborn dysfunction of the reproductive system) reasons? Are they not living individuals? In this context, some misapprehensions appeared concerning my cybernetic definition of a living individual. In order to answer this question, one should first realize that the psychical, social and cultural levels of reality that primarily determine our humanity cannot be completely, or even to a large extent, reduced to (or explained in terms of) the biological level. The psychical/social/cultural level possesses its own goals and mechanisms that frequently oppose/contradict the goals and mechanisms of biological individuals. Psycho-sociological mechanisms (motivations) can repress the biological imperative of reproduction as efficiently as inborn defects of the reproductive system. For this reason, people that do not have children (including the author of this book) can be treated from the biological point of view as defective cybernetic individuals, although they are, of course, alive—the great majority of their biological functions, including most negative feedbacks, are not disturbed. There is nothing odd in this statement—most people would agree that a person who must wear glasses (also including the author of this book) is a defective individual from the biological point of view. This is so because the superior purpose of the purely biological individual is to produce as many descendants as possible (in the Stone Age, a person with bad sight would not have survived and, therefore, would not have reproduced). However, everyone is primarily a psychical, social and cultural individual and a moral individual. After all, adopting genetically unrelated children constitutes a completely nonsensical (even harmful—supporting of "evolutionary rivals") act from the point of view of biological purposefulness and mechanisms of biological evolution. However, nature (as opposed to culture) is, in essence, amoral, and one should not apply the directives of the Darwinian "struggle for existence" to human societies and ethics.

Additionally, one should be aware that not only people who do not reproduce at all but also persons who do not produce the maximal possible amount of progeny (under given conditions) are defective from the biological point of view. A woman could potentially give birth during her life to around 20 children, perhaps even more. Every woman who does not do this evidently opposes the genetic-evolutionary imperative of propagation of her own genes (own identity) to a potentially large number of copies. Of course, in developed western societies, an obstacle on the way to intensive reproduction is the will to ensure a good start in life, prosperity, education *etc.*, to one's progeny, and also to have a convenient life, or alternatively sacrifice to the priesthood, fine arts, science or philosophy. In any case, it is debatable whether any people, for instance, some tribes, reproduce with the maximal possible speed in their given circumstances (even if so, this is because of high mortality at a young age, so this is not "effective" reproduction). Generally, this means that people are subject to not only biological but primarily to psychical, sociological and cultural motivations and conditionings (they are not only biological but first of all psychical, sociological and cultural individuals). This fact speaks against extreme forms of sociobiology (the problem of the divergence of biological and psychical motivations is more broadly discussed in Korzeniewski 2010).

The reasoning concerning people who do not reproduce can, to some extent, be applied to other biological individuals that cannot produce progeny, for example, to sterile inter-species cross-breeds such as the mule, which is the cross of a female horse and a male donkey. Again, they are defective living individuals. They are alive in the sense that they carry out most biological functions (and possess networks of negative feedbacks related to them) but not reproduction. However, they cannot realize their superior task and are, therefore, completely nonsensical from the biological point of view.

In a comparable way, we can deal with the problem of the parts (elements) of the human body, for instance, the liver and single cells, which of course, are not living (cybernetic) individuals, although they are alive, as they constitute sub-systems of a living individual containing integrated and relatively (although of course not completely) isolated complexes of purposeful negative feedbacks. The liver and single cells are only elements of a bigger system and are directed toward the self-copying of this system. The parts of the human body have no sense or significance after separation from the maternal system but can still perform some sets of functions with a clearly specified sense only in the functional context of the whole system. For instance, isolated human cells can be cultivated in laboratories and kept alive (they preserve their assembly of purposeful negative feedbacks) for a long time, but this procedure is completely senseless from the point of view of the living (cybernetic) individual.

Some people who have problems with understanding the cybernetic definition of a living individual argue that cells in the body of a multicellular organism can reproduce and take care of themselves and maintain internal homeostasis, and therefore fulfill this definition. Therefore, I would like to emphasize once more that firstly, according to this definition, a living individual is self-sufficient in the realization of its superior task (survival and reproduction); that is, it does not need any additional (external in relation to it) directed-on-something mechanisms (negative feedback and other regulatory mechanisms) that purposefully support it. Evidently, this does not take place in the case of single cells, which are supported by other cells and organs and, ultimately, by the whole organism. Second, the superior purpose of a living organism is reproduction. Again, cells in a multicellular organism, especially somatic cells, are eventually directed toward the reproduction of the whole organism and not of themselves. Cell divisions are under the strict control of the whole organism (usually, differentiated cells do not reproduce at all). Thirdly, particular cells cooperate within the body and do not compete with each other. Cells can sacrifice themselves for the good of the whole individual—for instance, they die ("commit suicide") due to apoptosis (programmed cell death) or defend the organism against infection (immunological system cells). Next, even more importantly, somatic cells die together with the death of the whole organism and, therefore, cannot transfer their identity to progeny. Cancer cells, which break free from the control of the whole organism and begin to realize their own selfish goal—their own reproduction and expansion—become, as discussed above, cybernetic individuals (they regain the freedom of the unicellular ancestors of multicellular organisms). Finally, the cybernetic individual comprises all negative feedbacks that are directed solely and exclusively toward its reproduction. Obviously, cells isolated from the whole organism do not possess *all* the negative feedbacks necessary for their maintenance and reproduction (imagine a nerve cell that tries to survive and reproduce outside the maternal organism), and many of their negative feedbacks serve other cells and the whole organism (a nerve cell evidently participates in the realization of the superior purpose of the whole organism, and not only their own selfish goals).

The cybernetic definition of a living individual can also be applied to the origin of life on our planet. Certainly, inanimate systems that were the direct ancestors of the first living organisms did not possess any regulatory mechanisms, especially negative feedbacks, compensating disturbances coming from the surroundings and supervising the purposeful functioning of the system. Such systems, which in my opinion were autocatalytic cycles of proteins and nucleic acids in objects resembling coacervates or proto-membrane-surrounded bubbles, could possess some passive properties allowing to buffer, and alleviate harmful influences of the environment, for example, organic compounds could be more stable and durable inside coacervate-like bubbles than in a water environment. However, as the discussed systems did not possess the possibility to record information about their structure and function or transfer this information to descendent systems, they could not create any purposeful, directed-on-something regulatory mechanisms directed toward preserving their integrality, survival and reproduction. This possibility appeared only when the first genetic code was shaped, coupling the previously separate autocatalytic cycles of proteins and nucleic acids into hypercycles. The first efficient *replicators* originated in this way, capable of quick and relatively faithful copying of their identity expressed in the sequence of nucleotides in RNA and amino acids in proteins. From this moment, proteins could undergo very quick evolution, adapting themselves to fulfilling such functions as enzymatic catalysis, forming structure, the transport of substances from and to surroundings and so on. Undoubtedly, one of the first functions of the self-specializing proteins was the maintenance of internal homeostasis, appropriately reacting to changes in external conditions, and the maximization of the rate of growth and divisions (this was very important in competition with other systems for building substances accessible in the surroundings). In this way, the first negative feedbacks were sustaining the superior positive feedback—the drive toward self-copying. In other words, the first dynamic *homeostats* were formed. Most probably, the origin of efficient replicators and homeostats was separated by a very short period (in the geological-evolutionary timescale). For this reason, identifying one of the above events with the moment of the emergence of life to a large degree seems arbitrary. Aside from this, all logical premises suggest that for a long time, the evolution, improvement and quantitative expansion of systems being both replicators and homeostats advanced in parallel and were mutually co-conditioned—an ample genetic record enabled the production of a more complex network of negative feedbacks, while this network, in turn, caused greater durability and more accurate copying of the genetic record. Personally, I prefer to regard systems that are not only efficient replicators but also homeostats as living individuals, at least because of the fact that many contemporary computer programs, which still rather inefficiently imitate the phenomenon of life, involve objects that are replicators but not homeostats (see below). Of course, living individuals defined as homeostats must also be replicators (purposefully directed toward self-copying), but the reverse relationship does not occur.

The origin of the first genetic code could have coincided almost perfectly with the emergence of the first negative feedback mechanism. What did this consist of? In my opinion, it was the possibly faithful copying of an RNA strand based on the complementarity of the pairs of bases C-G and A-U by some primitive proto-enzyme, the ancestor of RNA-dependent RNA polymerase still appearing in plants and viruses (including SARS-CoV-2) (and once present in LUCA ancestors) (de Farias et al. 2017). Its aim would have been preserving the "optimal" "assigned" value of the parameter—the sequence of nucleotides in RNA. According to our knowledge, RNA strands themselves cannot self-copy efficiently; in other words, they cannot synthesize a complementary RNA strand on the matrix of the maternal strand from free nucleotides available in the surroundings. Even polypeptide chains with an accidental sequence of amino acids could not be efficient in catalyzing this process. Only the origin of the primitive genetic code and proto-proteins, whose identity (the sequence of amino acids) was inherited from generation to generation and perfected due to evolution, could enable the faithful replication of relatively long RNA strands. Therefore, here the cybernetic aspect of the origin of life excellently intertwines with the genetic aspect—like all other aspects, they mutually co-conditioned each other.

Testing the cybernetic definition of the living individual

The network of negative feedbacks (and, more broadly, of regulatory mechanisms and information transfer and transformation pathways) univocally determines the identity of the cybernetic individual. In particular, this identity comprises such aspects as (1) the number of negative feedbacks, (2) parameters for which the value is regulated, (3) assigned (optimal) values of particular parameters in a given situation, (4) strength of steering, i.e., the rate of compensation of deviations of a given parameter value from the assigned value, and (5) relationships (connections) between a given negative feedback with other negative feedbacks within the whole network of negative feedbacks of the considered individual. As discussed below, the cybernetic identity of a living individual is strictly related to the ecological niche of this individual.

The proposed cybernetic definition of life (the living individual) fulfills all five criteria formulated in Subchapter 6.2 (Korzeniewski 2005):

(1) It is formulated within formal sciences: cybernetics and information theory.
(2) It uses only two primary concepts: negative feedback and positive feedback. Other properties of living individuals are high-level properties derived from the discussed definition and elementary features of the physical world (see below).
(3) It clearly distinguishes living individuals from other systems, including physical, inanimate, psychical and social systems, and biological systems that are "alive" but not constituting living individuals; such systems are situated either below living individuals in the hierarchy of biological systems, for instance, cells and organs in complex tissue organisms, or lie above them, for instance, populations, ecosystems and the whole biosphere (see Korzeniewski 2005, 2013a).
(4) It clearly and univocally specifies the unit of natural Darwinian selection = the subject of evolution.
(5) It defines the (amount of) biologically purposeful, aim-directed, "valuable" information associated with living systems (as opposed to devoid of any biological sense, indifferent, statistical thermodynamic information).

Generally, the cybernetic definition of a living individual reduces the number of elements (terms) to the minimum needed for the description of the essence of life, and as such, it seems to reach the core of the phenomenon of life. In my opinion, no other definition fulfills these criteria.

Confrontation of the cybernetic definition of the living individual with the real world

The definition considered here undoubtedly characterizes the living individual in a very general (for many people, too general), formal and abstract way. It seems to belong more to the Platonic world of pure ideas than the real physical world. Therefore, it can appear difficult to conceive how this definition refers to the present life on earth or to that which we would like to regard as life in general, for example, to various phenomena most probably existing on other planets in the universe. As with life on our planet, and most probably also extraterrestrial life, this possesses many features that are inseparably associated with it but that are not explicitly taken into account by the cybernetic definition of a living individual. It is possible to imagine very many theoretical kinds of life that would fulfill this definition; on the other hand, the real world imposes severe limitations on the possible physical realizations of the abstract directives contained in the discussed definition.

Life on earth, and probably any spontaneously-originated life existing in the physical reality, possesses many properties that do not result from the "pure" cybernetic definition of life, separated from the properties of the physical world. Living systems must be dissipative structures or, more generally, systems with low entropy far displaced from thermodynamic equilibrium, intensely dissipating energy and producing entropy. Most probably, spontaneously-originated life has to be based on organic compounds of carbon, although not necessarily exactly these that enter into the composition of terrestrial organisms. Living organisms have a complex, hierarchically-organized structure and a set of functions, reflecting the complexity and variability of their environment/ecological niche. They possess a record of identity (e.g., genetic record in DNA) separated from the executive machinery (e.g., proteins and other compounds). They mutate and evolve; their identity can change from generation to generation. They can accumulate (biologically purposeful) information over very long periods (due to which their fitness and the complexity of their structure and function can gradually increase). They are systems capable of supple adaptation to varying environmental conditions. Finally, they are systems equipped with a network of purposeful regulatory mechanisms (mainly negative feedbacks) constituting the cybernetic identity of the living individual. Therefore, real life in the physical universe, in particular the present life on earth, does not seem to have too many options with regard to its most general nature and properties.

Below, I intend to show that, in principle, all these features are generated, in a sense, automatically, when the abstract cybernetic definition of life (the living individual) becomes confronted with the real physical world. In other words, I suggest that these features are not primary but secondary—they can be derived from primary properties that form the cybernetic definition of a living individual.

Numerous properties of physical reality are crucial for the essence of life. First, the real world contains an enormous number of elements and aspects, such as various atoms and molecules of chemical compounds (and their properties resulting from quantum mechanics), concentrations of different substances, temperature, humidity, electromagnetic radiation of different wavelengths, physical forces (electromagnetism, gravity), mechanical properties (e.g., hardness), atmospheric and water currents, to enumerate only a few. Additionally, the physical environment (at least on the earth's surface) is characterized by huge differentiation in space and variability over time. Living organisms are forced to exploit their environment somehow, for example, as a source of matter, energy, shelter, and nesting places, and possess a complex of functions appropriate for this purpose. Additionally, every biological system is part of the physical world, and as such, it is built of the elements of this world and experiences various disturbances coming both from the surroundings and from its own interior. This fact constitutes a direct derivative of the high degree of differentiation of the real environment. It should be clearly emphasized that other individuals (of the same or other species) constitute an equally important (often much more important) component of the environment as its physical aspect. Various elements of the biosphere have a more or less direct impact on the ecological niche of a given individual. On the other hand, the "environment" of most computer models of some aspects of living systems and the environment of real experimental systems located in a test tube, container or cage, where conditions are strictly controlled, can be extremely simple.

The second property of the physical reality that is crucial for the essence of life is the fact that in the physical world, every system possessing a certain amount of information of any kind (especially biological information) must be a low-entropy system displaced from thermodynamic equilibrium (in particular—a dissipative structure) characterized by a large amount of thermodynamic information. In other words, every kind of information must be underlain by thermodynamic information (see Chapter 2).

The third property is that it seems as though among various chemical compounds, only organic compounds of carbon are characterized by a sufficiently high degree of complexity and diversity of structure and physicochemical properties to constitute the basis (primarily building material) of spontaneously-originated life (see Chapter 1).

The fourth property is the fact that it is not possible in the real world to store and copy information recorded on any physical carrier with infinite accuracy. Such storage and copying are disturbed by, for example, e.g., chaotic thermal movements and the collision of atoms and molecules (and frequently also by quantum phenomena). Generally, the discussed property results from the laws of thermodynamics (and quantum mechanics), especially from the second law saying that increasing entropy and information dissipation is unavoidable (Chapter 2). As a matter of fact, inanimate dissipative structures can locally decrease their entropy and increase the amount of thermodynamic information at the cost of an increase of entropy and a fall in the amount of information of the surroundings. However, the amount of information accumulated in this way is relatively small, lacking a purposeful, directed-on-something and long-lasting character.

Nevertheless, some mechanisms in the real world enable the preservation or even the spontaneous rise of information. Systems capable of doing this (equipped with the above-mentioned mechanisms) must fulfill two conditions: (1) they must copy information with a rate equal to (preservation of information) or greater than (possibility of increment of information) the rate of dissipation of information for thermodynamic reasons; (2) they must be able to recognize "profitable" (valuable, purposeful) information (due to this, systems possessing this kind of information are positively selected). Only such systems are capable of the spontaneous preservation, collection and accumulation of information over long periods (even, for example, an automatic spacecraft could collect and record scientific data only for some time, relatively short in the evolutionary scale, until it experiences technical failure) (see Chapter 4 and 5).

The fifth property is related to the fact that the environment of living organisms on the earth's surface has only a limited capacity. Moreover, particular environments such as lands, oceans (their depths and shallows), lakes, rivers, mountain ranges, valleys, *etc.*, possess much less capacity. Therefore, they accommodate only limited objects of finite dimensions, including living organisms.

The sixth property is that no function can exist in the real world without some structure underlying it. This is so because some physical carrier of function must exist, and any complex functions must be based on an appropriately structured (organized in space) assembly of elements (subsystems) that constitute the material realization of particular functions. Additionally, the more complex the set of functions is, the higher the required level of structural organization of their material carrier.

The seventh property relates to the fact that no complex three-dimensional physical system can be, for fundamental reasons, directly copied within three-dimensional space, in particular, without stopping its functioning. This occurs because copying a complex n-dimensional system into a descendant system in the way of the one-to-one assignment of elements in both systems corresponding to each other is only (theoretically) possible in an (n+1)-dimensional space. Therefore, it is easily possible in real three-dimensional space to copy in this way, for example, a sequence of letters in a text or a sequence of nucleotides in DNA (one-dimensional system), or to copy a woodcut or the arrangement of fonts in the printing press (two-dimensional system), but one can in no way copy directly a three-dimensional system, for instance, a car or a computer (therefore, cars and computers are built from the ground up, using two-dimensional blueprints (graphical schemes) and one-dimensional instructions (text).[18] Additionally, during copying, the elements being copied would not be able to fulfill their function, which would cause the necessity of cessation of the operation of the whole system during the copying of its key elements.

Let us now consider the consequences of the above fundamental properties of the real physical world, concerning the nature of the phenomenon of life existing in this world.

First of all, living individuals, whatever else they would be, must also be regulatory (cybernetic) systems, i.e., they must contain a complex of appropriate negative feedbacks and other regulatory mechanisms and pathways of information/signal transduction/transformation. In other words, they must be at least homeostatic systems—*homeostats*. This occurs because every system that aims to preserve its structural and functional identity and integrity for a long time in the physical reality must compensate for various disturbances from the surroundings and those appearing in its interior. Otherwise, these disturbances would destroy such a system sooner or later. Additionally, negative feedbacks seems necessary for controlling various purposeful processes directed toward correct embryonic development, cognitive functions and behavior, and ultimately survival and reproduction. Generally, to be an efficient replicator in the real (physical and biological) environment, the living individual must also be a precisely operating, broadly understood dynamic homeostat.

Every cybernetic individual must be a complex dissipative structure (or, if somebody prefers, a hierarchically-organized assembly of such structures) or, more generally, a system significantly displaced from thermodynamic equilibrium. This necessity results from, *inter alia*, the fact that some determined amount of biologically purposeful information is associated with the network of negative feedbacks constituting the identity of the cybernetic individual. This information cannot exist without the thermodynamic information underlying it. Every biological system is also physical and must be subject to physical laws. In fact, only a small fraction of the thermodynamic information associated with living organisms lies at the basis of biological information. Nevertheless, there is a strict relationship between the biological and thermodynamic information underlying it. Therefore, life could not exist in systems close to thermodynamic equilibrium, for example, in cold clouds of interstellar gas.

Confrontation of the cybernetic definition of the phenomenon of life (living individual) with the real world also implicates that spontaneously-

[18] Even a 3-D printer does not directly copy existing three-dimensional objects, but realizes the set of instructions recorded in its program. Additionally, such a printer can produce only static objects, but not dynamic systems.

originated life due to self-organization must be based on organic compounds as only known compounds that are sufficiently complex and diverse to constitute the building material of the first systems that can possess their own identity, inherit this identity, possess metabolism and evolve. The cybernetic definition of life suggests that the essence of life consists of a determined kind of complexity of biological systems rather than of the substance from which these systems are built. Therefore, there is no principal reason why every kind of life should be based on the same organic compounds. Theoretically, living individuals could be built of anything if they could carry out a determined set of functions. Nevertheless, the lowest level of the functioning of living individuals from the beginning was the molecular level, which remains the basic level today. On this level, only organic compounds can ensure a sufficiently rich and diverse set of building blocks and executive machinery to enable such a wide range of executed actions as the recording of information, the specific catalysis of various chemical reactions, isolation from the surroundings and so on. Spontaneously originating life had to start from the molecular level (although it currently contains a couple of higher levels, such as cellular and physiological). This is very important as contemporary devices constructed by humans, including robots and computers, function on much higher (more macroscopic) than the molecular level. It is not single chemical molecules that constitute their basic fundamental units but much larger elements such as microprocessors, chips or diodes based on metals, crystals and semiconductors. Such elements cannot spontaneously originate, while their shape, structure and functions, macroscopic from the molecular point of view, are not a direct derivative of only their building material but are first and foremost the action of the intellect and technology of their creators. Therefore, they cannot come into being due to self-organization.

The above considerations unavoidably lead us to the problem of the existence and nature of artificial life. Artificial living individuals can be imagined as virtual objects in the operational memory of a computer or as a kind of robots (built of metals, semiconductors, plastic *etc.*) capable of independent survival and reproduction in the physical environment (e.g., on some distant planet or moon). According to the cybernetic definition of life, to regard such individuals as alive, they would have to be able to self-copy and compensate for various disturbances coming from the (sufficiently complex and diverse) environment, for instance, to self-repair. Neither present computer models nor real robots fulfill this requirement; nevertheless, I do not see any principal reasons they could not do this in the future. However, for the reasons discussed above, such life cannot originate spontaneously—it must be constructed with the participation of some intellect. An excellent literary example of artificial life (although not completely logically consistent) is provided by Stanisław Lem in his novel *The Invincible* (1973). The problem of implications coming from the building material of artificial life for the nature of its individuals is discussed in more detail in Supplement 2 and in (Korzeniewski 2011).

Generally, the complexity of the environment (or, more strictly, the ecological niche) imposes (determines) the degree of complexity of living organisms (individuals). This is so because the more diverse and variable the environment (in particular, its aspects important for a given organism), the greater the number of (broadly understood) disturbances (variability in space and time) that takes place in it, and thus, the greater the number of complex properties of the environment that must be registered (perceived, in a very general sense of this term) and the greater the amount of adequate purposeful (in biological sense) actions that must be undertaken in order to react to these disturbances appropriately (Korzeniewski 2005). The regulatory, cybernetic complexity of living organisms implies (at least to some degree) their structural complexity, as some material elements of the system must implement particular negative feedbacks. Therefore, confronting the cybernetic definition of life with the real world imposes a determined structural and functional complexity on living individuals.

Admittedly, some creations that we classified as cybernetic individuals, such as viroids, viruses and transposons, are extremely simple, but as parasites, they cannot live independently—they are submerged in the rich networks of negative feedbacks of their hosts that do almost everything for them. Therefore, their environment (ecological niche), its aspects with which these beings must cope, is, in fact, very uncomplicated. In my opinion, there is a certain minimal degree of complexity of structure, function and regulatory mechanisms of free-living organisms that is evolutionarily stable. This degree can correspond to the degree of the organization represented by the simplest free-living prokaryotes (bacteria and archaea). It emerged soon (in the geological scale of time) after the origin of life on the earth, well over 3.5 billion years ago, and constituted at the same time the highest level of complexity for about two billion years until the appearance of eukaryotic (nuclear) organisms some 2-1.5 billion years ago. Nuclear and multicellular organisms occupied new, much more complex ecological niches in which an incomparably richer network of negative feedbacks was needed, and therefore a significantly greater complexity of structure and function was required. These needs were, in a sense, imposed on the organisms by the complexity of the elements of the physical (and biological) reality that were essential to their "style of living". At the same time, the reverse relationship also took place—for more complex organisms, more complicated elements and aspects of the environment became important (and therefore, the complexity of their ecological niche grew).

As discussed above, a concrete, such-and-not-another set of negative feedbacks constitutes the cybernetic identity of the cybernetic individual. The tasks of this identity are: (1) to control various functions of a living individual and maintain the stability of its internal conditions, and thus sustain the existence of this identity during the life of its "carrier" (the task of a homeostat); (2) to be transferred to a possibly great number of its progeny, i.e., descendant "carriers" of the identity (the task of a replicator). Apparently, these two tasks cannot be reconciled, as it is not possible to directly copy in the three-dimensional space of a three-dimensional object, especially when this object is just executing a complex set of functions necessary for its survival. However, this problem can be overcome if it is not a ready "adult" object that is copied but a sequence of instructions—a record of the procedures necessary for its self-construction, which is equivalent to the record of the identity of such an object. In this way, we build numerous copies of robots, cars and computers, and also living individuals behave (as they must). The important difference is that living individuals construct themselves using the instructions and set of "make yourself" tools in the fertilized egg cell.

An immanent property of all living organisms known to us (and, in my opinion, all real rather than virtual living organisms) is the separation of the set of instructions for the construction of the structure and function of an adult organism (in the form of genes and DNA fragments, enhancers and silencers participating in the regulation of their expression) from the executive machinery, which on the molecular level primarily has the form of proteins, and on the macroscopic level is, in fact, equivalent to the rest of the organism (apart from DNA). This machinery serves to carry out the embryonic development, growth, and functioning of an adult individual and to copy and transfer the record of the identity to progeny and implement it, while the record of the identity determines, using the executive machinery, the shape of new machinery that is created in developing descendant individuals and functions during the life of adult individuals. In this case, we again deal with a vicious circle involving the record of identity and executive machinery, so characteristic of living individuals. As already discussed, the transition from the one-dimensional identity record to the three-dimensional structure and function originally occurs when folding a newly synthesized protein chain with a given sequence of amino acids into a native conformation, which determines the structure and function of the protein. In turn, at higher levels of the organization, proteins determine, either directly or indirectly, the structure and functions of the cells and the whole organism. This double nature of proteins, which on the one hand represent a one-dimensional record of identity translated from DNA through the genetic code, and on the other hand, a three-dimensional structure and function, seems very characteristic for spontaneously-originated life, whose lowest, fundamental level of functioning is the molecular level. It is also worthy of note that the separation of the record of identity from the executive machinery enabled the storage of this record (in eukaryotes) in a safe place, far from the metabolic factory, and through this protecting it from serious damage, for example, by free radicals or harmful chemical substances. This is a subsequent reason why the realities of the physical world imposed the discussed property on cybernetic individuals.

At this point, it is worth emphasizing that it is not particular "selfish" genes (or their complexes) alone but the whole self-copying system, the cybernetic individual, that implements its complex structure and functions during its embryo- and ontogenesis. It is also certainly not the case that one gene is responsible for one property, but quite the contrary—all genes determine, more or less directly, to a larger or smaller degree, all phenotypic properties, although this is frequently to a tiny extent. It is frequently said that, for instance, a gene that encodes an enzyme which is an element of a metabolic pathway leading to the synthesis of a certain red pigment, is a gene "for the red color of flowers" because a mutation in this gene causes inactivation of the enzyme, a lack of the pigment and as a consequence, white flowers. However, thousands of other genes participate in the creation of this property. These are genes encoding other enzymes of the metabolic pathway, enzymes of other metabolic pathways producing energy in the form of ATP needed for the synthesis of the red pigment, morphogens controlling developing flowers (that must develop to possess any color at all), and so on. In principle, all genes have some impact here (e.g., a mutation in some "metabolic" gene could cause the plant, together with its flowers, to grow a little (or very) pigmy.

Furthermore, as a rule, particular genes influence very many different properties; for instance, a sublethal (harmful, but not leading to death) mutation in the gene for cytochrome oxidase, leading to the impairment of ATP production by oxidative phosphorylation, will cause a smaller or greater dysfunction of all parts of the organism, and ultimately a "weakness" and handicap of the whole individual. Of course, genes can "do" something through the proteins they encode, only taking into account the physical realities; for instance, chlorophyll is green and absorbs visible radiation, while the diffusion of morphogens causes forming of morphogen gradients. Therefore, developing the structure and function of an organism is an effect of the interaction of all elements of the genetic record, the executive machinery and the properties of the physical reality in general and a given concrete environment in particular. On the other hand, it is true that the huge majority of the structural-functional differences between a bacterium, a cauliflower and a rhinoceros result from the differences in their genetic record; if not the problems with compatibility, "putting" the genetic record of a cauliflower into the executive machinery in the egg cell of a rhinoceros would lead to developing a cauliflower and not a rhinoceros. In reality, this would not be possible because of the mentioned lack of compatibility between the genetic records and executive machinery (for instance, gene expression regulation systems) of evolutionarily distant organisms. However, in some cases (e.g., in the unicellular genus *Acetabularia*), the transplantation of a cell nucleus containing DNA to a cytoplasm artificially devoid of its own nucleus between two closely related organisms of different species leads to developing an organism with properties of the donor of the nucleus and not of the cytoplasm (Stambrook 2013).

Returning to the mainstream of this discussion, the primary characteristic property of life—biological evolution—emerges due to the confrontation of the cybernetic definition of a living individual with the realities of the physical reality. This was somewhat explained during the discussion on the essence of natural selection (see Fig. 4-8 and related discussion). Briefly speaking, for purely physical reasons, the record of the cybernetic identity of a living individual corresponding, as we remember, to a such-and-not-another network of negative feedbacks cannot be stored and copied without any errors. Mutations originating due to these errors cause the differentiation of individuals in a given population of some species—they will possess slightly different networks of regulatory mechanisms. Cybernetic individuals, which realize their superior positive feedback—reproduction will very quickly saturate their environment's limited capacity, constituting another obvious property of the real world. When this happens, a proportion of the population of descendant individuals must die before reaching the ability to reproduce. Only those individuals who are most efficient homeostats have the greatest probability to survive and reproduce; in other words, only those individuals whose networks of negative feedbacks will cope best with the diversity of the external (and internal) environment and have the highest probability of survival in a possibly good condition to the reproduction period have the possibility of leaving large numbers of progeny. Most mutations will negatively influence the chances of survival and reproduction or will not affect these chances. However, a few mutations will appear beneficial for the efficiency of the negative feedback networks and, therefore, will be consolidated in the population. The accumulation of many such profitable mutations will cause the cybernetic identity of a given individual (its network of negative feedbacks) to differ significantly from the identity of its ancestors. This is simply equivalent to the phenomenon of the evolution of natural selection. Therefore, biological evolution is not some irreducible and primary property—it can be derived as a logical consequence of the cybernetic definition of life confronted with elementary features of the real physical world.

Some definitions of life specify living systems as systems capable of supple adaptation to varying environmental conditions (Bedau 1998). Leaving aside the fact that such a definition concerns not only biological systems but also psychical and social systems, the ability for elastic adaptation is a direct derivative of biological evolution, which results from the cybernetic definition of life, as previously demonstrated.

One could question why just the cybernetic individual should be regarded as the subject of evolution (unit of selection). Only something purposeful in the biological (instrumental) sense can be the subject of evolution. The main known purposeful (directed-on-something) regulatory mechanism is negative feedback. However, negative feedback is not purposeful in isolation: it is purposeful only on account of the functional context of other negative feedbacks in the organism and, ultimately, on account of the realization of the superior positive feedback. Only the whole complex of inferior negative feedbacks subordinated to the superior positive feedback is self-sufficient in the realization of the master task of the subject of evolution—survival in a given environment and reproduction without the need for any help from outside or any other purposeful regulatory mechanisms to complete this task. Therefore, it seems reasonable to assume that only the whole system—a living individual defined within the cybernetic paradigm—constitutes the subject of evolution.

The amount of biologically purposeful information associated with the complexity of structure, function and the network of regulatory mechanisms can be maintained on a constant level or increase during long periods in the process of biological evolution. The theory of evolution does not presuppose that the complexity of organisms must grow due to evolution—after all, as discussed above, the general structure of prokaryotes most probably did not change significantly for over three billion years. On the other hand, a clear, at least during some periods, increase in complexity can be observed in many lines, for example, in the line leading from prokaryotic organisms through eukaryotic unicellular organisms to humans. Alternatively, at least, this could be an intuitive impression as no universal, commonly accepted definition of the complexity of living organisms related to their structure and function has been proposed. Also, in this regard, the cybernetic definition of life has much to offer because the (structural and functional) complexity of an organism (individual) can be equated with the above-defined amount of biologically-purposeful information, especially if one remembers that negative feedbacks control forming structure and function during embryonic development. Within this approach, it can be very sensibly claimed that a man is more complex than a bacterium or oyster, and in principle, because of the enormous amount of negative feedbacks associated with the functioning of the human brain, is more complex than any other living organism on the earth.

As previously mentioned, at the basis of every piece of information associated with structure and function, thermodynamic information must lie, which is inversely proportional to entropy. Therefore, increased biologically purposeful information in the process of evolution also involved increased thermodynamic information and a decrease of entropy. The capability for the preservation and cumulative accumulation of information and the fall of entropy in a given system over long periods (after all, even bacteria once originated from some simpler systems) is a unique property of living organisms, at least on organizational hierarchy levels lower than the psychical, social, cultural and civilizational level, which however are based on the biological level, without which they could not originate and exist. Inanimate dissipative structures do not possess this property, which can accumulate thermodynamic information only to a very limited degree over relatively short periods. According to the second law of thermodynamics, information in the real world undergoes quick dissipation in the absence of any mechanisms compensating for this process, and it cannot be preserved and accumulated for a long time. The only such mechanism is the fast copying of information and the possibility of selecting (for further copying) copies that contain the most correct (on account of some criteria) information. If the rate of copying is exactly equal to the rate of information dissipation, the amount of information will be preserved at best. If, however, the rate of copying exceeds the rate of dissipation, there occurs the possibility of the growth of the amount of information when due to accidental changes, some copies of information appear to be even more "correct" (in relation to some external criterion) than the original copies. Of course, in the living world, an exemplification of the described mechanism is the mechanism of natural selection. As demonstrated above, this results from the confrontation of the cybernetic definition of life with the real world. Therefore, the possibility of accumulating biologically-purposeful and the underlying thermodynamic information over long periods is also a derivative of this confrontation.

If someday we successfully construct artificial living organisms based on metals, semi-conductors, plastic, and so on (in Supplement 2, I discuss how such individuals could look), they would have at least partly different cybernetic identities than living individuals based on organic compounds known to us. This is at least because different negative feedbacks are needed in the case of finding, taking up and building bodies of metal or silicon, and others in the case of the assimilation of organic compounds or their synthesis and the production of proteins, sugars, fats and so on. Similarly, different sets of negative feedback loops must act during the construction of living individuals to produce ultimately different structures of these individuals. Even a small difference in the distribution of spots on a snail shell, if it did not originate due to environmental disturbances, is a derivative of slightly different sets of negative feedback controlling embryonic development, determining, for instance, different assigned concentrations of pigment in a given place on the shell. Therefore, the abstract and formally defined network of negative feedbacks determines a given individual's identity univocally when applied to the real world. In other words, the set of instructions in a given identity leads our physical reality to construct a strictly determined, such-and-not-another individual (of course, when we neglect stochastic processes and detailed environmental conditions).

To summarize, confronting the cybernetic definition of life with the properties of the real physical world leads to the conclusion that many conditions must be fulfilled for a certain physical system to become a living individual (a fundamental element of spontaneously-originated life). These conditions are as follows: (1) the system must be a dissipative structure significantly displaced from thermodynamic equilibrium; (2) the system must be built of organic compounds (at least in the case of spontaneously-originated life); (3) the system must possess a certain minimal degree of complexity of structure and function (proportional to the diversity and variability, and complexity, of its environment/ecological niche); (4) the system must have at its disposal a kind of metabolism that is the transformation of matter and energy; (5) the system must possess a record of identity separated from the executive machinery; (6) the system must be able to compensate for various disturbances coming from the environment (external and internal) and steer its own behavior and supervise its self-construction during morphogenesis. Once originated, such a system (its identity) will mutate, undergo natural selection and evolve. As a consequence, it will be able to accumulate information over long periods.

The cybernetic definition of life undoubtedly belongs to a large extent to the Platonic world of abstract concepts and ideas. However, when this definition is applied to physical reality, we obtain life as we know it very well from our everyday experiences. One of the great mysteries of our material world is that this world allows the realization of the cybernetic definition of life at all and, thus, the origin of the enormous diversity of living forms we observe around us. The reason for this state of affairs should be explained by a future complete physical theory of the universe, providing that such a theory is ever created.

Life on the surface of our planet astonishes us with its richness, diversity and the degree of organization of structure and function. Therefore, does the cybernetic definition of life, confronted with all this splendor, not seem extremely simple and sparse? Does it not lose the properties of life, which we intuitively associate with a bacterium, a maple or an octopus? Is it, to say it directly, an adequate definition of life?

A satisfactory spare and clear definition of some phenomenon should not be confused with an extensive, exhaustive picture description, quoting possibly many of its properties and characteristics. The role of a definition is completely different—first, a possibly simple and univocal separation of the defined phenomenon from all other phenomena (a logician would talk about a sharp separation of a set *A* from its complement *A'*), and second, reaching to the very core of the essence of this phenomenon. This is exactly what, in my opinion, the cybernetic definition of life does. On the other hand, the whole wealth of living forms emerges when this definition is applied to the real physical world, to the potential germs of structure (and function) virtually present in this world as laws of physics and chemistry (and organization!) discussed in the first chapters of this book.

The cybernetic definition of life (living individual) abstracts from accidental (although marvelous in its richness) features of terrestrial life and focuses on the essence of life in general and properties that differentiate life from all phenomena of inanimate nature. Of course, here, the process of abstraction requires some intellectual effort and resignation from innate intuition and acquired schemes and paradigms of thinking. However, this is, in principle, what the whole conceptual progress in science consists of.

CHAPTER 7

BIOLOGICAL INDIVIDUAL VS. PSYCHICAL INDIVIDUAL—ESSENCE OF SELF-CONSCIOUSNESS

A cybernetic model of the psychical individual

Two completely different sorts of "consciousness" should be clearly distinguished and separated as they are frequently confused: instrumental and psychical (Korzeniewski 2010, 2017a). Only the latter constitutes "true" consciousness. *Instrumental consciousness* represents something within something else, an object within a subject. For example, a consolidated picture of the fragment of the surroundings is represented within a camera; within a video camera, there is a moving picture; within a refrigerator, there is the internal temperature; within the metabolic-genetic network of a bacterium, there is the presence of lactose in the environment (the operon theory), while within the neural network of the frog's brain there is a fly flying by. The instrumental consciousness can be static (camera), dynamic (video camera) or regulatory, where the consciousness of something elicits some purposeful reaction (elevated internal temperature turns on the cooling device in a fridge, presence of lactose in the environment stimulates the production of enzymes decomposing this sugar in a bacterium, a fly flying by triggers the hunting instinct in the brain of a frog).

On the other hand, the "true" psychical consciousness constitutes a qualitatively different sphere of mental phenomena that is absent in most systems in the physical world, even very complex ones. While instrumental consciousness is quite easy to conceive, this is the psychical consciousness, whose nature and genesis require explanation. Below, only psychical consciousness is considered.

True, psychical consciousness alone cannot exist without self-consciousness (it does not make sense). In other words, the (psychical) consciousness appears together with self-consciousness and constitutes a part or an aspect of it (Korzeniewski 2010, 2020). Therefore, in the present work, psychical consciousness and self-consciousness are frequently referred to as (self-)consciousness. In my opinion, it must be underlain by a certain very special relationship between the organization of elements entering into the composition of some dynamic complex real systems.

I previously proposed (Korzeniewski 2010, 2013a, 2015, 2017a, 2020) what the emergence of (self-)consciousness, i.e., the subject or *psychical individual*, due to a certain specific pattern of the neurophysiological activity of the human brain consists in. Below, I summarize my opinions on the neural basis of (self-)consciousness, and in particular, I present my cybernetic model of the psychical individual as opposed to the biological individual.

The psychical individual can be identified with conscious "I." The neural network in the human brain, which underlies the psyche and self-consciousness, is composed of neural cells, i.e., neurons. They transfer neural signals through specialized appendages—dendrites and axons—connected by synapses. The function of a given neuron is partly determined by its internal properties, which is how it transfers signals received at the input (from synapses situated on dendrites and the cell body) into signals on the output (the axon and synapses located on it, connecting it with the dendrites and or cell bodies of other neurons). However, the role of a given neuron in the whole neural network is primarily determined by its functional context, in other words, which other neurons it is connected with and in which (functional) way. The role of these other neurons is, in turn, determined by their own internal properties and external connections, and ultimately, the function of all neurons is "defined" not only by all other neurons in the neural networks in the brain, but also by the receptors and effectors (mainly muscles) to which this network is "plugged." Therefore, particular neurons and their complexes "mean" by connotation, the functional mutual relations between them and the mentioned receptors and effectors.

Signals coming from appropriately stimulated receptors (the sense organs: eyes, ears, nose, skin receptors of touch, warmth, coldness and pain, receptors of "internal feeling," *etc.*) are processed and integrated into sensory data ("sensory pictures") of different modalities (visual, auditory, olfactory, tactile *etc.*). For example, in the case of the sense of vision, the varying in time stimulation of particular photosensitive cells (rods and cones) by photons (quanta of electromagnetic radiation of a given wavelength within the visual range), analogous to pixels in a video camera, is integrated by the neural network in the brain into spots (in fact, this happens already in the retina), and next into lines/contours/bars of various inclination, colors and movement (left-right, top-down, approaching-receding), simple and more complex shapes, individual objects, discrete causal events, and finally, for example, the faces of particular persons. The so-called *integrative structures* in the brain are responsible for this (Korzeniewski 2010, 2017a).

A repeating coincidence in time of the stimulation of different sets of integrative structures undergoes consolidation (organization) into the neural structures corresponding to particular units, "objects," and "facts" of the external world that have been called *primary (lower-level) associative structures* (Korzeniewski 2010, 2017a). *Secondary (higher-level) associative structures* corresponding to general categories, rules, laws, principles and ideas are formed on a higher level through repeating the stimulation of sets of lower-level associative structures characterized by temporary or situational coincidence. Spontaneous, autonomous activity of the brain (mainly prefrontal cortex), generally understood as thinking processes, participate in this process. There are extracted repeating patterns of the stimulation of various neural circuits and analogies within lower-level associative structures. Particular associative structures of lower and higher levels mean, like single neurons, by *connotation*, mutual relations between them and the remaining associative structures.[1]

Associative structures can correspond to certain patterns of behavior (e.g., the choice of the plant or insect to eat or the decision to follow such-and-not-another hunting strategy) or to some category of objects or regularities present in the external world (e.g., the recognition of particular objects, knowledge of laws and relationships governing the physical and biological environment, the anticipation of future events). They also interconnect stimuli and sensory pictures from different senses (vision, hearing, smell) and combine them into one integral whole. For example, due to the accumulation of experiences, there can emerge an associative structure corresponding to a leopard (or a more obscure associative structure corresponding to a large ground predator), a directive saying "do not eat plants with big round leaves with red spots or you will get poisoned," or an action plan stating "it is easiest to hunt an antelope in the rainy season by lurking in bushes covered with dense foliage." On the neuronal level, this process occurs through the formation or enhancement of neural connections

[1] The described kind of learning and memory record formation concerns *semantic memory* that is associated with understanding "what is what" (recognition of objects, processes and laws ruling the world). The prefrontal cortex and (especially anterior) temporal lobe participate in the consolidation of this sort of memory. Acquiring the *episodic memory*, concerning particular single events from one's life, is mediated by the hippocampus that links various aspects of a given episode recorded in different fragments of the brain cortex together (Banich & Compton 2018).

(through the creation of new synapses or lowering the sensitivity threshold of existing synapses) between "paths," "circuits," or "local sub-networks" in the neuronal network in the brain that: (1) represent appropriate properties/aspects of the world extracted from the sets of signals generated by receptors (and processed by integrative structures) or (2) are responsible for the initiation of appropriate behaviors (or inhibition of undesirable behaviors).

Equally important for the brain is the removal or weakening of inappropriate connections (through liquidation or elevation of the sensitivity threshold of synapses), which lead to the incorrect identification of objects or inadequate behaviors (from the point of view of fitness).

An associative structure constitutes a particular "circuit", local sub-network, neuronal "object" in the brain where neural impulses can easily flow and are "willingly" transferred; this allows the activation of the whole structure by a relatively weak stimulation. For instance, perceiving small fragments of a leopard running through high grass easily activates the associative structure: "a leopard" (Korzeniewski 2010).

The process of the formation/enhancement and removal/weakening of associative structures is supervised by the motivational system—inborn instincts/drives (e.g., survival instinct, sexual drive, the instinct of satisfying hunger, the instinct of avoiding injuries, illnesses and poisoning, cognitive instinct)—plus the reward system. This system forms and enhances associative structures that are correlated (temporarily or situationally) with the satisfying of instinctive needs and therefore stimulates the "reward center" in a positive way while it removes/weakens those associative structures that are correlated with a lack of satisfaction of instinctive needs and therefore negatively stimulate the discussed center (Korzeniewski 2010, 2017a). The reward system is primarily based on the neurotransmitter dopamine, which is released by the ends of axons of neurons of the dopaminergic system at the moment of its positive stimulation, predominantly in the prefrontal cortex and the hippocampus, and there it diffuses to neighboring neurons. The cell bodies of the neurons of the dopaminergic system are primarily situated in the ventral tegmental area (tegmentum) in the midbrain (Banich and Compton 2018, 375-376). If a given behavior leads to a realization of some instinctive needs/drives, such as satisfying hunger, the associative structure underlying such behavior is enhanced/consolidated. The reward system causes dopamine release, which diffuses to surrounding neurons and synapses, strengthening those connections (synapses) that were just activated and thus leads to the consolidation/enhancement of appropriate neural circuits/associative structures. The same happens when some fuzzy outline recognized as a leopard actually appears to be a leopard. On the other hand, the poisoning that follows eating an insect with a characteristic pattern of coloration, the lack of success in hunting after choosing a particular hunting strategy for a given recognized kind of prey, or the incorrect identification of the prey will result in a weakening of the associative structure underlying such a behavior or object recognition. The instinct/drives system plus the reward system causes the whole neural network of the brain to become *intentional* (Korzeniewski 2010, 2017a).

Psychologically, integrative structures correspond to impressions, while the lower- and higher-level associative structures correspond to primary and secondary concepts (Korzeniewski 2010, 2015, 1017a), while the positive/negative activation of the reward/punishment system corresponds to pleasure/displeasure (pain).

The discussed mechanisms cause the whole network of associative structures to be purposeful and *intentional* (in the operative sense) and not directionless and accidental. This network becomes a certain *representation of* (some aspects of) *the external world* that primarily serves the superior goal of the living individual: survival and reproduction (fitness maximization). Certainly, the neural network in the brain constitutes a part of the whole negative feedbacks network constituting the cybernetic identity of the living individual.

I previously proposed that self-consciousness emerges as an aspect, by-product and or consequence of the functioning of the neural system (the ABC theory of self-consciousness; Korzeniewski 2020) when the "cognitive center" in the brain becomes recurrently directed on itself, it receives and processes signals both from receptors (through the sensory cortex) and from itself (Korzeniewski 2010, 2015, 2017a, 2020). This center constitutes a complex "neural circuit," which compares data received from receptors (integrated by the sensory cortex into sensory pictures) with memory records, coordinates various functions of the central nervous system, directs one's attention, plans future actions and undertakes decisions concerning the stimulation of effectors (muscles). This center confronts signals from the surroundings with the existing neural network containing memory records and a picture of the world shaped within this network. At the same time, it uses the newly arriving integrated sensory data to develop further and extend this network (among others, to form new memory records). On the neural network level, its action consists of activating, forming and modifying associative structures of lower and higher levels. A fundamental role in its functioning is played by the autonomous spontaneous activity of the cerebral cortex, the thinking processes. The cognitive center is situated in the prefrontal cortex, although it uses sensory pictures and memory records located in (dispersed over) the frontal, parietal, temporal and occipital cortices. In animals devoid of the subjective psyche and (self-)consciousness, it only receives and processes signals from receptors through the sensory cortex. In the self-conscious brain also receives signals from itself (Korzeniewski 2010, 2020). Therefore, in agreement with the presented conception, the essence of the psychical individual is based on the relation of *self-reference*. In other words, the cognitive center in the self-conscious human brain is *directed at itself*, creating a representation (model) of itself within itself. It is just this very special type of dynamic complexity that, in my opinion, is equivalent to the subjective psyche and (self-)consciousness. Just the appearance of the relation of self-directing on itself or self-applicability, which constitutes a completely new quality, specific kind of functional (operational) organization of the neural network, led in the course of human evolution to the emergence of self-consciousness (Korzeniewski 2020).

In other words, the emergence of (self-)consciousness within the neural network underlying the discussed cognitive center is based on the directing of a part of the system's inputs on the system itself, meaning the self-recognition of the processes occurring in the cognitive center. Apart from the picture of the external world, the cognitive center also creates its own picture within itself. As mentioned above, according to my conception, psychical consciousness is strictly associated with self-consciousness, as the former cannot exist without the latter.

The general difference between the self-conscious brain and the brain devoid of self-consciousness is depicted in Fig. 7-1.

The inseparable relationship between (self-)consciousness and the kind of dynamic complexity of the neural network underlying it has far-reaching consequences on, for example, the problem of zombies, frequently dealt with and analyzed within philosophy. Zombies that refer to people that are materially (in all details of their structure and function) identical to normal people (including the system of neural connections in the brain) but are devoid of (self-)consciousness, cannot exist, and are even nonsensical. Because when an appropriate kind of complexity of neural network in the human brain comes into being, it *must* generate self-consciousness and a subjective psyche (Korzeniewski 2015).

brain devoid of self-consciousness

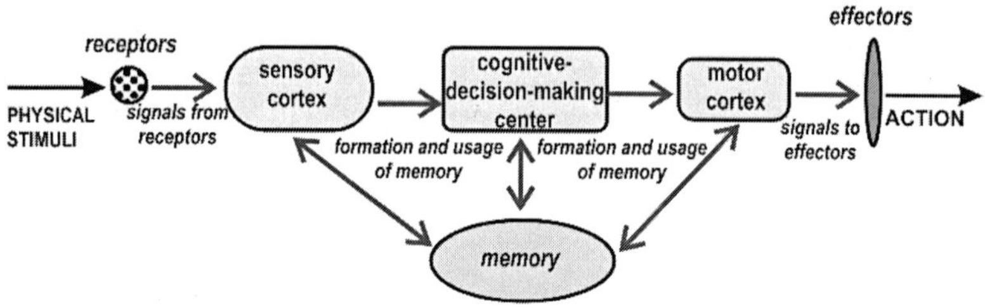

brain endowed with self-consciousness

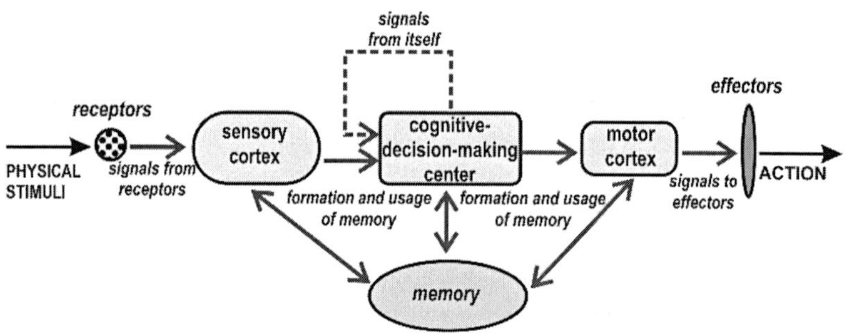

Fig. 7-1. Difference between the brain devoid of self-consciousness and the brain endowed with self-consciousness. In the brain, the sensory cortex receives and integrates signals coming from receptors – e.g., the eyes (occipital and parietal lobe) and ears (temporal lobe) – and transfers complex "sensory pictures" to the cognitive-decisive center ("operational memory") being the seat of processes of planning and data analysis (prefrontal lobe). This center sends signals concerning the undertaken decisions about the initiation of various actions to the premotor and motor cortex (back part of the frontal cortex), which in turn transcribes them to the stimulation of particular effectors: muscles and their groups. Their action is coordinated by the cerebellum. The sensory cortex, cognitive-decisive center and motor cortex on the one hand participate in the formation of episodic and semantic memories entering into the composition of the picture of the external world (parietal, temporal and frontal lobe), while on the other hand, they use the already existing memory records to recognize objects, understand rules and execute current actions. In the self-conscious brain, the cognitive-decisive center receives signals not only from the sensory and associative cortex, but also from itself; it creates its own representation within itself and undergoes recurrent directing on itself. As a result of this, the subjective psychic sphere and self-consciousness appear, while the processes of data analysis, planning and decision making adopt the form of thoughts.

Formal similarities between the biological individual and the psychical individual

The essence of the psychical individual can be summarized as follows: (1) the neural network (comprised of a system of dynamic connections between neurons), especially its cognitive-decisive part, was formed as a kind of *representation of the external world*—the functional "knots" of this network roughly correspond to various aspects, objects and properties of the physical reality; (2) the elements of this network—neurons and their groups—mean by co-conditioning, *connotation*—particular neurons constitute the functional context for other neurons, and ultimately, only the whole neuronal network (plus receptors and effectors) determines the functional sense of all elements it is composed of; (3) this network is *intentionally* (in the operational sense) directed towards the survival and reproduction of its "carrier", which is the living individual equipped with it, and also on itself, its own self-representation, the creation of its own picture within itself; (4) psychical consciousness and self-consciousness emerge due to the *self-directing* on itself of the "cognitive center" in the human brain. This center, which in (most) animals that are devoid of self-consciousness receives processed stimuli coming, more or less directly, from the sense organs and the sensory part of the brain, in humans began to receive and process stimuli also coming from itself. As a consequence, it developed a map of itself within itself. All these properties characterize the cybernetic psychical individual.

It is difficult not to notice an analogy between the conditions necessary for the emergence of the cybernetic psychical individual (self-consciousness) from the cooperation of neurons and the emergence of the cybernetic biological individual from the cooperation of negative feedbacks, underlain on a lower level by interactions between atoms. From an abstract formal point of view, these individuals are strictly analogous. Therefore, one can be tempted to formulate a universal definition of an individual (Korzeniewski 2013a). Namely, an individual (biological or psychical) is a system comprising a dynamic network of elements (negative feedbacks or neurons) that possesses the following common properties: (1) it is *intentional* (in the instrumental, operational sense); (2) its elements have meaning (have sense) through *connotation* (mutual relations between them); (3) it contains an instrumental *representation of* (some aspects of) *the world*; (4) it is self-referent, *recurrently directed on itself* (its own reproduction or self-representation). Life and self-consciousness exhibit deep, formal, structural similarities. The kind of complexity characterizing both biological and psychical individuals is responsible for the "special", "distinguished" status of the phenomenon of life and (self-)consciousness.

Therefore, the phenomena of life (living individual) and self-consciousness (psychical individual) originated when some instrumentally *intentional* dynamic network of elements which mean by *connotation*, constituting a *representation* of some aspects *of the external world,* became *directed on itself*. Some "individuality" features can also be seen in various social, economic and political systems (Korzeniewski 2013a).

Does complexity exist objectively?

The human brain, like all parts/properties of humans serving some biological purposes, originated due to evolution. Certainly, it has not developed for practicing science or philosophy but for efficiently hunting mammoths and participation in complex social behaviors. Despite our enormous achievements in the domain of natural sciences, technology and medicine, we are not universal cognitive machines. The structure and functioning of the human brain, the manner of its integration and processing of signals and the discrete nature of language have a profound impact on shaping our picture of the world (Korzeniewski 2010, 2013b, 2015, 2017a).

Our brain, and through this, the mind, has inborn predispositions to create the idea (concept) of substance and matter (this statement also concerns the "spiritual substance", that is, the soul) and of objects built of material substance that is in general "palpable things." On the other hand, the human brain is less efficient in "immediate" and "visible" perception, conceptual conceiving of the phenomenon of complexity and complicated dynamic structure, not only organized in space but also "happening in time." Complexity and structure are usually intuitively treated as phenomena secondary in relation to matter/substance, space and time and are therefore more "subjective" and less real than the latter, being situated rather inside our minds than in the real external world. This questions the very fact of the "real", "objective" existence of the phenomenon of life and (self-)consciousness, underlain after all by the special kind of complexity of structure and function discussed above, constituting the essence of the biological and psychical individual. Nevertheless, one could ask the basic question of whether this secondary character constitutes a "true" property of the world or is simply a derivative of the functioning of our brain/mind working in a such-and-not-another way. Do space, time, and matter exist in a concrete, undeniable, or even somewhat absolute way, while complexity and structure are only elusive, vague and, to some extent, accidental properties of solid and concrete matter "suspended" in well-grounded space and time? Or maybe the "real" state of affairs is the opposite, and just the complexity and structure of matter in space and time (and mutual relations between them) constitute more fundamental and primary aspects of the world than matter, space and time themselves?

Physics is undoubtedly the most fundamental natural science dealing with the broadly-understood material world. However, does physics seek to reveal the immanent essence and substance of the external world? Does it investigate what is "actually" broadly understood as matter? Contemporary physics treats elementary particles as points of zero dimensions or vibrating strings of zero thickness. However, what then constitutes the "substance" of elementary particles? "Something" point-like, zero-dimensional? "Something" having the form of a one-dimensional string? What would such a string be made of? Does such a particle simply constitute a place in space with the greatest intensity of some physical field (interaction, force), such as electricity or gravity, as discussed in Subchapter 1.6? In such a case, what constitutes the "substance" of such a field?

Elementary particles propagate in space in the form of a wave, and the probability of finding a particle in a given point of space is described by a so-called wave function (this probability is proportional to the square of this function, although we have no idea why) (see Chapter 1). However, again, what constitutes the "substance" of the probability wave? Moreover, what happens with this substance at the moment of the so-called collapse of the wave function? We do not perceive matter in its immanence, as a Kantian "thing in itself" (noumenon), as naïve realism might suggest, but through the prism of our brain/mind (co-created through the integration and association of lower and higher levels of signals coming from sense organs—see above). The properties of matter, in particular, and the so-called external world on the physical level are generally approached by us in theoretical physics by mathematical constructs. Physics offers only a formal mathematical description of "reality." It does not care, as it cannot, what kind of "substance" fills these constructs in the real world. Mathematics, lying at the basis of physics and being the source of its greatest successes, simply ignores the problem of the real content of the world. Therefore, we cannot decide whether, for example, an electron and the wave function associated with it, orbital in an atom or curved space-time, are only abstract mathematical concepts or real existing physical objects/phenomena. We do not even know whether such a distinction is legitimate at all.

Furthermore, because mathematics constitutes a creation of the human mind (although some believe that it exists in an absolute way in the world of Platonic ideas), we cannot demarcate any reliable border between the content of our psyche and the "substance" of the physical world. Examples of the devouring subjectivization of physics include the commonly known Schrödinger's cat paradox (my approach to and solution of this paradox are proposed elsewhere: Korzeniewski 2016) and the distinguishing of macroscopic states in thermodynamics (see Subchapter 2.3). Therefore, the contradistinction of the supposedly "objective" physics and the "subjective" because operating with concepts of (special kind of) complexity, purposefulness and instrumental intentionality biology and neurophysiology (including the theory of self-consciousness) sounds like a misunderstanding. Admittedly, a physical system can be much better described by mathematics than biological or psychical systems, but this has nothing to do with the objectivity of their existence but only testifies to the greater complexity of the latter. Therefore, uncertainty, relativity, relationality and subjectivity will always remain unavoidable elements of our picture of the world.

As it is discussed in the next subchapter, all "components" of the universe (matter, space, various kinds of elementary particles) mutually co-condition each other and, therefore, can exist only *in relation* to each other; no absolute, independent or "objective" status can be attributed to them. On the other hand, various aspects of the real world emerge as properties of systems characterized by an appropriate degree and type of complexity of structure and function. This complexity exists in a way that is no less or no more "objective" (or, if somebody prefers, "subjective") than space, time and matter. This is only the way of organizing the human brain that gives preference to the latter.

To summarize, the apparently objective phenomena and aspects of the external reality, including matter (e.g., elementary particles), space, time, discrete individual objects and events, or causality, are not objective. During the development of modern physics, especially theoretical physics, they are becoming increasingly more subjective products of our mind, concepts associated rather with psychical phenomena than with real "things in themselves." Various apparently subjective properties of physical, chemical, biological and psychical systems equivalent to their special kind of complexity are no more and no less objective/subjective than the elementary particles and atoms they are built of, the space in which they form various structures, and the time in which they "happen."

Let us consider a few levels of complexity occurring in the real world. The properties of atomic nuclei (e.g., stability/instability) are equivalent to the complexity of the system of protons and neutrons entering into their composition. Certainly, the nucleus does not behave as a simple "sum" of protons and neutrons, even its mass is smaller than the sum of the masses of "free" nucleons; in order to smash the nucleus, it is necessary to supply a large amount of energy equivalent to mass; this energy is released during thermonuclear synthesis.

The properties of atoms of particular chemical elements (size, valence, electronegativity, energetic levels of orbitals, absorption and emission spectrum and other physicochemical properties) are equivalent to the complexity of the system: the nucleus plus electrons, in particular of the complexity of the set of electron orbitals of different shapes (s, p, d, f), completely or partly filled by electrons, especially external orbitals that can participate in forming chemical bonds. The properties of elements as substances (e.g., color, specific mass, state of matter at a given temperature, reactivity) result from the properties of single atoms and complex interactions between them. For instance, it is not trivial that sulfur as a substance is yellow, hard and brittle, as single sulfur atoms are neither "yellow" nor "brittle."

The physicochemical properties of water molecules (e.g., polarity) are derivatives of the spatial arrangement of one oxygen atom and two hydrogen atoms, particularly the shape and properties of their molecular orbitals. The properties of water as a substance (liquidity at room temperature, the temperature of ice melting and vapor condensation, transparency, dielectric constant, and refractive index) are equivalent to the dynamic spatial organization of a huge number of water molecules and the mutual interactions between them. The fact that water is wet does not result simply from the chemical composition of its molecules. The same concerns more complex molecules and substances.

Similarly, the properties of yet more complex molecules, in particular organic polymers, and chiefly linear heteropolymers (proteins and nucleic acids), are equivalent to the complexity of the assemblies (appropriately arranged in space and mutually connected by bonds) of atoms and monomers of which they are composed. This primarily concerns the molecules constituting the (molecular) basis of life and responsible for the identity of particular individuals: DNA, RNA and proteins, lipids, polysaccharides, *etc*. DNA adopts the form of a stiff double helix, but RNA is much more elastic and flexible, and because of this, it can arrange itself into single- and double-strand segments, including loops, and form various three-dimensional spatial structures; proteins can adopt very diverse spatial forms and fulfill various functions, which results from the complexity of the folded strings of monomers they are composed of. The structure (frequently very complicated), function and properties of a given protein are equivalent to appropriate interactions of amino acids connected in a such-and-not-another sequence and characterized by such-and-not-other physicochemical properties. Looking at the one-dimensional sequence of amino acids in a polypeptide chain, we cannot say in which three-dimensional structure (conformation) it will fold itself and what function it will fulfill in the cell (although we can try to simulate this using appropriate computer models, so this is a problem of our knowledge, and not of some intrinsic uncertainty of how the chain will fold).

An analogous situation occurs in the case of structures of a higher order, including complexes of macromolecules, such as ribosomes, chromosomes, cytoskeleton or lipid-protein membranes composed of proteins, nucleic acids, lipids and or carbohydrates. Their structure, function, and "macroscopic appearance" are a derivative of the complexity of the elements entering their composition. On a yet higher level, this also refers to cellular organelles, e.g., mitochondria.

The characteristic properties, structure and function of the whole living cell are equivalent to an appropriate complexity (functional organization) of the assembly of its components: lipid-protein membranes, cytosol, cytoskeleton, organelles, ribosomes, chromosomes and so on.

Living organisms as whole systems, both uni- and multicellular, possess (are equipped with) a special kind of dynamic complexity with many aspects: genetic, structural, biochemical, enzymatic, thermodynamic, energetic, reproductive, *etc*. One such aspect, which was especially emphasized in Chapter 6, is the cybernetic aspect (regulatory, information transfer/transformation), which can be identified with an appropriately-organized complex of inferior negative feedbacks (more widely: regulatory and signal transduction mechanisms) directed at the realization of the superior positive feedback—reproduction—the copying of own identity in a potentially large number of possibly similar copies. In this way, we come to the cybernetic living individual whose essence and characteristic properties are equivalent to the complexity of the assembly of the regulatory mechanisms and the processes of transfer, transformation and possible integration of information in a purposeful manner directed at its self-copying. These, in turn, are assemblies of atoms and molecules appropriately organized in space and time.

Finally, subjective psyche and self-consciousness (the psychical individual) are, in my opinion, equivalent to the kind of functional complexity of the neural network in the human brain (Korzeniewski 2010, 2013a, 1015, 2017a), as discussed in Subchapters 7.1 and 7.2.

Generally speaking, there is no reason to treat various properties of real systems equivalent to the (type of the) complexity of assembly of their elements on different levels of a hierarchy of dynamic organization of matter in space and time as less "objective", "fundamental" or primary, than matter, space and time themselves. This is particularly because the latter undergoes an advancing subjectivization within modern physics (especially in quantum mechanics and relativity theory). Biological and psychical (self-conscious) systems are characterized by a high degree and special kind of complexity, which is characteristic and constitutes their essence. Complexity exists in a possibly objective way. Just a special kind of complexity determines the core of the nature of life and self-consciousness.

On the essence of physical reality

The progress of contemporary physics allows one to speculate about what the emergence of the universe from nothingness, which most probably occurred about 13.7 billion years ago, could have consisted of. More and more premises allow us to suppose that particular elements and aspects of the physical reality do not constitute independent, autonomous beings, but their existences mutually condition each other. For example, the universe's total energy is probably zero, as the positive energy associated with matter (mass) exactly counterbalances the negative energy associated with the recession (shifting away) from each other of masses in space in the gravitational field (Penrose 2002; Barrow 2013) . Therefore, space and matter constitute, in a sense, the *yin* and *yang* of the universe. Suppose the energy of "nothingness" (whatever this could mean) and of the whole universe is exactly zero because of the reciprocal "canceling" of the energy of matter and space with their opposite signs. In that case, the question arises of what, in fact, actually originated and appeared together with the emergence of the universe from non-existence. My answer to this question (Korzeniewski 2013b, 2015) is the *information* about the mutual distribution of matter and space (please recall that, just after the Big Bang, entropy was minimal, and therefore, thermodynamic information was maximal). Somebody religious could therefore say that "the word became flesh."

Similarly, elementary particles cannot exist without anti-particles, fermions without bosons, leptons without hadrons, matter without interactions (forces), *etc*. Matter (mass and energy) curves space-time, while the space-time curvature, equivalent to gravity, acts on matter. Therefore, perhaps all these elements of the universe emerged through the "delamination" of nothingness into opposite aspects/elements, and because of this, their existences mutually co-condition each other until the present today, creating a kind of network of reciprocal relations. These relations constitute a mapping (representation) of the fundamental mathematical laws governing the physical world. Furthermore, much indicates that all elementary particles of the universe that were (as quanta of energy not yet "formed" as particles) in strict contact with each other at its dawn, yet before the inflation phase, are closely "correlated" through quantum entanglement (the so-called non-locality of quantum mechanics; Baggott 2010; Al Khalili 2012). One can speculate (as I do) that the universe was generated (or generated itself) by the multiple recurrent application of a certain mathematical procedure (analogously, e.g., being a fractal, the Mandelbrot set is generated). This possibility seems very attractive philosophically, at least for me. This would be a subsequent (after the living individual and self-consciousness) example of the relation of self-applicability (self-reference), a recurrent directing on itself of a system composed of co-conditioning each other, meaning by "connotation" elements. Is such a system characterized by intentionality? Well, please address this question to theologists.

Therefore, it cannot be ruled out that the emergence of the physical level of reality (the universe) from nonentity was analogous to the emergence of the biological level from the physical level (of living individuals from atoms) and of the psychical level from the biological level (of self-consciousness from the neural network). This could speak for the unity of the universe as a hierarchical system, in which higher levels of the organizational hierarchy are generated in a certain determined way, identical for all main levels, based on the lower levels.

CHAPTER 8

CONCLUSIONS

Here ends our journey leading from atoms to living individuals. In realizing the main theme of this book, I tried to investigate how a SYSTEM, characterized by a specific kind of complexity, emerges from the appropriately-organized assembly of elements. At this moment, it can, at last, be said with whole responsibility: a living individual is, in some very important sense, nothing more than a complex and specifically organized dynamic configuration of atoms. If I dared to formulate such a statement at the beginning of this book before presenting the entirety of the above route, it could seem to be primitive reductionism, not too distant from the claim that apart from atoms, nothing real actually exists. Nevertheless, one should decidedly oppose such reductionism. First, as shown in Chapter 1, it is not only atoms but also the elementary particles they are built of that do not exist fully objectively. Contemporary physics more and more decidedly heads to the conclusion that an absolute, completely autonomous existence is apparently not even attributed to such fundamental elements of the physical reality as space, time and matter, but also wave functions, elementary particles, orbitals and forces—various aspects of the world seem to mutually co-condition each-other (see above and Korzeniewski 2013b, 2015), and the creations of the human mind become indistinguishable from the physical objects and phenomena, located "out there" on the side of the real world. The (proper kind of) complexity underlying the essence of the living individual (and psychical individual) exists at least as objectively as the mentioned phenomena. Second, when we analyze consecutive levels of the hierarchy of the structure of living organisms—atoms, inorganic compounds, simple organic compounds, complex (polymeric) organic compounds, assemblies of organic compounds of different sizes (proteins with many sub-units, lipid-protein membranes, chromosomes, ribosomes, cytoskeleton), organelles (e.g., cellular nucleus, mitochondria), cells of many (even hundreds of) types, tissues, organs, systems of organs, and finally entire individuals—then after all, *somewhere*, on some level of the hierarchy, the system becomes alive, and this is not caused by the appearance of some mysterious factor or force, but simply by an increase of the degree of (a certain kind of) complexity of the system. Whether ribosomes, mitochondria, or only the whole cells are already alive can be discussed. If we assume the latter, it is, in principle, possible (only present technical limitations are an obstacle) to assemble the living cell (initially a prokaryotic one) from its still inanimate components.

Moreover, this would mean bringing into life a previously dead system. Generally, the phenomenon of life constitutes an as objective aspect of the world, as various elements of the physical world: space, time, matter, energy, forces, elementary particles and so on. In other words, the complexity of a SYSTEM that is dynamic, organized in space, "happening" in time and composed of atoms is no more, no less real, than the very atoms of which this system is built, the space in which these atoms are dislocated, and the time in which the functioning of the whole system takes place. This is the manner how stimuli from the external world are integrated by our brain that causes us to intuitively perceive space, time and atoms as much more concrete, real or even "palpable" than the specific dynamic complexity underlying the biological (and psychical) cybernetic individual.

In this book, I have chosen a slightly different approach to deriving life from inanimate atoms, emphasizing thermodynamic, evolutionary and cybernetic (related to the internal purposefulness of living individuals) aspects. First, I have demonstrated how it is possible at all in our universe to form the structure on different levels of hierarchy; how atoms can join into less or more complex molecules of chemical compounds; how some of them—organic compounds—can form polymeric compounds, their complexes and yet greater structures. Second, I have argued that despite potential predispositions to form complex structures from atoms, the probability of the spontaneous origin of such structures instantaneously, in one step, without the intervention of some external forces, is very (astronomically) low, and such structures by themselves are very nondurable because of the second law of thermodynamics referring to the tendency for the increase of entropy. In the face of this, complex objects/phenomena can originate spontaneously only as dynamic dissipative structures driven by the intensive dissipation of energy and the production of entropy in systems significantly displaced from thermodynamic equilibrium. Only thermodynamic gradients supply the external driving force and the thermodynamic information necessary for the spontaneous origin and sustaining of the complexity of physical systems. Third, I have explained that dissipative structures devoid of appropriate regulatory homeostatic mechanisms directed at achieving a certain goal can be characterized only by their limited complexity and a small amount of information. In order to preserve and potentially elevate the amount of information associated with complex structures, the information must be copied with a rate equal to or higher than the rate of its decay caused by entropy increase, and for further copying, only the correct (with regard to external criteria determining the chances of its "carriers" for survival and reproduction) copies of information should be selected. In living organisms, this is ensured by natural selection. Fourth, I have shown that the structures associated with living organisms could appear spontaneously and reach a high degree of complexity because life originated and evolved gradually, step by step. The increase in complexity was frequently achieved according to the principle "first multiply, then differentiate," which describes the origin of many copies of a given subsystem of structure/function followed by the evolution of particular copies in different directions. The basis of evolution and the growth of the complexity of the structure/function of living cells was modifications of the gene expression regulation cascade inside the cell, while for multicellular organisms, it was also a modification of the cascade of morphogenetic processes occurring during embryonic and individual development. Fifth, I was presenting the argument that the fundamental, decidedly distinguished level in the hierarchy of organization of the phenomenon of life is the living individual and that this individual can be best (in the most objective and parsimonious way, in terms of formal sciences) defined within the cybernetic paradigm. According to the cybernetic definition, the living individual is a system controlling (through a rich network of purposeful regulatory mechanisms, primarily negative feedbacks) its individual development, current functioning, adequate behavior in response to the environmental stimuli and the constancy of internal conditions, directed exclusively toward copying its own identity (reproduction), being at the same time self-sufficient in the realization of this task. Speaking very briefly, the living individual is simply a dynamic homeostatic replicator.

The organized and complex dynamic configurations of atoms create a SYSTEM called the living individual, which on the macroscopic level, seems to represent a completely new quality. Additionally, the special organization of atoms in the system plays a key role here. For the cybernetic individual to come into being, the co-conditioning network of negative feedbacks (regulatory mechanisms), as a functional reflection (representation) of varying properties of the diverse environment that are important for the organism, must undergo the intentional directing on its own reproduction (positive feedback). Such a definition of the living individual is undoubtedly very abstract and belongs to a large extent to the Platonic world of ideal concepts rather than to the tangible reality we deal with every day. However, when we analyze which properties a given system would have to fulfill in order to implement this definition in the real physical world, we will discover with surprise that from our considerations, there emerges a bacterium, an oak, a bat and the uncountable diversity of other forms of life existing on the earth and most probably on numerous planets in the universe surrounding us.

At first look, it seems that an impassable abyss separates the living individual from the atom, as these entities differ from each other so drastically, both in their essence and their degree of complexity. However, on the road taken by us, leading from the valleys of the simplicity of atoms to the peaks of the complexity of the living individual, we did not meet any cleft that was impossible to cross, and we made the subsequent steps securely and safely. Our journey was not disturbed by any unresolvable mysteries or supernatural forces necessary to inspire life into a pellet of inanimate atoms. The climb through subsequent levels of complication was a rather interesting trip targeted at admiring landscapes and not the back-breaking passage through impassable walls. We did not need any specialist climbing equipment—elementary knowledge of physics, chemistry, biology and cybernetics was sufficient for us. We started from atoms and advanced cautiously, thus patiently overcoming subsequent obstacles, and at last, we stepped imperceptibly to the top of the realm of complexity called the living individual. Moreover, when we turned and looked back, we saw that the abyss appearing before our eyes was apparent because we had in our memory the path leading through it, winding among the fissures which led us to the top. Furthermore, even though from the perspective of the peak, an optical illusion again shows us an impassable precipice separating us from atoms lying in a dark valley, we are not so easily deceived.

Finally, after traveling the whole way, we are ready to answer the supreme question asked in the introduction of this book: ***what purpose does the frog serve?*** Astonishingly many people do not know the correct answer to this question, although an attentive reader certainly already conjectures it.

One of the possibilities that seems to arise is that a frog's supreme purpose is maintaining its structure's integrity and function in the surrounding environment, simply speaking, its survival, the continuation of its existence in a (relatively) intact state. After all, the heart, brain, liver and kidneys sustain the liveliness and functioning of all other organs in a living organism (e.g., our frog), and therefore the result—the functioning and survival of the whole organism in good condition. The matter seems similar in the case of a car[1]—all its parts and sub-assemblies are subordinated to the efficient functioning of the whole while providing, for example, resistance to rusting or mechanical deformation to maintain its structure and functioning over time. However, is it the car's supreme task to preserve itself and avoid disintegration over (tens of) years? Of course, as discussed in the introduction, it is not the car's main "purpose"; for the car to move efficiently, it must maintain its structural and functional integrity. This fact suggests that the same concerns living organisms—for them to perform their superior task, whatever it would be, it is necessary, at least for some time, to prevent their bodies from losing the integrity of structure and function. However, the realization of this task always fails sooner or later. After all, ultimately, every living organism (including the frog) dies for "natural" reasons (senility), which happens rather rarely in nature or due to influences of the environment (predators, parasites, climatic factors, random events).

So maybe frogs serve the whole ecosystem in which they live, for example, as food for storks? Perhaps this could be regarded as true in common-sense reasoning. However, this statement is evidently false from the biological-evolutionary point of view. The frog certainly is not purposefully directed toward being eaten by storks. It does not strive to become their food. Just the opposite—the protective coloration of a frog impedes it from being found by storks, and if necessary, it escapes from a predator. In the course of biological evolution, frogs have specialized in performing this function for hundreds of millions of years. What would we say about a car that escapes from its owner? Would we state that it is purposefully directed at being his/her vehicle? Certainly, from the frog's point of view, its superior purpose is not to be eaten by a stork.

Therefore, the goal of the frog is neither its possibly long survival nor its functioning as a useful element of the ecosystem. As the frog is a (cybernetic) living individual, its whole structure and function, its complex of all purposeful regulatory mechanisms (including negative feedbacks) ensuring internal homeostasis, supervising embryonic development and current functioning (including behavior) of the organism are directed toward one superior purpose: reproduction, that is copying itself, its own identity, in a potentially large number of copies that are possibly similar to

[1] There are also very important differences. The car is not a dissipative structure far displaced from thermodynamic equilibrium. It does not possess cybernetic regulatory mechanisms (especially negative feedbacks), including self-repair mechanisms, purposefully directed on the maintenance of internal homeostasis. Through its "body" a continuous stream of matter, energy and information does not flow. It does not carry out self-construction from a relatively simple germ. Finally, it is not able to grow and change its form during its "life."

the original. Therefore, the answer to the question: "What purpose does the frog serve?" appears to be elementarily simple: "***The purpose of the frog is to produce frogs***" and nothing else. It is not like a car, for which the objective is to provide people with a possibly quick, easy and convenient locomotion, or like a robot that strictly executes the orders of its creators. Such a superior purpose in the form of the imperative of the ruthless copying of self-identity can seem senseless, soulless or even repellent from the point of view of the human psyche. However, one who does not understand this purpose does not understand the essence of the phenomenon of life and living individuals. They do not understand the essence of the frog.

SUPPLEMENT 1

DOES THE FLOW OF TIME REALLY OCCUR?

Thermodynamics explains to us the origin of the (thermodynamic) arrow of time, which is the immanent asymmetry of the time dimension in macroscopic systems composed of many elements. However, the arrow of time does not imply the real *flow* of time, which is inseparably associated with the concept of the *present*. Because if time actually flows from the past to the future, then the present indicates the point on the time axis in which the physical world currently "finds itself". This is like with an audio cassette tape containing the record of some music (unfortunately, this analogy does not work equally well with a CD disc)—one can univocally state at which point the tape is being read by the cassette player and, therefore, which fragment of the music is being played. However, if one accepts the determinism of the physical reality, then the four-dimensional space-time filling our universe can be treated, if we look at it from outside, as a congealed, consolidated, "frozen" block, which does not happen in time (as it contains time within it), but simply *is* (lasts timelessly, whatever this could mean). In such an approach, all the time moments exist "at the same time" simultaneously, while the flow of time, the past, present and future, is only an illusion of our mind. In other words, our tape is not played at all—both it and the music recorded on it constitute a certain determined being existing at once in all its fragments from the beginning to the end, and it only seems to the human consciousness, for not entirely clear (but rather unimportant) reasons, that there is now only being played the determined, this-and-not-another fragment, following the previous fragment and preceding the next fragment. It is astonishing how many physicists willingly accept such a vision of time (Convey and Highfield 1991).

However, this vision does not explain one elementarily simple thing: why (it seems to us that) we find ourselves (together with the whole physical reality) just *in this-and-not-another* point in time. In other words, this is the problem of *anchoring in the present*, between the past and the future, so that the point of anchoring monotonously moves from the past to the future (Korzeniewski 2016). Therefore, the flow of time can be reduced within this approach to the shifting border (the present) between the past and the future.

Because nobody actually knows what the flow of time consists of, I dared to present my own conception of it (Korzeniewski 2016). I call it the "freezing of reality." As mentioned before, the conception of the congealed and stiffened space-time seen from outside is only possible under the assumption of a full determinism of the laws of physics. However, if the contemporary interpretation of quantum mechanics is right, the world is governed by indeterministic, probabilistic laws. Therefore, while the past is (already) congealed, certain and unchangeable (these are the most elementary attributes of the past), there are multiple chances for various possible events in the future, which can only be determined stochastically. The transition from the uncertain, probabilistic future to the certain, congealed past happens in the present.

Furthermore, just this transition, occurring at speed equal to the speed of the flow of time (that is, half-jokingly, with the speed of one second per second), corresponds, in my opinion, to the movement of the point of the present along the axis of time. This somewhat resembles the phase transition during the freezing of water. The liquid phase (water), with the chaotic movement of molecules behaving, to a large extent, in an unpredictable way, passes to the solid phase (ice) with molecules "congealed" in determined places within the crystal network. This transition spreads in water from some crystallization centers in supercooled water. Similarly, according to my conception, the present is a translocating "front of the wave" associated with the "phase transition" of an uncertain future into a congealed past. The front of the phase transition can be located only in one point on the axis of time at the same time (the present), univocally dividing time into the "solid", "congealed" part (the past) and the "liquid", dynamic part (the future). At the same time, this front can move only in one direction, fixing the arrow of time. This "freezing of reality" naturally determines our "anchoring in the present" and the phenomenon of the real flow of time.

All great physical theories (Newton's mechanics, Maxwell's electrodynamics, Einstein's relativity and quantum mechanics with their many forbears) not only do not take into account the real flow of time but are, in their essence, symmetrical, reversible concerning time—their mathematical formalism does not distinguish the future from the past and does not involve the arrow of time (Convey and Highfield 1991). The only postulated process in dynamic physics, which is irreversible in time, undetermined and implicating the real flow of time, is the reduction (collapse or decoherence) of the wave function appearing within quantum mechanics (Baggott 2010; Al-Khalili 2012). A one-time reduction of the whole probability wave due to a sudden violent interaction with the surroundings is the collapse of the wave function. On the other hand, a gradual reduction from a greater to a smaller number of quantum states due to more subtle interactions with the surroundings is the decoherence of the wave function (Al-Khalili & McFadden 2014). However, the reduction of the wave function is not a part of the "core" of the mathematical and conceptual apparatus of quantum mechanics—it was added *ad hoc* to explain the observed behavior of physical systems.

On the other hand, presently it is the only chance to explain the phenomenon of the arrow of time and the real flow of time. We can call it the quantum flow of time.[1] Additionally, the cosmological arrow (and flow) of time associated with the evolution of the universe, especially with regard to its expansion, requires explanation.

Even if I am right on this point (which is, of course, not certain), the principal question remains: what is the relationship between the direction of the flow of time defined above and the thermodynamic arrow of time? I do not know the answer to this, but I think the solution to this fascinating problem constitutes one of the greatest challenges of contemporary science. It is probably not a matter of chance that the universe (and the time dimension that emerged at its beginning) started from a state of extremely low entropy, and therefore, out of necessity, this entropy increases with the flow of time. What can seem intuitively surprising is that this primeval low entropy was associated with a homogeneous distribution of matter in the gravitational field, while the far future of the universe is matter concentration in still greater and greater (due to accumulation of the surrounding matter and joining with each other) black holes—objects with the highest entropy. If the entropy of the universe from the very beginning had adopted a maximal value, nothing would happen in it. We do not know, however, whether the initial state resulted from necessity (laws of physics) or chance. Certainly, the uniform distribution of matter/energy was partly the result of the inflation phase just after the Big Bang, in which an enormous expansion of space took place. The problem of the reality of the flow of time is discussed in more detail elsewhere (Korzeniewski 2016).

[1] Even in the absence of the collapse/decoherence of the wavefunction, there are some asymmetries (arrows of time) in quantum mechanics resulting from initial conditions. For instance, the wave of matter associated with an electron emitted from some point source propagates in space in all directions (like waves in water propagate concentrically outwards from the point where a pebble was thrown into the water), while the opposite process, which is the convergence of the wave of matter into a single point, does not result in reality.

SUPPLEMENT 2

ARTIFICIAL LIFE BASED ON SUPRA-MOLECULAR LEVEL

Natural life based on molecular elements/machines

An extremely distinctive property of naturally spontaneously originated life in comparison to various devices, machines, robots, computers, engines and mechanisms built by people is the fact that the structure, and especially the function of living individuals, is based on ("begins" from, is anchored in) the molecular level of physical reality. This means that single (macro)molecules of organic compounds, primarily proteins but also lipids, carbohydrates and nucleic acids (RNA, DNA), lie at the basis of the spatial forms adopted by life and all processes occurring in these forms. Living individuals' structural molecular bricks (cells) comprise such objects as the cytoskeleton, lipid-protein membranes and various proteins and other compounds and their complexes suspended in the cytosolic water phase. Molecular machines that "realize" the whole metabolism are protein enzymes (composed of one, a few, several or even tens of polypeptide subunits, frequently also containing non-protein cofactors). An elegant example of a molecular machine is ATP synthase, one part of which rotates in relation to another part, which is driven by the dissipation of the proton gradient through an appropriate lipid-protein membrane and coupled with ATP synthesis. There also exist greater (more macroscopic, but still molecular) machines, for example, flagella and cilia (driven by molecular motors), which move single cells in water or translocate water outside the cells of multicellular organisms; actomyosin filaments responsible for muscle contraction; ribosomes (composed of two big protein subunits and three or four strands of ribosomal RNA, rRNA) carrying out protein synthesis; karyokinetic spindles engaged in chromosome separation to descendant nuclei and cells during cell divisions; kinesins and dyneins transporting various objects along microtubules *etc*. Generally, just (macro)molecules constitute the most fundamental units of structure and especially the function of biological systems. Briefly speaking, the molecular level is the lowest structural-functional level of natural, spontaneously originated life. Nevertheless, this conclusion does not at all imply a primitive reductionism, ignoring various emergent properties appearing on the cellular, physiological and organismal levels.

The fact that natural living individuals are based on molecular machines implies their numerous secondary properties. First, a continuous stream of matter, energy and thermodynamic information flows through biological systems. At the same time, their form remains preserved or at least changes much more slowly than the exchange of their component substances occurs. In other words, living individuals are dissipative structures organized in space and time, which can maintain an almost unchanged structure, but unceasingly uptake and excrete building material, driven by energy dissipation (the increase of the entropy of the surroundings). The atoms entering our bodies' composition are removed from them over the weeks/months and are replaced by new atoms taken up as food (some exceptions here are the mineral components of teeth and bones). Proteins and other organic compounds are damaged (e.g., by free radicals or simply due to thermal movements of molecules) and become continuously replaced with newly synthesized compounds. Additionally, many cells undergo exchange, for example, epidermal or blood cells. Generally, molecular machines not only underlie the structure and functioning of living individuals but are continuously synthesized and decomposed (which is, of course, also carried out by molecular machines).

Another property characteristic of living individuals based on molecular machines is the ability to repair various cases of damage without the need to exchange large subunits (modules, parts) of an organism. Put simply, living individuals do not use spare parts (subunits), without which it would be difficult to repair, for example, a car. Already existing elements form new, regenerated elements. This is possible as the repair/regeneration occurs more or less directly on the molecular level.

Damage to the double DNA helix is repaired, while damaged proteins are decomposed and replaced with newly synthesized proteins. The same concerns other organic compounds of the cell. Heat shock proteins help proteins denatured because of high temperature or other stress-generating factors adopt native (physiological) conformation. Similarly, higher-level structures or whole cells are repaired or replaced. The repair/regeneration also occurs on the physiological level—examples include the healing of wounds, bone restoration (knitting), fighting off parasites and infections and regeneration of lost limbs in some amphibians. In all these cases, the repair/regeneration occurs more or less directly on the molecular level (it is based on, for example, protein synthesis and the action of pre-existing proteins).

Descendant living individuals self-develop (self-construct) themselves during embryonic and individual development. Their parents do not build them by adding subsequent parts/elements. Instead, the structure and function of their bodies are formed gradually through growth and the differentiation of specific, relatively simple "germs." In particular, in multicellular organisms, embryonic development usually starts from fertilized egg cells (zygotes) and consists of the division, the possible increase of size and the differentiation of cells. The entire process is principally endogenous; parental living individuals do not participate (with small, rather unimportant exceptions). Of course, in mammals, for example, the mother supports the fetus and young individual (she feeds it and ensures its protection and gaining of experiences), but this does not change the essence of the problem. Again, such an independent, autonomous development of the initial "germ" is possible because of the molecular background of its functioning —the key role is played here by, for example, gene expression regulation, the production of morphogen gradients, the mutual support of various cell elements in the synthesis of new elements, and so on.

Artificial life based on supra-molecular elements/machines

Numerous existing or imagined systems/objects/phenomena treated sometimes as "artificial life", both real—robots—and virtual, existing only in the operational memory of a computer, are not in fact "alive" at all as they do not fulfill the cybernetic definition of a living individual—they do not possess purposeful regulatory mechanisms directed at their survival or, first and foremost, at their reproduction.

Here I would like to limit the concept of "artificial living individuals" to some real systems left alone by some intelligent creators, for example, on the surface of some distant planet. Such systems would have to be completely self-dependent in their existence and reproduction. At the same time, I do not mean here some variation on the theme of our terrestrial life, for instance, some cells built of slightly different organic compounds, having slightly different genetic codes *etc*. I rather mean principally different creations built of metals, plastics, silicon, glass, semiconductors, fullerenes, *etc*. Such systems would have to execute a set of various functions, including the following: 1. finding and exploiting various substances (metal ores, silicon, carbon, sulfur, different kinds of rocks and so on) needed for building the components of their bodies (or hunt for other forms of artificial life in order to gain these substances); 2. transforming these "primeval" raw materials into secondary substances directly necessary for the production of these components (e.g., smelt metals from ores, carry out various chemical syntheses, producing plastic or glass); 3. producing the building components/elements of their bodies (e.g., silicon plates, wires, capacitors, semi-conductors, appropriately-shaped metal elements); 4. producing elements/parts of higher order (e.g., microprocessors, chips, batteries/accumulators, electrical circuits, engines, information storage media); 5. detecting defects/damage to their bodies; 6. exchanging/repairing incorrectly working parts/elements of their bodies; 7. constructing/producing the whole bodies of their descendants; 8. preserving, copying and implementing their "genetic record," i.e., a set of instructions leading to the creation of such-and-not-another structure and complex of functions in descendant individuals; 9. receiving various stimuli from the surroundings; 10. responding appropriately to these stimuli through suitable behavior; 11. exploiting available sources of energy (electromagnetic radiation, nuclear processes, chemical reactions, atmospheric electric discharges) for the construction of their bodies and functioning; 12. maintaining internal homeostasis and compensating for various disturbances from the surroundings; 13. regulating, steering and integrating all the above functions; 14. of course, all these functions/goals/tasks would be subordinated to the copying of such artificial living individuals, of their identity to a potentially large number of copies, i.e., their reproduction in agreement with the cybernetic definition of a living individual.

Natural living individuals, being "components" of spontaneously-originated life, cope with the above tasks without any problems—these are tasks for which they originated and evolved (please do not interpret this sentence in a teleological way). The discussed functions are implemented in them, more or less directly, by molecular machines possessing the character of complex organic compounds.

And what about "artificial individuals"? A problem appears here. According to our knowledge, only organic carbon compounds are complex and diverse enough to constitute molecular elements/machines underlying the functioning of natural living individuals. Therefore, the basic, "lowest" level of the functioning of artificial living individuals based on other substances would have to be the supra-molecular level, as it takes place in contemporary robots and computers (their most fundamental elements, e.g., microprocessors, cables, diodes, microchips) are huge (from the molecular point of view) complexes of atoms and molecules artificially shaped during the technological process. Nevertheless, these molecules are usually much simpler than complex organic macromolecules. This fact has enormous consequences.

In artificial "supra-molecular" individuals, a continuous flow of matter and energy through their bodies does not occur—their parts/elements/subunits remain unchanged, and the matter (substance) they are composed of is not exchanged with the environment (therefore, of course, artificial living individuals are not dissipative structures). Alternatively, more precisely, in these elements/parts/subunits, there is a slow continuous increase of entropy, which can lead to various defects after some time. Of course, artificial living individuals can also be damaged much more violently due to the influence of external factors. In both cases, they can become partly or completely non-functional ("sick", "wounded = damaged," or "dead"). They are incapable of self-repair or self-regeneration as they do not possess any internal mechanisms responsible for this (please imagine a self-repairing microchip). Instead, the spoiled or damaged elements/parts/subunits must be repaired from outside, by some external devices/mechanisms, or simply be exchanged for appropriate spare parts (it is much easier to exchange a defective microprocessor than to repair or "regenerate" it). This problem becomes even more acute in the case of the production of progeny. An artificial "supra-molecular" individual cannot self-develop from a relatively simple "germ" produced by parental individuals, as, again, there are no molecular mechanisms that could be responsible for this (this would be unthinkable); any self-developing supra-molecular robot or computer cannot exist. Instead, a descendant "adult" (a fully shaped artificial individual) would have to be constructed by parental individuals, assembled step by step by the addition of consecutive elements/parts/subunits, like computers and robots are constructed by people (or by other robots/devices/machines created by people). For similar reasons, artificial living individuals would not be able to grow spontaneously; at best, they could be extended by adding new parts but not by enlarging the parts already in their composition. There are several other implications of the supra-molecular nature of artificial living individuals. For example, while the energy in the form of ATP is produced in every living cell of natural living individuals, in artificial individuals, there would probably have to be one (or a few) central generator(s) of energy (e.g., in the form of electricity).

All this leads to the main problem—is it at all possible to construct and make functional an artificial living individual constituting one structurally-integrated whole? While every natural (spontaneously-originated) individual—and to a large extent, every cell—realizes its "life functions" without any problems, it would be extremely difficult to imagine an artificial individual, a kind of robot, which would be able to simultaneously: find, intake and appropriately transform its building material; produce appropriate elements/parts/subunits; appropriately assemble them during the construction of progeny; detect damage in its body and repair or exchange spoiled elements (and produce appropriate spare parts); produce energy in a commonly available form (e.g., electricity); maintain internal homeostasis, appropriately react to signals/disturbances from the surroundings; implement various cognitive functions and so on. It seems that such a structurally-integrated artificial individual would have to be a chimera of various mines, smelters, chemical factories, power plants, a combination of different kinds of robots executing various tasks, a cognitive device, an information processing center (something resembling a universal computer), a complex of appropriate engines, and so on. It seems obvious that such a "monstrous" creation would be completely non-functional and, in fact, even impossible to imagine. It is worth realizing that the enumerated tasks are implemented within the human, technical civilization by hundreds and thousands of various devices/units/engines/factories *etc*., which are usually separated spatially from each other, and the whole must be supervised by people. Does this mean that artificial "supra-molecular" individuals (and therefore "truly" artificial life) cannot exist at all?

In my opinion, the answer to this question is negative, and the solution to the discussed problem is something I call a dispersed living individual. The dispersed individual is a complex of particular subunits ("sub-individuals") that are functionally integrated (also cybernetically, through a network of negative feedbacks) but not structurally. Some images of how such a dispersed individual could look are offered by the above-mentioned human technological civilization (of course, devoid of people as supervisors). In my opinion, artificial life not based on molecular machines would have to possess a similar (as to the order of magnitude) degree of complexity. Of course, the artificial living individual composed of hundreds of separate mines, factories, computers, robots, and so on seems monstrous, but everything indicates that there is no other choice if it would have to be self-sufficient concerning survival and reproduction on some distant planet.

Dispersed living individuals are not plucked out of thin air. We are very familiar with them within our natural terrestrial life. Of course, I mean colonies of (eu)social insects—ants, termites, wasps and bees (and mammals—hairless moles). Also, in this case, the whole cybernetic individual (the colony) is divided into separate units—particular insects (or rodents)—that are integrated only functionally but not structurally (leaving aside the fact that most of them live in well-separated nests). Insects belonging to different casts (the queen and fertile males, sterile workers and soldiers of different sizes) specialize in fulfilling various functions. Nevertheless, while in the case of natural terrestrial "molecular" life, being the dispersed individual constitutes an interesting exception, this seems an absolute necessity for artificial "supra-molecular" living individuals.

As mentioned, the interesting idea of dispersed individuals, although for understandable reasons not quite precise from the formal point of view, was presented by Stanisław Lem in his intellectually fertile (and very engrossing) novel *The Invincible* (1973). This topic is discussed in more detail elsewhere (Korzeniewski 2011).

REFERENCES AND SUGGESTIONS FOR FURTHER READING

In order to understand the content of this book, an elementary knowledge of biology and other natural sciences, in particular evolution theory, biochemistry, genetics and developmental biology, at the secondary education level is needed. Many handbooks offer an appropriate range of knowledge.

Some selected more detailed literature is given below. Most of these positions are cited explicitly in the text, and others are suggestions for further reading for an interested reader.

Finally, at present, the Internet offers countless resources of information.

Adami, Christoph. 1998. *Introduction to Artificial Life.* Santa Clara: Springer-Verlag.

Adami, Christoph. 2002. "What is Complexity?" *BioEssays* 24: 1085-1094.

Alberts, Bruce, Dennis Bray, Alexander Johnson, Julian Lewis, Martin Raff, Keith Roberts, and Peter Walter. 1998. *Essential Cell Biology. An Introduction to the Molecular Biology of the Cell.* New York & London: Garland Publishing.

Al-Khalili, Jim. 2012. *Quantum: A Guide for the Perplexed.* Weidenfeld & Nicolson.

Al-Khalili, Jim, and Johnjoe McFadden. 2014. *Life on the Edge. The Coming of Age of Quantum Biology*. Bantam Press.

Ashby, W. Ross. 1962. *Introduction to Cybernetics.* London: Methuen.

Atkins, Peter, Julio de Paula. 2002. *Atkins' Physical Chemistry.* New York: W.H. Freeman and Company.

Baggott, Jim. 2013. *Farewell to Reality: How Modern Physics has Betrayed the Search for Scientific Truth.* Constable & Robinson.

Baggott, Jim. 2015. *Origins: The Scientific Story of Creation.* Oxford: Oxford University Press.

Baker, Joanne. 2007. *50 Physics Ideas you Really Need to Know.* Quercus.

Banich, Marie T., and Rebecca J. Compton. 2018. *Cognitive Neuroscience.* Cambridge: Cambridge University Press.

Barrack, Jeffrey E. and Ronald R. Breaker. 2007. "The Power of Riboswitches." *Scientific American* 296, no. 1: 50-57.

Barrow, John D. 1994. *The Origin of the Universe*. Basic Books.

Barski, G. and F. Cornefert-Jensen. 1966. "Cytogenetic Study of Sticker Venereal Sarcoma in European Dogs." *Journal of National Cancer Institute* 37, no. 6: 787-797.

Beatty, J. Thomas, Jörg Overmann, Michael T. Lince, Ann K. Manske, Andrew S. Lang, Robert E. Blankenship, Cindy L. Van Dover, Tracey A. Martinson, F. Gerald Plumley. 2005. "An obligately photosynthetic bacterial anaerobe from a deep-sea hydrothermal vent." *Proceedings of the National Academy of Sciences USA* 102, no. 26: 9306-93010.

Bedau, Mark A. 1998. "Four Puzzles about Life." *Artificial life* 4, no. 2: 125-140.

Berner, Robert A., Antonio C. Lasaga, and Robert M. Garrels. 1983. "The Carbonate-Silicate Geochemical Cycle and its Effect on Atmospheric Carbon Dioxide over the past 100 Million Years." *American Journal of Science*, 283, no. 7: 641–683.

Bougourd, Susan M., and R. Neil Jones. 1997. "B Chromosomes: a Physiological Enigma." *New Phytologist* 137: 43-54.

Brooks, Daniel, and E.O. Wiley. 1986. *Evolution as Entropy. Toward a Unified Theory of Biology.* Chicago & London: University of Chicago Press.

Brown, Terry. 2012. *Introduction to Genetics: A Molecular Approach.* New York & London: Garland Science.

Bryson, Bill. 2016. *A Short History of Nearly Everything.* Broadway Books.

Cairns-Smith, A. Graham. 1996. *Evolving the Mind. On the Nature of Matter and the Origin of Consciousness.* Cambridge: Cambridge University Press.

Carroll, Sean B., Benjamin Prud'homme, and Nicolas Gompel. 2008. "Regulating Evolution: How Gene Switches Make Life." *Scientific American* 298, no. 5: 60–67.

Cavalier-Smith, Thomas. 2001. "Obcells as Proto-Organisms: Membrane Heredity, Lithophosphorylation, and the Origins of the Genetic Code, the First Cells, and Photosynthesis." *Journal of Molecular Evolution* 53: 555–595.

Chance, Britton, and G.R. Williams. 1956. "The Respiratory Chain and Oxidative Phosphorylation." *Advances in Enzymology* 17: 65-134.

Chandrasekhar, B. S. 1998. *Why Things Are the Way They Are.* Cambridge: Cambridge University Press.

Copley, Shelley D., Eric Smith, and Harold J. Morowitz. 2007. "The Origin of the RNA World: Co-Evolution of Genes and Metabolism." *Bioorganic Chemistry* 35, no. 66: 430-43.

Coveney, Peter, and Roger Highfield. 1991. *The Arrow of Time*. London: Flamingo.

Coveney, Peter, and Roger Highfield. 1996. *Frontiers of Complexity. The Search for Order in a Chaotic World.* Ballantine Books.

Cryer, P. E. 1991. "Regulation of Glucose Metabolism in Man." *Journal of Internal Medicine Supplement* 735: 31-39.

Crick, Francis. 1995. *The Astonishing Hypothesis: The Scientific Search for the Soul.* Scribner.

Darwin, Charles. 1859. *The Origin of Species by Means of Natural Selection, or the Preservation of Favored Races in the Struggle for Life.* New York: Modern Library.

Davies, Paul. 1996. *About Time. Einstein's Unfinished Revolution.* Simon & Schuster.

Dawkins, Richard. 1995. *River out of Eden: A Darwinian View of Life.* Basic Books.

Dawkins, Richard. 1976. *The Selfish Gene.* Oxford: Oxford University Press.

Dawkins, Richard. 1996. *Climbing Mount Improbable.* W. W. Norton & Company.

Difference Between. 2020. "Microstate and Macrostate." Accessed May 27, 2020. https://www.differencebetween.com/difference-between-microstate-and-macrostate/

Doolittle, W. Ford, and Carmen Sapienza. 1980. "Selfish Genes, the Phenotype Paradigm and Genome Evolution." *Nature* 284: 601-603.

Deamer, David W. 2019. *Assembling Life: How Can Life Begin on Earth and Other Habitable Planets.* Oxford: Oxford University Press.

Deamer, David W. 2020. *Origin of Life. What Everyone Needs to Know.* Oxford: Oxford University Press.

de Farias, Savio T., Ariosvaldo P. dos Santos Junior, Thais G. Rêgo, and Marco V. José. 2017. "Origin and Evolution of RNA-Dependent RNA Polymerase." *Frontiers in Genetics* 8: 125.

DeLaurier, April. 2018. "Evolution and development of the fish jaw skeleton." *WIREs Developmental Biology* 8, no. 2: e337.

Dunbar, Robin. 1998. *Grooming, Gossip and Evolution of Language.* Cambridge MA: Harvard University Press.

Dunbar, Robin. 2005. *The Human Story. A New History of Mankind's evolution.* Gardeners Books.

Dunbar, Robin. 2016. *Human Evolution: Our Brains and Behavior*. Oxford: Oxford University Press.

Eigen, Manfred, and Peter Schuster. 1979. *The Hypercycle: a Principle of Natural Self-Organization*. Berlin: Springer.

Eldredge, Niles, and Stephen Jay Gould. 1972. "Punctuated Equilibria: an Alternative to Phyletic Gradualism." In: *Models in paleobiology*, edited by Thomas J. M. Schopf, 82-115. San Francisco: Freeman, Cooper and Co.

Emmeche, Claus. 1992. "Life is an Abstract Phenomenon: is Artificial Life Possible?" In: *Toward a Practice of Autonomous Systems. Proceedings of the First European Conference on Artificial Life*, edited by Francisco J. Varen and Paul Bourgine, 466-474. Cambridge, MA: MIT Press.

Emmeche, Claus. 1997. "Aspects of Complexity in Life and Science." *Philosophica* 59, no. 1: 41-68.

Faller, James E., and Alan H. Cook. 2019. "Gravity." *Encyclopædia Britannica*, Inc., 10 Apr. 2019.

Fastowsky, David E., and David B. Weishampel. 2016. *Dinosaurs. A Concise Natural History.* Cambridge: Cambridge University Press.

Futuyma, Douglas J. 2013. *Evolution.* Sinauer Associates.

Gánti, Tibor. 1975. "Organisation of Chemical Reactions into Dividing and Metabolizing Units: the Chemotons." *BioSystems* 7, no. 1: 189-195.

Gell-Mann, Murray. 1995. *The Quark and the Jaguar. Adventures in the Simple and the Complex.* Griffin.

Gilbert, Scott F. 2010. *Developmental Biology.* Sunderland, Sinauer Associates.

Gould, Stephen Jay. 1990. *Wonderful Life: The Burgess Shale and the Nature of History.* W. W. Norton & Company.

Gould, Stephen Jay. 1994. "The Evolution of Life on earth". *Scientific American* 271, no. 4: 62-69.

Greene, Brian. 2005. *The Fabric of the Cosmos: Space, Time, and the Texture of Reality.* Vintage.

Greene, Brian. 2010. *The Elegant Universe: Superstrings, Hidden Dimensions, and the Quest for the Ultimate Theory.* W. W. Norton & Company.

Griesemer, James. 2000. "The Units of Evolutionary Transition." *Selection* 1: 67-80.

Haldane, John B. S. 1929. "The Origin of Life." *Rationalist Annual* 148: 3-10.

Hall, Brian K. 1999. *Evolutionary Developmental Biology*. London: Chapman & Hall.

Harari, Yuval N. 2015. *Sapiens: A Brief History of Humankind.* Harper.

Harris, Hugh M.B., and Colin Hill. 2021. "A Place for Viruses on the Tree of Life." *Frontiers in Microbiology* 11: 604048.

Hawking, Stephen. 1988. *A brief history of time: From big bang to black holes.* New York: Bantam.

Heylighen, Francis. 1999. “The Growth of Structural and Functional Complexity During Evolution.” In: *The Evolution of Complexity*, edited by Francis Heylighen, J. Bollen and A. Riegler, 17-44. Dordrecht: Kluwer Academic Publishers.

Hill, Richard W., and Gordon A. Wyse. *Animal Physiology*. HarperCollins Publishers.

Hoffmeyer, Jesper. 1997. “Biosemiotics: toward a New Synthesis in Biology.” *European Journal for Semiotic Studies* 9, no. 2: 355-376.

Hokkanen, J. E. 1999. “Visual Simulations, Artificial Animals and Virtual Ecosystems.” *Journal of Experimental Biology* 202, no. 23: 3477-3484.

Holland, Heinrich D. 2006. "The oxygenation of the atmosphere and oceans". *Philosophical Transactions of the Royal Society: Biological Sciences* 361: 903–915.

Hopkins, William G. 1995. *Introduction to Plant Physiology.* New York: John Wiley & Sons.

Hug, Laura A., Brett J. Baker, Karthik Anantharaman, Christopher T. Brown, Alexander J. Probst, Cindy J. Castelle, Cristina N. Butterfield, Alex W. Hernsdorf, Yuki Amano, Kotaro Ise, Yohey Suzuki, Natasha Dudek, David A. Relman, Kari M. Finstad, Ronald Amundson, Brian C. Thomas, and Jillian F. Banfield. 2016. “A New View of the Tree of Life.” *Nature Microbiology* 1: 16048.

Jacob, François, and Jacques Monod. 1961. “On the Regulation of Gene Activity.” *Cold Spring Harbor Symposia on Quantitative Biology* 26: 193-211.

Jones, R. N. 1975. “B-Chromosome System in Flowering Plants and Animal Species.” *International Review of Cytology* 40: 1-100.

Kado, Clarence I. 1998. “Origin and Evolution of Plasmids.” *Antonie van Leeuvenhoek* 73: 117-126.

Kardong, Kenneth V. 2002. *Vertebrates. Comparative Anatomy, Function, Evolution.* McGraw-Hill.

Kauffman, Stuart A. 1986. “Autocatalytic sets of proteins.” Journal of Theoretical Biology 119, no. 1: 1-24.

Keymeulen, Didier, Masaya Iwata, Yasuo Kunioshi, and Tetsuya Higuchi. 1998. “Online Evolution for a Self-Adopting Robotic Navigation System Using Evolvable Hardware.” *Artificial Life* 4, no. 4: 359-393.

Kingsley, David M. 2009. “From Atoms to Traits.” *Scientific American* 300, no. 1: 52-59.

Knoll, Andrew H. 2021. *A Brief History of Earth: Four Billion Years in Eight Chapters.* Harper Collins.

Koga, Yosuke, and Hiroyuki Morii. 2007. "Biosynthesis of Ether-Type Polar Lipids in Archaea and Evolutionary Considerations." *Microbiology and Molecular Biology Reviews* 71, no. 1: 97-120.

Korzeniewski, Bernard. 2001. "Cybernetic Formulation of the Definition of Life." *Journal of Theoretical Biology* 209, no. 3: 275-286.

Korzeniewski, Bernard. 2005. "Confrontation of the Cybernetic Definition of a Living Individual with the Real World." *Acta Biotheoretica* 53, no. 1: 1-28.

Korzeniewski, Bernard. 2010. *From Neurons to Self-Consciousness. How the Brain Generates the Mind*. Amherst: Humanity Books.

Korzeniewski, Bernard. 2011. "Artificial Cybernetic Living Individuals Based on Supramolecular-Level Organization as Dispersed Individuals." *Artificial Life* 17, no. 1: 51-67.

Korzeniewski, Bernard. 2013a. "Formal Similarities Between Cybernetic Definition of Life and Cybernetic Model of Self-Consciousness: Universal Definition/Model of Individual." *Open Journal of Philosophy* 3, no. 2: 314-328.

Korzeniewski, Bernard. 2013b. "Magic of Language." *Open Journal of Philosophy* 3, no. 4: 455-465.

Korzeniewski, Bernard. 2015. "Mind-Body Problem: Does Complexity Exist Objectively?" *Open Journal of Philosophy* 5, no. 6: 351-364.

Korzeniewski, Bernard. 2016. "Freezing of Reality: Is Flow of Time Real?" *Open Journal of Philosophy* 6, no. 3: 256-264.

Korzeniewski, Bernard. 2017a. "Does Matter Matter? Should We Mind the Mind? Can Philosophy be Reduced to Neurophysiology?" *Open Journal of Philosophy* 7, no. 3: 265-328.

Korzeniewski, Bernard. 2017b. "Regulation of Oxidative Phosphorylation through Each-Step Activation: Evidence from Computer Modeling." *Progress in Biophysics and Molecular Biology* 125: 1-23.

Korzeniewski, Bernard. 2020. "Self-Consciousness as Product of Biological Evolution." *Journal of Consciousness Studies* 27, no. 7-8: 50-76.

Knoll, Andrew H. 2021. *A Brief History of Earth: Four Billion Years in Eight Chapters.* Custom House.

Koshland, Daniel E Jr. 2002. "The Seven Pillars of Life." *Science* 295, no. 5563: 2215-2216.

Krem, Maxwell M. and Di Cera Enrico. 2001. "Molecular markers of serine protease evolution." *EMBO Journal* 20: 3036-3045.

LabXchange. 2022. "Microstates, Macrostates, and Entropy." Accessed June 12, 2022.
https://www.labxchange.org/library/items/lb:LabXchange:ac117f1a:html:1

Lane, Nick. 2010. *Life Ascending. The Ten Greatest Inventions of Evolution.* W. W. Norton & Company.
Lane, Nick. 2015. T*he Vital Question. Why is Life the Way It Is?* Profile Books.
Leakey, Richard. 1996. *The Origin of Humankind.* Basic Books.
Lederman, Leon, and Dick Teresi. 2006. *The God Particle. If the Universe is the Answer, What Is the Question?* Mariner Books.
Lem, Stanisław. 1973. *The Invincible*. San Francisco: The Seabury Press.
Lem, Stanisław. 2002. *Solaris.* Mariner.
Lem, Stanisław. 2020. *His Master's Voice.* The MIT Press.
Lincoln, Tracey A., and Gerald F. Joyce. 2009. "Self-sustained Replication of an RNA Enzyme." Science 323: 1229-1232.
Lipson, Hod, and Jordan Pollack. 2000. "Automatic Design and Manufacture of Robotic Life Forms." *Nature* 406, no. 6799: 974-978.
Lloyd, Elisabeth A. 1988. *The Structure and Confirmation of Evolutionary Theory.* New York: Greenwood Press.
Lovelock, James. 1979. *Gaia, A New Look at Life on earth*. New York: Oxford University Press.
Macdougall, J. D. 1996. *A Short History of Planet earth. Mountains, Mammals, Fire and Ice.* New York: John Wiley & Sons.
Mackie, G.O., P.R. Pugh, and J.E. Purcell. 1988. "Siphonophore Biology." *Advances in Marine Biology* 24: 97-262.
Macphail, Euan M. 1998. *The Evolution of Consciousness.* Oxford: Oxford University Press.
March, Jerry. 1992. *Organic Chemistry: Reactions, Mechanisms, and Structure.* New York: John Wiley & Sons.
Margulis, Lynn. 1981. *Symbiosis in the Cell Evolution.* San Francisco: W.H. Freeman.
Marks, Dawn B., Marks, Allan D., and Smith, Colleen M. 1996. Basic *Medical Biochemistry. A Clinical Approach.* Baltimore: Williams & Wilkins.
Martin, B.R. 1987. *Metabolic Regulation. A Molecular Approach.* Oxford: Blackwell Scientific Publications.
Mattick, John S. 2004. "The Hidden Genetic Program of Complex Organisms." *Scientific American* 291, no. 4: 60-77.
Maynard Smith, John. 1970. "Time in the Evolutionary Process." *Studium Generale* 23, no. 3: 266-272.
Maynard Smith J., and Eörs Szathmáry. 1995. *The Major Transitions in Evolution*. Oxford: W.H. Freeman.
McShea, Daniel W. 1991. "Complexity and Evolution: What Everybody Knows." *Biology and Philosophy* 6: 303-324.

Miglino, Orazio, Henrik Hautop Lund, and Stefano Nolfi. 1995. "Evolving Mobile Robots in Simulated and Real Environments." *Artificial Life* 2, no. 4: 417-434.

Molles, Manuel C. Jr. 1999. *Ecology: Concepts and Applications.* McGraw-Hill.

Muller, H. J. 1935. "Life." *Science* 121, no. 3132: 1-9.

Muller, H. J. 1966. "The Gene Material as the Initiator and Organizing Basis of Life." *American Naturalist* 100, no. 915: 493-517.

Murray, Robert K., David A. Bender, Kathleen M. Botham, Peter J. Kennelly, Victor W. Rodwell, and P. Anthony Weil. 2012. *Harper's Illustrated Biochemistry*. New York: McGraw Hill.

Nakashima, T. 1987. "Metabolism of proteinoid microspheres." *Topics in Current Chemistry* 139: 58-81.

Nicholls, David G., and Stuart J. Ferguson. 2002. *Bioenergetics 3.* Academic Press.

Nilsson, Dan E., and Susanne Pelger. 1994. "A Pessimistic Estimate of the Time Required for an Eye to Evolve." *Proceedings: Biological Sciences* 256, no. 1345: 53-58.

Noble, Dennis. 2006. *The Music of Life.* Oxford: Oxford University Press.

Nur, Uzi, John H. Werren, Danna G. Eickbush, William D. Burke, and Thomas H. Eickbush. 1988. "A "selfish" B Chromosome that Enhances its Transmission by Elimination of the Paternal Genome." *Science* 240, no. 4851: 512-514.

Oparin, Aleksandr I. 1938. *The Origin of Life.* New York: Dover Publications.

Orgel, Leslie E., and Francis H. C. Crick. 1980. "Selfish DNA: the Ultimate Parasite." *Nature* 284, no. 5757: 604-607.

Orr, H. Allen. 2009. "Testing Natural Selection with Genetics." *Scientific American* 300, no. 1: 44-51.

Oyama, Susan. 1985. The Ontogeny of Information. Cambridge: Cambridge University Press.

Ozaki, Kazumi, and Christopher T. Reinhard. 2021. "The Future Lifespan of Earth's Oxygenated Atmosphere." *Nature Geoscience* 14: 138–142.

Ozima, Minoru, Jun Korenaga, and Qing-Zhu Yin. 2012. *The Earth. Its Birth and Growth.* Cambridge: Cambridge University Press.

Patrick, Graham. 2000. *Instant Notes, Organic Chemistry.* BIOS Scientific.

Paul, Natasha, and Gerald F. Joyce. 2002. "A self-replicating ligase ribozyme." Proceedings of the National Academy of Science USA 99, no. 20: 12733-12740.

Penrose, Roger. 2002. *The Emperor's New Mind: Concerning Computers, Minds, and the Laws of Physics.* Oxford: Oxford University Press.

Pollard, Katherine S. 2009. *What Makes Us Human? Scientific American* 300, no. 5: 44-49.

Postgate, John. 1994. *The Outer Reaches of Life.* Cambridge: Cambridge University Press.

Powers, William T. 1973. *Behaviour: The Control of Perception.* Chicago: Adline.

Prigogine, Ilia. 1980. *From Being to Becoming: Time and Complexity in Physical Sciences.* San Francisco: W.H. Freeman.

Prigogine, Ilia, and Isabelle Stengers. 1984. *Order out of Chaos. Man's New Dialogue with Nature.* London: Heinemann.

Prusiner, Stanley B. 1998. "Prions." *Proceedings of the National Academy of Science USA* 95, no. 23: 13363-13383.

Ray, Thomas S. 1992. "An Approach to the Synthesis of Life." In: *Artificial Life II: Proceedings of an Interdisciplinary Workshop on the Synthesis and Stimulation of Living Systems (Santa Fe Institute Studies in the Sciences and Complexity, vol. 10)*, edited by Christopher G. Langton, Charles Taylor, J. Doyne Farmer, and Steen Rasmussen., 371-408. Reading, Addison-Wesley.

Reece, Jane B., Lisa A. Urry, Michael L. Cain, Steven A. Wasserman, Peter V. Minorsky, and Robert B. Jacksomn. 2011. *Campbell Biology.* Boston: Pearson.

Ricardo, Alonso, Jack W. Szostak. 2009. "The Origin of Life on Earth." *Scientific American* 301, no. 3: 54-61.

Rinke, Christian, Patrick Schwientek, Alexander Sczyrba, Natalia N. Ivanova, Iain J. Anderson, Jan-Fang Cheng, Aaron Darling, Stephanie Malfatti, Brandon K. Swan, Esther A. Gies, Jeremy A. Dodsworth, Brian P. Hedlund, George Tsiamis, Stefan M. Sievert, Wen-Tso Liu, Jonathan A. Eisen, Steven J. Hallam, Nikos C. Kyrpides, Ramunas Stepanauskas, Edward M. Rubin, Philip Hugenholtz, and Tanja Woyke. 2013. "Insights into the phylogeny and coding potential of microbial dark matter." *Nature* 499: 431–437.

Schleif, Robert. 1993. *Genetics and Molecular Biology.* Baltimore and London: The Johns Hopkins University Press.

Schrödinger, Erwin. 1992. *What is Life?* Cambridge: Press Syndicate of the University of Cambridge.

Shannon, Claude E. 1949. The Mathematical Theory of Communication. Urbana: University of Illinois Press.

Shapiro, Robert. 2007. "A Simpler Origin for Life." *Scientific American* 296, no. 6: 46-53.

Slack, Jonathan. 2001. *Essential Developmental Biology.* Blackwell.

Smith, Eric, and Harold J. Morowitz. 2016. *The Origin and Nature of Life on earth. The Emergence of the Fourth Geosphere.* Cambridge: Cambridge University Press.

Smokers, Iris B.A. Merlijn H. I. van Haren, Tiemei Lu, and Evan Spruijt. 2022. "Complex Coacervation and Compartmentalized Conversion of Prebiotically Relevant Metabolites." *ChemSystemsChem* 4: e202200004.

Smolin, Lee. 1999. *Life of the Cosmos.* Oxford: Oxford University Press.

Solomons, T. W. Graham , Craig B. Fryhle, and Scott A. Snyder. 2017. *Solomons' Organic Chemistry.* Wiley.

Sparrow, Giles. 2016. *50 Ideas You Really Need to Know: Astronomy.* Quercus.

Stambrook, Peter J. 2013. "Nuclear transfer between two species of Acetabularia." *Experimental Biology and Medicine* 238, no. 3: 257-8.

Steward, Ian, and Jack Cohen J. 1997. *Figments of Reality. The Evolution of the Curious Mind.* Cambridge: Cambridge University Press.

Stringer, Chris, and Robin McKie. 1998. *African Exodus. The Origins of Modern Humanity.* Holt Paperbacks.

Ślesak, Ireneusz, Halina Ślesak, and Jerzy Kruk J. 2012." Oxygen and Hydrogen Peroxide in the Early Evolution of Life on earth. In Silico Comparative Analysis of Biochemical Pathways." *Astrobiology* 12, no. 8: 775-784.

Toyama, Brandon H., and Martin W. Hetzer .2013) "Protein homeostasis: live long, won't prosper". *Nature Reviews. Molecular Cell Biology* 14, no. 1: 55–61.

Tsokolov, Serhiy A. 2009. "Why Is the Definition of Life So Elusive? Epistemological Considerations." *Astrobiology* 9, no. 4: 401-412.

Twyman, Richard M. 2001. *Instant Notes, Developmental Biology.* BIOS Scientific.

Umbarger, H. E. 1978. "Amino Acid Biosynthesis and its Regulation." *Annual Review of Biochemistry* 47: 533-606.

van Kranandonk Martin J., David W. Deamer, and Tara Djokic. 2017. "Life Springs." *Scientific American* 317, no. 2: 28-35.

Voet, Donald, Judith G. Voet, and Charlotte W. Pratt. 1999. *Fundamentals of Biochemistry*. New York: John Wiley & Sons.

von Bertalanffy, Ludwig. 1968. *General Systems Theory.* New York: George Braziller.

Walsh, D. M. 2015. *Organisms, Agency, and Evolution.* Cambridge: Cambridge University Press.

Wächterhäuser, Günter. 2007. "On the chemistry and evolution of the pioneer organisms." *Chemical Biodiversity* 4, no. 4: 584-602.

Weinberg, Steven. 1993. The First Three Minutes: A Modern View of the Origin of the Universe. Basic Books.

Weiss, Madeline C., Filipa L. Sousa, Natalia Mrnjavac, Sinje Neukirchen, Mayo Roettger, Shijulal Nelson-Sathi, and William F. Martin. 2016. "The Phylogeny and Habitat of the Last Universal Common Ancestor." *Nature Microbiology* 1: 16116.

Wiener, Norbert. 1948. *Cybernetics, or Control and Communication in the Animal and the Machine.* Cambridge MA: MIT Press.

Williams, George C. 1997. *The Pony Fish's Glow and Other Clues to Plan and Purpose in Nature.* Basic Books.

Wilson, Edward O. 1971. *The Insect Societies.* Cambridge, MA: Harvard University Press.

Wilson, Edward O. 1975. *Sociobiology: The New Synthesis.* Cambridge, MA: Harvard University Press.

Wolpert, Lewis. 1981. "Positional information and pattern formation." *Philosophical Transactions of Royal Society London* 295: 441-50.

Woods, Gregory M., Lauren J. Howson, Gabriella K. Brown, Cesar Tovar, Alexandre Kreiss, Lynn M. Corcoran, and A. Bruce Lyons. 2015. "Immunology of a Transmissible Cancer Spreading among Tasmanian Devils." *J Immunol* 195, no. 1: 23–29.

Wynne-Edwards, V.C. 1962. *Animal Dispersion in Relation to Social Behaviour*. Edinburgh: Oliver & Boyd.

Yarlett, Nigel, and Johannes H. P. Hackstein. 2005. "Hydrogenosomes: One Organelle, Multiple Origins." *BioScience* 55, no. 8: 657–668.

Zimmermann, W. 1952. "Main Results of the 'Telome Theory'." *The Palaeobotanist* 1: 456-470

INDEX